The Evolving Sphere of Food Security

For Tina and Hamid -
With many
thanks for your
leadership -

The Evolving Sphere of Food Security

Edited by Rosamond L. Naylor

OXFORD
UNIVERSITY PRESS

OXFORD
UNIVERSITY PRESS

Oxford University Press is a department of the University of Oxford.
It furthers the University's objective of excellence in research, scholarship,
and education by publishing worldwide.

Oxford New York
Auckland Cape Town Dar es Salaam Hong Kong Karachi
Kuala Lumpur Madrid Melbourne Mexico City Nairobi
New Delhi Shanghai Taipei Toronto

With offices in
Argentina Austria Brazil Chile Czech Republic France Greece
Guatemala Hungary Italy Japan Poland Portugal Singapore
South Korea Switzerland Thailand Turkey Ukraine Vietnam

Published in the United States of America by
Oxford University Press
198 Madison Avenue, New York, NY 10016

Library of Congress Cataloging-in-Publication Data
The evolving sphere of food security / edited by Rosamond L. Naylor.
pages cm
Includes bibliographical references and index.
ISBN 978–0–19–935405–4 (hardcover : alk. paper)
ISBN: 978–0–19–935406–1 (pbk. : alk. paper)
1. Food security. I. Naylor, Rosamond.
HD9000.5.E96 2014
338.1'9—dc23
2014024467

9 8 7 6 5 4 3 2 1
Printed in the United States of America
on acid-free paper

For Wally and Nicki

CONTENTS

PART FOUR Agriculture's Dependence on Resources and the Environment

PART FIVE Food in a National and International Security Context

FOREWORD

From my childhood days growing up in Ghana, through my years working at the UN, to my more recent efforts to help promote a green revolution in Africa, I have been deeply concerned by hunger and malnutrition among us. The fact that over 800 million people lack basic energy for a productive life, and that up to 3 billion suffer from protein and micronutrient deficiencies is an unconscionable moral failing of our global community.

In November 2011, I was invited to give a keynote speech at Stanford University for the launch of its Center on Food Security and the Environment. I urged the students and faculty in the audience to get involved in solving the global problems of hunger, climate change, and complacency in the face of human suffering. I wanted answers to tough questions: Why did the Green Revolution pass Africa by? Why do hundreds of millions of people go hungry in countries with rapid economic growth such as China and India? Why do one in five children in the United States go to school every day without adequate nutrition? And I reminded them that global food insecurity should not be interpreted as a problem of "*those* chronically malnourished people." Rather, it is *our* collective problem, a stumbling block to a fairer and more secure world.

The Evolving Sphere of Food Security is a response to the challenge that I laid down that day. Uniquely, it connects food to other key challenges of human well-being and security. Any society that fails to build on the potential of all its members will always be at a disadvantage. Food and nutrition insecurity reflects and contributes to this unmet potential. It affects everything from the health of an unborn child, to the educational attainment of a given society, to the political stability and economic progress of the world. Food, water, health, energy, climate, rights to land, suffrage, and education are all linked in the daily lives of people around the world.

In this book, experts from many fields have pooled their knowledge to understand and interpret this complex web of development. Even the concept of development can be misleading if it fails to include a focus on resources, climate, and the environment. As the book highlights, many low-income farm communities are in desperate need of water for irrigation and fertilizers to improve crop yields, and these resources will play a critical role in helping them to adapt to climate change. As countries break out of poverty and their food demands expand, they need to be especially careful to manage fertilizers, water, and other farm inputs in ways that minimize environmental damage and resource depletion. The interdisciplinary perspective of the book enables the authors to identify technological and management strategies that can help reduce the long-term stresses of agricultural growth, while improving food and nutrition security and safeguarding the environment.

There is no easy fix for persistent hunger, and no single strategy can be deployed in all locations. Improving food and nutrition security is a matter of good governance, rigorous science, water and soil management, innovation, investment, health, vision, courage, leadership, and human compassion. It also depends on us putting in place the right policies to address poverty and inequality, promote gender equality, strengthen land rights, and accelerate wider economic and social development. With such complexity at so many levels, it is sometimes easier to search for simple answers and walk away from the deeper domains of history, politics, and human and natural resource constraints that keep people hungry. *The Evolving Sphere of Food Security* provides a guide for understanding the root causes and trade-offs that governments and communities face as they attempt to put nutritious food on the plates of all.

This book deserves a wide audience, especially among college and university students. As today's youth become of age, we will rely upon them more and more to solve problems like food and nutrition insecurity. My dialogues with young leaders from all corners of the world, as well as the enthusiasm of the students who attended my lecture at Stanford, convince me that they are eager to do so. I hope this book informs their vision and their efforts to alleviate hunger, end poverty, and promote sustainable development.

Kofi Annan
Founding Chairman of the Alliance for a
Green Revolution in Africa (AGRA)
Former Secretary-General of the United Nations

ACKNOWLEDGMENTS

This volume is the result of many years of collaborative research and teaching at Stanford University, and I owe sincere thanks to the faculty members and students who helped bring it to fruition. The faculty who contributed chapters took time to attend informal workshops as we designed the volume, to edit each other's work, and to re-draft their own work to emphasize the real-world aspects of their research and the connectivity among sections. I learned an enormous amount from all of them in the process. I owe special thanks to Thomas Hayden, professional editor and Lecturer in Environmental Communication in the School of Earth Sciences at Stanford. Tom worked tirelessly with me and the other authors to bring out personal stories, to make the prose more accessible to a wide audience, and to integrate the chapters. His editing skills, vision, and patience are exceptional. I would also like to thank two special Research Assistants, Natalie Johnson and Karrah Phillips, who provided support at all stages of the volume, from substantive research and data collection to formatting and editing. They are both extremely talented, and I suspect we will see books under their own names in the future.

The motivation for this volume emerged from an annual symposium series at Stanford University called *Connecting the Dots: Food, Energy, Water, and Climate*, funded by the TomKat Center for Sustainable Energy. After the first symposium I was urged by Kai Lee of the Packard Foundation to publish our interdisciplinary perspective on food security for student and informed lay audiences worldwide. His suggestion resonated with me because I have devoted my career to working at the interface between disciplines. I am grateful to Kai, and to Kat Taylor and Stacey Bent (co-Founder and Program Manager, respectively, of the TomKat Center), for pushing me towards this end.

The volume would not have been possible without the establishment and support of the Center on Food Security and the Environment (FSE) at Stanford, which I founded and have directed since 2006. Several individuals and institutions have provided core support for FSE, including Steve and Roberta Denning, Lawrence and Tricia Kemp, Whitney and Betty MacMillan, Jeff and Tricia Raikes, Julie Wrigley, Alison and Geoff Rusack, Richard and Gloria Kushel, Zachery Nelson and Elizabeth Horn, the Cargill Foundation, and the Bill and Melinda Gates Foundation. As a constituent Center of Stanford's Freeman Spogli Institute for International Studies and the Stanford Woods Institute for the Environment, FSE has also benefitted from the backing of

Chip Blacker, Gerhard Casper, Tino Cuéllar, Jeff Koseff, and Buzz Thompson. I am grateful to FSE's administrative team—Lori McVay, Mary Smith, Ashley Dean, Laura Seaman, and Rita Robinson—for their day-to-day efforts surrounding this volume, and to Donald Kennedy and Bill Burke, two important academic members of FSE, for their substantive input. Finally, I would like to thank Scott Parris and Cathryn Vaulman from Oxford University Press for their excellent editorial and publishing assistance, as well as David Battisti and a dozen anonymous reviewers whose insightful critiques improved the quality of the volume.

On a personal note, I am extremely fortunate to have the family support of Wally and Nicki Naylor, and the friendship of Wally Falcon. They have remained by my side throughout my non-traditional academic career, and have supported me through the thick and thin of this volume.

Roz Naylor
March 7, 2014

LIST OF CONTRIBUTORS

Eran Bendavid is an Assistant Professor of Medicine in the Division of General Internal Medicine, and a Research Fellow at the Center on Health Policy Research within the Freeman Spogli Institute for International Studies at Stanford University.

Jennifer A. Burney is an Assistant Professor of Science, Technology, Engineering, and Policy in the School of International Relations and Pacific Studies at the University of California (San Diego) and a Visiting Scholar at the Center on Food Security and the Environment at Stanford University.

Mariano-Florentino Cuéllar is the Stanley Morrison Professor of Law and the Director of the Freeman Spogli Institute for International Studies at Stanford University.

Jennifer Davis is an Associate Professor in the Department of Civil & Environmental Engineering at Stanford University and the Higgins-Magid Faculty Senior Fellow at the Stanford Woods Institute for the Environment, where she directs the Program on Water, Health, and Development.

Walter P. Falcon is the Helen Farnsworth Professor of International Agricultural Policy in the Department of Economics (Emeritus), and a Senior Fellow at the Center on Food Security and the Environment within the Freeman Spogli Institute for International Studies and the Woods Institute for the Environment at Stanford University.

Christopher B. Field is the Founding Director of the Carnegie Institution's Department of Global Ecology, Professor of Biology, Professor of Environmental Earth System Science, Faculty Director of Stanford's Jasper Ridge Biological Preserve, and Senior Fellow of the Stanford Woods Institute for the Environment, the Precourt Institute for Energy, and the Freeman Spogli Institute for International Studies at Stanford University.

Jikun Huang is the Founder and Director of the Center for Chinese Agricultural Policy, Professor at the Institute of Geographic Sciences and Natural Resources Research within the Chinese Academy of Sciences, and an affiliated scholar at the Center on Food Security and the Environment at Stanford University.

Eric F. Lambin is the George and Setsuko Ishiyama Provostial Professor in Environmental Earth System Science, a Senior Fellow at Stanford Woods Institute for the Environment, and an affiliated faculty member of the Center on Food Security and the Environment at Stanford University. He is also a Professor in the Department of Geography at the University of Louvain in Louvain-la-Neuve, Belgium.

David Lazarus is the inaugural Food and International Security Fellow at the Center on Food Security and the Environment and the Center on International Security and Cooperation within the Freeman Spogli Institute for International Studies at Stanford University.

David B. Lobell is an Associate Professor in the Department of Environmental Earth System Science, Associate Director of the Center on Food Security and the Environment, and Senior Fellow at the Freeman Spogli Institute for International Studies and the Woods Institute for the Environment at Stanford University.

Pamela A. Matson is the Chester Naramore Dean of the School of Earth Sciences, a Senior Fellow in the Freeman Spogli Institute for International Studies and the Woods Institute for the Environment, and an affiliated faculty member in the Center for Food Security and the Environment at Stanford University.

Rosamond L. Naylor is the Director of the Center on Food Security and the Environment and the William Wrigley Senior Fellow at the Freeman Spogli Institute for International Studies and the Woods Institute for the Environment at Stanford University. She is also a Professor of Environmental Earth System Science and a Professor (by courtesy) of Economics at Stanford University.

Amy J. Pickering is a Research Associate at the Stanford Woods Institute for the Environment at Stanford University, and an affiliated researcher at the International Center for Diarrheal Diseases Research in Bangladesh.

Scott Rozelle is the Helen F. Farnsworth Senior Fellow, and co-Director and Principal Investigator of the Rural Education Action Program (REAP) within the Center on Food Security and the Environment at the Freeman Spogli Institute for International Studies at Stanford University.

Ximena Rueda is an Assistant Professor of Geography at the Universidad de los Andes in Colombia and a Visiting Scholar at the Center on Food Security and the Environment at Stanford University.

Whitney L. Smith is a visiting Research Fellow at the Center on Food Security and the Environment within the Freeman Spogli Institute for International

Studies and the Stanford Woods Institute for the Environment at Stanford University.

Stephen John Stedman is a Senior Fellow at the Center on Democracy, Development, and the Rule of Law (CDDRL), a Professor of Political Science (by courtesy), and an affiliated faculty member at the Center for International Security and Cooperation and the Center on Food Security and the Environment at the Freeman Spogli Institute for International Studies at Stanford University.

Johan Swinnen is a Professor of Development Economics at the University of Leuven (KUL) in Belgium and a Visiting Professor at the Center on Food Security and the Environment at Stanford University.

Barton H. Thompson, Jr. is the Perry L. McCarty Director of the Stanford Woods Institute for the Environment, Robert E. Paradise Professor of Natural Resources Law, and Senior Fellow (by courtesy) at the Freeman Spogli Institute for International Studies at Stanford University.

Peter M. Vitousek is the Clifford G. Morrison Professor in Population and Resource Studies in the Department of Biological Sciences at Stanford University. He is also a Senior Fellow in the Freeman Spogli Institute for International Studies and the Stanford Woods Institute for the Environment, and an affiliated faculty member in the Center on Food Security and the Environment at Stanford.

Xiaobing Wang is a Research Associate in the Center for Chinese Agricultural Policy at the Chinese Academy of Sciences and a member of the research team of the Rural Education Action Program within China (REAP-China), an affiliate of REAP-Stanford and the Center on Food Security and the Environment at Stanford University.

The Evolving Sphere of Food Security

PART ONE

Introduction

1

The Many Faces of Food Security

Rosamond L. Naylor

Hunger is an intensely human experience. Hundreds of millions of men, women, and children go to bed each night with empty stomachs and wake up each morning knowing that food will be scarce, if it can be found at all. For many families, chronic hunger means the disgrace of eating spoiled or stolen food, or the despair of battling relentless cycles of infectious disease. Other forms of malnutrition are more subtle but equally damaging to the quality of life; they often entail persistent lethargy, poor concentration, low achievement in school and at work, or long-term physical and mental disabilities. Most professionals who have spent their careers in the food and agricultural fields have seen the devastating effects of cyclones, floods, droughts, and heat waves on rural communities, or the nutritional and economic impacts of a sudden drop in crop yields from agricultural pests or disease outbreaks. They have observed the unrest in urban areas when food prices spike, and the humiliation of families who have lost their jobs or their land and have been forced into refugee camps, soup kitchens, or welfare programs. Many policy analysts have also witnessed politicians wrestling over subsidy programs related to food and agriculture, often (but not always) with humanitarian intentions. Such events occur in every country of the world, and they often go unnoticed by the general public.

The primary objective of this volume, therefore, is to explore the many faces and facets of food insecurity—their symptoms, their roots, and their possible remedies—through the lens of professionals who have worked on these issues from a wide variety of disciplinary angles.[1] There are two main perspectives that distinguish this book from other volumes on food security. The first is that food security, in its broadest form, is tied to security of many other kinds: energy, water, health, climate, the environment, and national security (Figure 1.1). As a result, we take our analysis beyond narrow definitions and

[1] This book serves as a companion to a more conventional volume on food security in the twenty-first century based largely on economic development analyses. See Falcon and Naylor (2014).

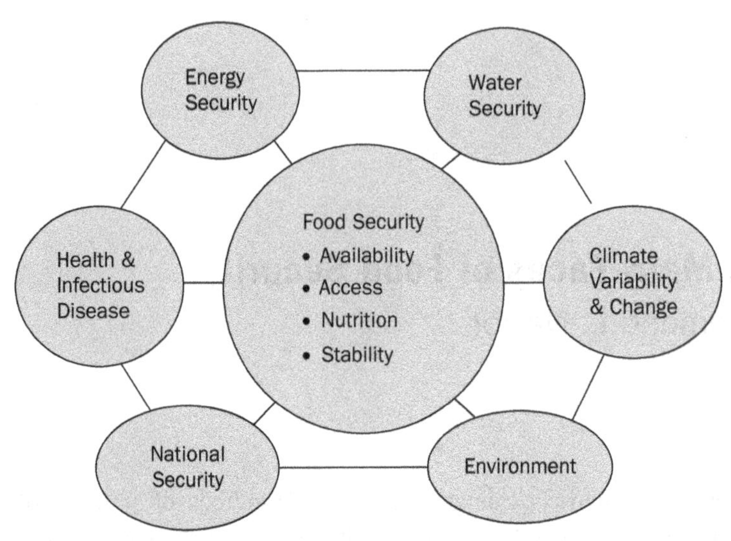

FIGURE 1.1 Connecting the Dots Between Food, Energy, Water, Climate, Environment, Health, and National Security

seek strategies to improve food security that look beyond farmers' fields, and beyond the traditional domains of agriculture. Enabling the production and distribution of adequate food is crucial, of course. But solving food insecurity in various locations may also involve investing in renewable energy, adapting to climate change, managing HIV/AIDS, creating viable water markets, or establishing an equitable system of property rights for land.

Examples of solutions that are complementary to, but not directly within, agriculture's realm can be found all over the world. I recently visited the International Crops Research Institute for the Semi-Arid Tropics (ICRISAT) in Niamey, Niger, with a research group from Stanford (see Chapter 6 by Burney). As we drove along a road overlooking the Niger River, we passed village after village with dry fields, emaciated goats, and children standing naked with the distended bellies typical of *kwashiorkor* (severe protein deficiency). We were told that malnutrition associated with inadequate diets, limited access to water, and diarrheal diseases kills at least one in seven children under the age of five in this region. Although a wide river was flowing nearby, it was clear that no human or other forms of energy were available to lift the water to the farm plots. The temperature was 46C (115F) that day in early March, prior to the main growing season in the Sahel, along the Sahara Desert's southern boundary. It would likely cool down to 34C when the rains finally arrived, but climate projections suggest that later this century growing season temperatures in the Sahel will exceed even the warmest seasons on record (Battisti and Naylor 2009). As the region warms, soil moisture will also decline. In order to improve food security in Niger, therefore, there is an urgent need for investments in decentralized energy systems that can support household water availability, small-scale irrigation, and diversified cropping systems—particularly in the face of global climate change (Burney et al. 2013).

The U.S. corn economy presents a very different illustration. Almost everyone who has lived or worked in America's Corn Belt is familiar with the coffee shops where farmers gather to discuss the two main issues typically on their minds: prices and weather.[2] Both have been highly variable during the past decade, so the conversation has been lively. Corn plays an oversized role in the U.S. agricultural economy: it occupies the greatest harvested area; it is the country's leading export crop; and it receives the largest value of farm subsidies (see Chapter 4 by Cuéllar and colleagues). Corn is used in the production of multiple products, including various foods, livestock feeds, sweeteners (high-fructose corn syrup), and more recently, ethanol. Ethanol production in the United States was promoted heavily through Energy Policy Act of 2005 and the Energy Independence and Security Act of 2007, with the goals of improving domestic energy security (by weaning the United States off foreign crude oil sources), strengthening national security (by reducing dependency on rogue governments for oil to fuel the domestic economy), mitigating climate change (by advancing renewable fuels that have lower greenhouse gas emissions than crude oil), and revitalizing the rural economy through greater demand for agricultural products. By 2010, ethanol accounted for over 40 percent of domestic corn use—exceeding the amount used in livestock feeds for the first time on record—and many observers became increasingly worried about the effect of U.S. biofuel policies on international grain prices and food security.[3] The connections between biofuels, climate, agriculture, and food security are discussed in greater detail later in this volume (see Chapters 4 and 9). The main point here is that energy policy, not agricultural policy per se, was largely responsible for the marked price volatility in global grain markets during the half-dozen years spanning 2006 to 2012 (Figure 1.2). The impacts of biofuel-influenced price spikes have been felt around the world, and have been particularly detrimental for less-developed, food-importing countries. In Guatemala, for example, the poorest households are often farmers and yet still net purchasers of corn. Higher global corn prices translate into less food for home consumption, or less disposable income for other needs such as health care, education, or clothing.

The U.S. example underscores the second distinguishing perspective of this volume: Because the world food system is tightly linked, major changes in agricultural production, consumption, and trade in one area typically have important spillover effects for food security elsewhere. As a result, policy actions in one country or region can impact food security far away, and for

---

[2] For some local flavor of the U.S. Corn Belt, see the chronicles by Walter Falcon at: http://foodsecurity.stanford.edu/news/stuck_in_the_mud_stanfords_scholarly_farmer_on_the_soggy_fortunes_of_midwest_growers_20130716/.

[3] As reviewed in Naylor and Falcon (2011) and Naylor (2014).

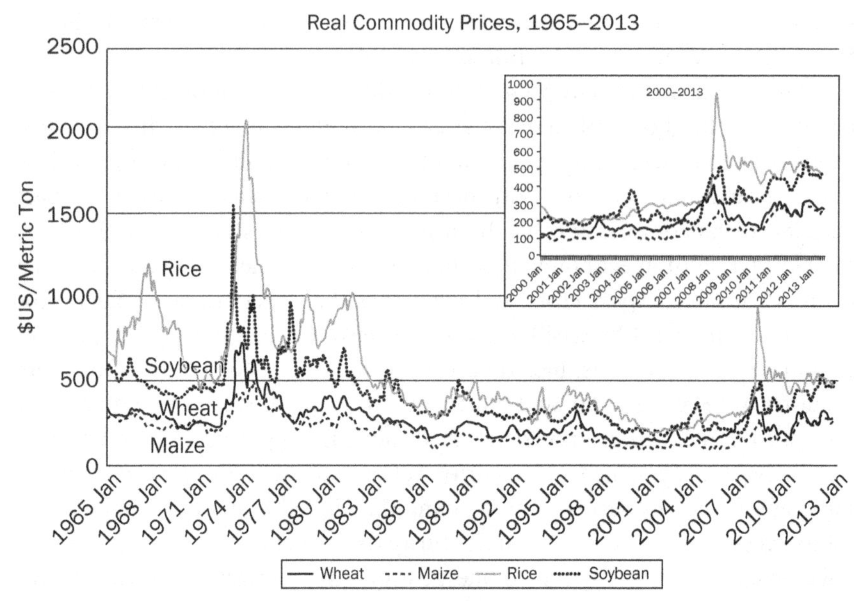

FIGURE 1.2 Historical Pattern of Agricultural Prices Nominal prices deflated using US GDP deflator.
Source: IMF e-Library, accessed 9 August 2013.

years or decades to come. Students who focus on one country at one point in time will therefore gain only a partial understanding of the dynamics of global food security. There are almost always historical precedents to national food and agricultural policies (as discussed in Chapters 2 through 5), or to domestic laws and institutions governing land (Chapter 8) or water (Chapter 11). Understanding the food situation anywhere in the world means understanding the historical development of these policies and institutions, and their spillovers in time and space.

Looking around the globe, agricultural trade causes shocks in crop production or consumption in one location to reverberate across regions and sectors. The effect is further propagated by substitutions in food, feed, and fuel markets. For instance, in 2010 an extreme heat wave decimated wheat yields in one of Russia's main grain belts, causing global wheat prices to spike.[4] Grain yields in parts of Eastern Europe were also damaged by severe heat and drought. Prices continued to rise when Russia, which a year earlier had supplied 17 percent of the global wheat market, announced a one-year ban on wheat exports in order to secure enough grain at low prices for domestic food and feed use. Russia's policy, in response to the devastating climate event, also caused prices for corn (which substitutes in feed) and rice (which substitutes in food) to rise in international markets (Figure 1.2).

---

[4] For a short review of this event, see Naylor (2011).

Economic and nutritional spillovers from food and agricultural policies often occur in both time and space. As discussed extensively in Chapters 4 and 5, agricultural policies implemented in the United States and European Union decades ago in the twentieth century were designed to support grain production and rural incomes at a time when agriculture contributed significantly to those regions' economic output and employment. These policies inadvertently led to the build-up of large grain surpluses and to the decline of world grain prices (on trend) until the post-2006 price spikes (Figure 1.2). Low prices in world markets made it difficult for developing countries to compete internationally in grain exports, and many governments, particularly in sub-Saharan Africa, found it less costly in the short-run to import cheap grain rather than invest in domestic agricultural production. Largely as a result, rural incomes stagnated and food security deteriorated in several of these countries (World Bank 2007). Grain surpluses were also used to support the development of the livestock and high fructose corn syrup industries, which have played an insidious role in rising obesity rates, especially in the United States. Such policies are currently under active contention, and the world is undergoing major demographic, economic, and climatic shifts. The dynamics of global agricultural prices, dietary preferences, and food security in the coming decades thus remain open to debate.

What Is Food Security?

In its most basic form, "food security" means having adequate supplies of affordable food throughout the year to ensure a healthy and productive life (see Glossary). It can be measured at the household level, with a focus on the distribution of food and work effort within households, or at community, regional, national, and global scales. The concept of food security encompasses four key elements: availability (a physical measure of food supplies), access (an economic measure of the income needed to purchase food), utilization (a nutritional measure of calorie, protein, and micro-nutrient intake and absorption), and stability (measures through time of variations in food prices, supplies or access arising from adverse weather, pests and pathogens, or political and economic instability). Because food security has many dimensions, it can be an elusive goal—challenging enough to measure, let alone achieve, particularly for poor households and less-developed countries.[5]

On the global scale, farmers produce plenty of food to feed the world's seven billion-plus people. Total production of grains more than doubled during the past half-century, from less than 1 billion metric tons (mt) in the mid-1960s to over 2.3 billion mt in 2012 (FAO, 2013a). Global supplies of

---

[5] For an excellent discussion of food security measurements, see Headey and Ecker (2013); see also FAO/WFP/IFAD 2012.

other agricultural commodities, such as vegetables, fruits, and oilseed crops, also increased during this time period. The growth in grain output alone has allowed the average availability of calories per capita to rise, despite a more than doubling of the world's human population. As a measure of food security, however, this calculation is flawed.[6] It assumes that calories are all that matter; that physical distribution and purchasing power are sufficient; that crop losses are minimal; that everyone consumes grains directly as food instead of consuming them indirectly through meat, dairy, and aquaculture products; and that food crops are used for human consumption and not for transportation fuel or other industrial purposes. In reality, not one of these assumptions holds. Food insecurity lies in the gaps between "what could be" and "what is."And "what could be"—the potential of agricultural systems to feed the world with adequate nutrition—will become increasingly constrained during the twenty-first century by global climate change and natural resource scarcity as the human population expands to eight, nine, or even ten billion.

For now, however, the biggest food security problem is economic access: most people living day-in and day-out with hunger are forced to do so because they cannot afford adequate diets. Economic access to food is typically measured by calculating the income needed to purchase sufficient calories for an active working life (assuming a given share of income spent on food), and then determining how many people fall above or below that income threshold (FAO/WFP/IFAD 2012). Based on measurements of extreme poverty and near poverty, it is reasonable to conclude that fully one-third of the world's population lives in chronic hunger or is highly vulnerable to food insecurity if their incomes fall or if food prices rise (Box 1.1).

The concept of economic access to food is complicated by the fact that most rural households are both producers and consumers of food, and for the lowest-income bracket, food often represents a major expense, as well as a critical source of income. The majority of the world's rural poor depend on crop or livestock production for part or all of their livelihoods, and a large share of their assets—human and non-human (e.g., land)—are tied up in agriculture. In most sub-Saharan African countries, over half of the economically active population earns its income from agriculture, and the share is closer to 90 percent for some countries, such as Niger, Rwanda, and Burkina Faso (UNDP 2012). In these countries, food accounts for roughly three-quarters of household

---

[6] Calorie availability is commonly calculated at the national scale from the United Nations Food and Agricultural Organization (FAO) Food Balance Sheet series; the data show per capita dietary energy supply (DES), which indicates the number of calories available for direct consumption based on total crop production accounting for (net) trade, minus waste and storage (FAO 2013b). The FAO estimated that 870 million people were undernourished based on measures of dietary energy supply during the 2010–2012 period. For further detail on measurements, see FAO/WFP/IFAD (2012).

BOX 1.1

Poverty and Economic Access to Food[1]

Poverty measurements are typically used to calculate economic access to food. In 2010, the World Bank used purchasing power parity (PPP) metrics of $1.25/day per capita (cap) and $2.00/cap/day to gauge the extent of extreme poverty and near-poverty, respectively, around the world.[2] The extreme poverty threshold of $1.25/cap/day was set according to the mean national poverty line for the world's poorest 20 countries. The near poverty threshold of $2/cap/day represented the mean national poverty line for all developing countries.

The good news is that global rates of extreme poverty fell dramatically over the previous three decades, from 52 percent in 1981 to 20 percent in 2010. Despite these gains, however, 1.2 billion people still lived in extreme poverty in 2010—over one-fifth of the population living in the developing world at that time. The rate of extreme poverty in sub-Saharan Africa fell below 50 percent for the first time in 2008, but remained stubbornly high at 48.5 percent in 2010. Based on these recent trends, it is likely that sub-Saharan Africa will have a disproportionately high rate of extreme poverty in 2015.

As the extreme poverty rate dropped, the number of people living in near-poverty (between $1.25/day and $2/day) increased by over 80 percent during this same 30-year period—from 650 million in 1981 to 1.2 billion in 2010. That means that in 2010, almost 35 percent of the people in the world lived on $2 or less per day (compared to an average of nearly $130/day in the United States). *In other words, in 2010 one-third of the world's population lived in chronic hunger or remained highly vulnerable to food insecurity if their incomes were to fall or if food prices were to rise.*

[1] The data used in this box were taken from: The World Bank update on poverty statistics: http://web.worldbank.org/WBSITE/EXTERNAL/TOPICS/EXTPOVERTY/EXTPA/0 ,,contentMDK:20040961—menuP K:435040—pagePK:148956—piPK:216618—theSitePK:430367—isCURL:Y,00.html; the United Nations Population Division on population figures: http://esa.un.org/wpp/unpp/panel population.htm.; and the International Monetary Fund on U.S. per capita income: http://www.imf.org/external/pubs/ft/weo/2010/02/weodata/index.aspx. All web links accessed August 7, 2013. As of May 2014 when this volume went to press, the World Bank was considering a significant increase in its global poverty line to reflect the rapid expansion of developing country economies and associated changes in purchasing power parity (PPP).

[2] The PPP is used to compare purchasing power across countries based on adjustments in foreign exchange rates and relative rates of inflation—essentially, it is a tool to compare apples to apples (not apples to oranges).

expenses for the poorest fifth of the population.[7] As a result, economic access to food for the rural poor depends on earnings from agriculture and on the cost of food, both of which can vary widely (not always in the same direction) with crop failures, shifts in demand, and changes in land or labor productivity.

Stability is therefore an extremely important dimension of food security, but represented poorly by most metrics (Headey and Ecker 2013). The lack of stability in food production and prices can be especially harmful to households living in poverty or at the brink of poverty. A sudden jump in food prices

[7] The proportion of income spent on food varies widely among poor households throughout the world (Banerjee and Duflow 2008, 2011). Poor households may spend a sizeable share of their income on ceremonies (e.g., weddings, funerals) and other non-food items, even when the nutritional needs of all family members are not fully met. More typically, household expenditures cover basic calorie needs and vary more widely with respect to other food and non-food expenditures depending on income constraints.

or drop in agricultural production due to erratic market conditions, extreme weather, or pest or disease infestations often translates into fewer meals per week or smaller portions per meal, with the brunt of the cuts commonly faced by women and girls in the household.

Food availability and economic access are both measures of potential calorie intake. Calories are only part of the story, however. The nutritional aspect—*what* people eat beyond the calories contained in food—is arguably more important for the long-run physical and mental health potential of the population in question (as reviewed by Leathers and Foster 2009). Severe malnutrition can lead to *wasting* (low weight-for-height, often associated with loss of fat and muscle tissue caused by acute starvation) and *stunting* (low height-for-age, often leading to permanent physical and mental underdevelopment as a result of chronic hunger). Childhood malnutrition is ultimately responsible for about one-third of all deaths of children under the age of five (WHO 2012). In 2010, an estimated 100 million children were underweight (weight-for-age)—a condition commonly considered to be the largest risk factor associated with infectious disease in the developing world (FAO/WFP/IFAD 2012).

The most basic forms of malnutrition are calorie and protein deficiencies, which are generally solved by increasing the amount of food that people eat. But infants and young children have relatively high protein requirements for growth, which cannot be met by the relatively protein-poor staple grains alone—small children simply can't consume the bulk that would be needed even if it is available. Similarly, a diet of starchy staples (yams, sweet potatoes, cassava) may provide sufficient calories but often results in severe protein deficiencies known as *kwashiorkor* and *marasmus* that cause stunting. Pregnant and lactating women also have higher protein needs, and if they lack sufficient protein, stunting can occur in fetuses and nursing infants.[8]

The difference between sufficient calories and adequate nutrition became clear to me during my first field research visit to Central Java, Indonesia, in 1985. Indonesia had made exceptional progress in promoting agricultural development and rural income growth during the previous 15 years, as discussed by Falcon in Chapter 2. The growth in rice productivity was impressive, and rural poverty rates had fallen dramatically. Indonesia had transitioned from being the world's largest importer of rice to being an occasional exporter. During this period of relative prosperity, I visited the home of a farmer I had interviewed earlier that day. She lived with her elderly parents, her husband, and her two children in a one-room house with dirt floors. Although the family had a small sewing business and owned a quarter-hectare of land in a booming rural area of Java, they appeared to live on the brink of extreme poverty. As I sat at their table, I noticed her 14-year-old son eating on a bench outside.

[8] For a thorough review of maternal and child malnutrition causes and treatments, see *The Lancet* 2013 series: http://www.thelancet.com/series/maternal-and-child-nutrition. Accessed August 7, 2013.

His frail form made him appear to be closer to 10 years of age, and he ate his large bowl of rice listlessly, slightly slumped over. I asked if he was sick, and his mother replied that he was fine, just tired, as he was on most days. The boy was visibly malnourished, probably from lack of protein and micronutrients—starchy calories alone were not sufficient for his development and hard work in the field. He was not one of those skeletal forms that appear on the six o'clock news when a serious famine erupts someplace in the world. Yet the sight of this 14-year-old boy, eating so slowly and absently, has never left my mind.

Micronutrient deficiency, often referred to as "hidden hunger," afflicts roughly one-third of the global population (FAO/WFP/IFAD 2012). It is generally caused by a combination of low diet diversity and lack of micronutrient and mineral supplements or enriched foodstuffs. Although many micronutrients are required in the diet, there are three main deficiencies worldwide (Leathers and Foster 2009). Vitamin A deficiency is associated with night blindness and increased mortality from respiratory and gastrointestinal disease. Iodine deficiency is associated with goiter and a reduction in mental abilities. Iron deficiency is associated with anemia that leads to a decline in work productivity, diminished ability to concentrate and learn, increased susceptibility to infection, and greater risk of death during pregnancy and childbirth. Together, these three micronutrient deficiencies impact 2 to 3 billion people worldwide, and result in illness, reduced learning ability, lost productivity, and early death. As summarized in the *2012 State of Food Insecurity* report, hidden hunger inevitably leads to "a tragic loss of human potential" (FAO/WFP/IFAD 2012, p. 23).

In the lexicon of food security, hidden hunger comprises one element of utilization, or the intake of nutrition. Another component is "secondary malnutrition," the inability to digest or absorb nutrients due to diarrhea, respiratory illnesses, intestinal parasites, measles, HIV/AIDS, and other diseases. Like hidden hunger, it can lead to long-term impairment of physical and cognitive development. Tackling secondary malnutrition is a difficult task, because it often involves addressing not just diet deficiencies, but also infectious disease, sanitation problems, and access to clean drinking water.

A final category of malnutrition that affects rich and poor countries alike is over-nutrition. Globally, the number of overweight people—estimated at over 1.4 billion in 2012—exceeds the number of calorie-deficient people (FAO/WFP/IFAD 2012). Health problems related to overweight and obesity, such as diabetes, hypertension, and heart disease, are more widely seen in high-income countries and result from diets rich in calories, saturated fats, salt, and sugar. Even the poorest countries in sub-Saharan Africa and South Asia face rising obesity problems, however, with urbanization, more sedentary lifestyles, and consumption of processed foods that are high in calories and low in nutrients. The United Nations Development Program (UNDP) reports that in a survey of recent mothers in 31 sub-Saharan African countries since 2000, more women were overweight than underweight (UNDP 2012). In fourteen of these countries, over one-fifth of the women surveyed were overweight and more than 5 percent were obese. The connection between undernourishment and obesity is now considered by many experts to be a life-cycle phenomenon (FAO/WFP/IFAD 2012; Black et al. 2013). Children who are born undernourished and remain so, from the womb through infancy, have a greater propensity to become obese if fed high-calorie diets later in life than do children who have not suffered from early malnutrition. How this "nutrition transition" will play out in the developing world in future decades remains unclear. Nonetheless, the double burden of under- and over-nutrition underscores a delicate balance—between inadequacy and excess, and between quality and bulk.

Connecting the Dots

Societies at all stages of development pay a large and enduring price for malnutrition in any of its forms. The resulting disease burdens, loss of human productivity, and long-term physical and mental disabilities place heavy fiscal burdens on governments and often perpetuate income inequalities within a country's population. At the same time, policies and programs aimed at improving farm incomes, agricultural productivity, or nutrition have their own set of costs. They can be expensive, taxing on the environment and natural resources, and biased toward wealthier producers or regions that have more political influence. The path that individual countries choose to follow while meeting basic food needs—as well as satisfying demands for high-value agricultural products and crop-based biofuels—raises two key questions with respect to global food security. First, how do the priorities and challenges of achieving food security change over time as countries develop economically? And second, how do the policies used to promote food security in one country affect nutrition, food access, natural resources, and national security in other countries? This volume establishes a framework for answering these critical questions by tracing four key areas of the food security field: (1) the political economy of food and agriculture; (2) challenges for the poorest billion; (3) agriculture's dependence on resources and the environment; and (4) food security in a national and international security context. A major objective of this book is to connect these areas in a way that tells an integrated story about human lives, resource use, and the policy process—a story about global food security.

THE POLITICAL ECONOMY OF FOOD AND AGRICULTURE

Agricultural policy, when viewed from a global perspective, presents a basic paradox. In industrialized countries, where agriculture typically comprises less than 5 percent of GDP and national employment, government support for food and agriculture tends to be extremely high (e.g., in the form of producer subsidies, crop insurance schemes, nutrition support programs, and trade protection). By contrast, in less-developed countries, where agriculture often comprises over one-third of GDP and engages more than half of the labor force, the agricultural sector is often taxed rather than subsidized, either directly (e.g., export taxes) or indirectly (e.g., via exchange rate or interest rate management). This paradox is easy to understand in terms of the financial resources available to governments: rich countries can afford producer subsidy programs and safety nets for consumers, while less-developed countries have few options but to tax agriculture in order to earn public sector revenues. But basic rationality suggests that it would make more sense for governments in industrialized countries to shift their financial support to other sectors of the economy that are more important for GDP and employment. Similarly, it appears risky for less-developed countries to create disincentives for agricultural producers when food and agriculture are so crucial for economic growth, employment, and nutrition. Resolving this food policy paradox requires a sense of balance by countries at varying stages of development—a balance between supporting the agricultural sector and letting it decline. And few countries with strong farm economies seem very willing to permit the latter to occur.

In virtually all cases, policymakers keep a close eye on their political supporters so they can retain power and accomplish their broader goals. In many less-developed countries, urban constituents have a stronger political voice than do farmers—and when food prices spike, it is usually the urban working class, not the rural poor, who take to the streets in protest. Once in the streets, the discontent of protesters often spreads, sometimes resulting in short-term appeasement, policy changes, or even the ousting of governments, as was the case in Tunisia in 2011, setting in motion the "Arab Spring."[9] In richer countries, farmers, and the agribusinesses that have developed around them, typically demonstrate more political clout than in poorer countries. Agriculture's paradoxically powerful role in wealthy nations has its origins in earlier periods when the sector was more important for employment, incomes, and economic growth, and is actively protected by vested interests. Maintaining strong support for agriculture and agribusiness is made easier politically when it is tied to government spending on nutrition programs and other consumer safety nets, as it is in the United States and European Union (discussed further in Chapters 4 and 5).

Although wide variation exists in the policies and politics of countries throughout the world, virtually all nations undergo a similar process of structural transformation as they move from less-developed to more-developed

[9] Thomas Friedman provides an interesting perspective on the connections between food prices, political unrest in the Middle East, and climate change: http://www.nytimes.com/2012/04/08/opinion/sunday/friedman-the-other-arab-spring.html. Accessed August 9, 2013.

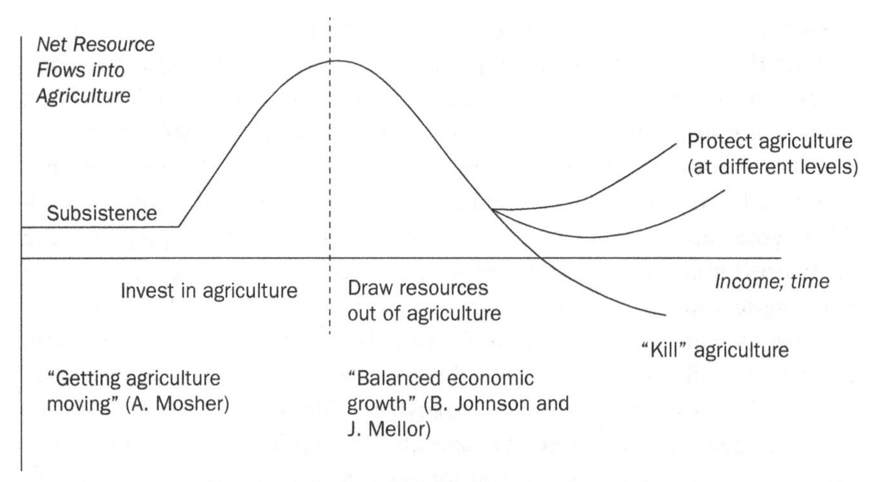

FIGURE 1.3 Structural Transformation and the Political Economy of Agricultural Development

status. This process is defined by the declining relative role of agriculture in employment and GDP as their economies expand.[10] There are three main stages of structural transformation that characterize the political economy of agricultural development, as illustrated in Figure 1.3.

The first stage is what Arthur Mosher, one of the world's most famous agricultural development specialists of the twentieth century, described as "getting agriculture moving" (Mosher 1966). The idea here is that in order to develop an agricultural surplus that will foster increased rural incomes, economic growth, and food security, poor agrarian economies need to invest in agriculture—specifically in appropriate technology (e.g., seeds and capital equipment that are in line with their natural and human resources), as well as in infrastructure, farm inputs, and supply chains. From the 1960s into the mid-1980s, widespread adoption of Green Revolution technologies—especially high-yielding seed varieties that performed particularly well with fertilizer supplements and irrigation—occurred in many parts of Asia and Latin America (Evenson and Gollin 2003; Pingali 2014). The seed varieties, developed mainly for wheat and rice, were scale neutral (meaning they could produce high yields when grown on either small or large plots of land) and were made available through public agricultural research institutions. High-yielding seed technologies were also introduced for maize, primarily in the form of hybrid seeds that were controlled largely by the private sector. The process of agricultural intensification via the Green Revolution resulted in a tripling of grain production in Asia and Latin America between 1965 and 1985 (FAO 2013a).

---

[10] Robust empirical measures of structural transformation are seen in time series data as individual countries become richer, and in cross-country comparisons between low-, middle-, and high-income countries at any point in time. For a deeper discussion on structural transformation, see the papers by O. Badiane and C. P. Timmer in Falcon and Naylor (2014).

According to Mosher's theory, and to development theory more generally, widespread investment in agricultural technology and irrigation infrastructure boosts the demand for unskilled labor and improves labor productivity, leading to significant growth in rural wages and incomes. This theory has played out well in practice. Significant growth in agricultural productivity enabled two of world's most populous countries, China and Indonesia, to reduce their domestic poverty rates from around 80 percent in the 1960s to less than 15 percent in the 1980s (World Bank 2007). Together, their successes caused global poverty numbers to fall precipitously during the 20-year period. Walter Falcon describes Indonesia's experience in Chapter 2, drawing on his long-term involvement as a policy advisor in the country. In the late 1960s, when Falcon began his work, Indonesia was ending a prolonged period of civil conflict. It had among the world's highest rates of population growth and poverty, despite having ample natural and mineral resources. Falcon discusses the types of policy decisions, trade-offs, and leadership that were needed to achieve a sustained process of rural economic growth in Indonesia, including the support of family planning services by a largely Muslim government. Had an aggressive agricultural and rural development strategy not been pursued, a large share of its population might have remained among the world's poorest people today.

The second stage of structural transformation occurs when sustained agricultural growth is able to contribute to broader economic development—a period often referred to as one of "balanced economic growth," following the early development theory of Johnson and Mellor (1961). During this stage, the expansion in agriculture and non-agricultural sectors is supported by surplus food production, increased land and labor productivity, rising rural incomes, growth in demand for non-agricultural goods and services, and higher exports and savings. Investments in agriculture remain positive and serve as an engine of growth to the rest of the economy, even as labor and savings are drawn out of agriculture and into manufacturing and services. China offers one of the best examples of this middle stage of structural transformation: it has achieved phenomenal growth in agricultural land and labor productivity since the early 1970s; surplus food production has largely supported the calorie requirements of its massive population (over 1.3 billion in 2013); sustained economic growth (as measured in GDP) has surpassed all other countries in recent decades; and it has experienced the largest human migration—from rural to urban areas—ever witnessed on the face of the earth.

This development process, while laudable in many respects, also has its drawbacks and limitations. In China, the income gap between rural and urban sectors has expanded, land and water resources have become increasingly constrained, environmental damage from agricultural and non-agricultural activities has escalated, and food insecurity remains a serious problem in many rural areas. In Chapter 3, Scott Rozelle and his colleagues discuss how the food security challenges faced by China have changed during the course of its economic development. Having achieved sufficient calorie availability for most of its population, the country now confronts a deeper problem of micronutrient deficiencies, which may be compromising its rural education system and its ability to move into a more advanced stage of economic development.

Rozelle's insights from the field, where he and his colleagues have conducted more rural surveys than any other development experts in China or elsewhere on record, bring the problem of hidden hunger to the fore. Anyone who spends time in the Chinese countryside knows that agricultural villages are disproportionately populated by elderly adults and young children—grandparents and their grandchildren, essentially. Many middle-age parents work in the cities, and the older children are often sent to boarding schools for secondary education. During Rozelle's earlier visits to these schools, he was struck by the fact that the students would stay inside and nap during recess. Anemia was endemic in the schools, particularly in the western provinces, and it was sapping the students' energy and seriously constraining their ability to learn. The critical issue for China in the twenty-first century is how to manage its increasing income disparities, which are likely made worse by hidden hunger. This issue is not only a threat to food security—it is also a threat to the country's social order, and hence to its national security. Other large emerging economies, such as India, Brazil, and South Africa, face similar problems as they grapple with enduring hunger and poverty in the face of rising wealth.

The importance of agriculture for food availability, economic access to food, and nutritional intake in all countries justifies the large investments that most governments make in the agriculture and food sectors. These investments often represent major sunk costs, as seen, for example, in the construction of irrigation infrastructure, the conversion of land from swamps to fields, and the establishment of agricultural universities, research institutions, and extension services. The success of these investments supports growth throughout the economy over time, but also leads to a decline in the relative value of agriculture compared to manufacturing and services, because household spending on food versus non-food items tends to fall as incomes rise.[11] Rising food supplies stemming from agricultural investments, matched with a relative decline in the demand for food, leads to surplus crop production and falling farm prices, unless countries can export their agricultural products or use them increasingly in the production of feeds, fructose, fuels, and other industrial products.

These challenges represent the third stage in structural transformation. Countries at this stage have typically invested economically *and* politically in their agricultural sectors, and the political bodies that emerge as a result are rarely willing to give up much support for agriculture. In Chapter 4, Mariano-Florentino Cuéllar and colleagues describe the U.S. experience as it has evolved

[11] This process is also known as Engel's Law, defined by a declining income elasticity of demand for food in aggregate as incomes rise. See Timmer et al. (1983).

from an agrarian to an industrial society. Despite the diminished role that agriculture now plays in the U.S. economy, food and agricultural legislation, particularly through the Farm Bill, remains a dominant priority for the federal government, worth tens of billions of dollars in public spending each year. The authors' collective experience, in locations from the White House to a farmhouse in Springville, Iowa, underscores the point that improving global food security requires more than a focus on changing domestic food and agricultural policies—it also requires changes in the political process itself.

A comparable assessment for the European Union is presented by Johan Swinnen in Chapter 5. The European Union differs dramatically from the United States in that it is comprised of 28 countries, each with its own agricultural goals, different demographic and food security challenges, and varying financial resources to bring to the table. Coming from Belgium—the de facto capital of the EU (Brussels)—and having advised the European Commission on agriculture during the critical period when Eastern Europe joined the European Union, Swinnen describes the dramatic development of the European Union's agricultural policy. What links the United States, the European Union, and other industrialized countries at this stage in structural transformation is their competition in global agricultural markets. There is arguably no issue more contentious in international trade negotiations than agriculture, as discussed in these chapters. The powerful G7 countries fret, in public discussion, about global food security. But most of their policy focus around this topic tends to be on domestic producers and consumers—that is, on the national political economy of food and agriculture. As discussed later in this volume by Stephen Stedman (Chapter 13), the concept of food security as an international security issue has not yet taken hold at the upper reaches of the security establishment or for the G7.

CHALLENGES FOR THE POOREST BILLION

Many practitioners enter the food and agriculture policy arena with a sense of optimism. But no one lasts long without coming to appreciate that failures are at least as common as successes. Failures occur at all scales and for different reasons. For example, the introduction of treadle pumps for human-powered irrigation, targeted toward poor farmers, may be disadopted because they require too much time for the benefit they provide, or because individuals lack the needed strength due to chronic infectious disease. In locations where treadle pumps work, the program may be taken over by more wealthy households in the village who can better afford to pay for the pumps up front (Gugerty and Kremer 2004, 2008). At the national scale, a program designed to employ rural residents to build large-scale irrigation infrastructure in exchange for food (a classic "work-for-food" program) may fail due to competition for funds among different government agencies engaged in the program. Such failures are numerous and often go unreported in the academic literature, whereas successes are almost always well documented.

The Green Revolution represents a development program that has worked extremely well in many countries, particularly in Asia and Latin America, but has had more limited success in sub-Saharan Africa, or "SSA" (Evenson and Gollin 2003). There are several plausible reasons why it has not always worked in SSA. The agricultural landscape is heterogeneous and fractured, making it difficult to scale irrigation, seed, mechanization, and fertilizer technologies. Less than 5 percent of SSA's agricultural land is irrigated, and soils in many areas are old and highly weathered. Civil conflict, corruption, and poor governance have curtailed investments in agricultural research and extension programs, and have led to an inconsistent mix of food and agricultural policies that have largely prevented the emergence of a strong private sector (Jayne et al. 2006, 2014). Without a strong private sector, supply chains for farm inputs and outputs are not sufficient to support growth in rural incomes for many poor smallholders. In addition, rural financial markets, roads, and other transportation infrastructure have not been widely developed. At the household scale, poor water quality and sanitation have contributed to chronic diarrhea and other infectious diseases, reducing labor productivity. The existence of the tsetse fly, which has prevented the use of animal power for plowing in broad regions of the continent, has also constrained agricultural labor productivity growth. Although the economic situation in SSA has improved markedly since the turn of the twenty-first century, persistent poverty remains a problem in many rural areas (UNDP 2012).

Where large rural development programs falter, there is often room for smaller-scale interventions that solve local constraints on water, energy, and food. In Chapter 6, Jennifer Burney describes the vicious cycle of poverty that arises when energy and water are not available for year-round agricultural production. A physicist by training, Burney lived in Benin, West Africa, in the mid-2000s, where she designed and introduced an innovative solar-powered drip irrigation system (requiring no batteries or other purchased inputs for operation) for use in vegetable production by women in poor households.[12] Subsequent economic evaluations by Burney and her colleagues of this technology and other small-scale, distributed irrigation systems revealed significant returns on investment and major improvements in food availability, household nutrition, and economic access to food (Burney et al. 2010, 2013; Burney and Naylor 2011).

---

[12] This solar-powered drip irrigation system was provided to women's cooperatives (representing the poorest households in the targeted villages) for use in small-scale vegetable gardens with help from the Solar Electric Light Fund (SELF) and the International Crops Research Institute for the Semi-Arid Tropics (ICRISAT). For further detail see http://foodsecurity.stanford.edu/research/agricultural_innovation/. Accessed August 21, 2013.

There are other equally effective interventions that can improve food security, but that lie outside of agriculture's direct domain. Jenna Davis and colleagues explore in Chapter 7 the connections that exist between domestic water use (for drinking and washing), productive water use (for livestock or vegetable crops), the incidence and progression of infectious disease (diarrhea, HIV/AIDS), and food security at the household scale in western Kenya. Based on their diverse field experiences, the authors show how programs to address HIV/AIDS or to improve the quality of drinking water in households may be more useful for solving hunger in certain locations—particularly hunger stemming from secondary malnutrition—than programs to enhance staple crop production.

A development strategy of implementing one intervention at a time, then evaluating its impacts before proceeding further, has a certain appeal given the vast resources wasted on grand schemes that have failed in the past, particularly in SSA.[13] Persistent poverty cannot be eliminated, however, by one-shot interventions that improve the welfare of individuals at the margin but keep them poor (Easterly 2009). Finding some middle ground that enables localized solutions to be scaled up, and that engages extremely poor *and* near-poor households in order to pave the way for upward income mobility, is critical for sustained development and rural income growth (Barrett and Carter 2013).

The debate over small-scale interventions versus transformative solutions is especially contentious when it comes to property rights and the allocation of land to smallholders versus large agricultural producers in sub-Saharan Africa.[14] There has been widespread media focus on land acquisitions in Africa since the mid-2000s. Many observers view these acquisitions as land grabs by wealthy governments (e.g., Saudi Arabia, China) and national elites; others see these deals as constructive investments in agriculture that have been long overdue. Whitney Smith and I explore the tricky issue of land rights and rural development in sub-Saharan Africa in depth in Chapter 8, drawing on our combined expertise in the legal and economics fields. Our analysis highlights

[13] Attacking poverty through marginal interventions, and evaluating each intervention before scaling up to the next level, is a development approach that has been championed by Esther Duflo, Abhijit Banerjee, and their colleagues at the Abdul Latif Jameel Poverty Action Lab (J-PAL). Their approach uses randomized, controlled experiments in selected communities to measure outcomes, which avoids large problems of endogeneity and poor-quality data that typically plague traditional micro- and macro-economic modeling exercises in developing country contexts. For further information, see Banerjee and Duflo (2011) and http://www.povertyactionlab.org/. Accessed August 21, 2013.

[14] Paul Collier, one of the world's most prominent development experts, advocates for a major transition toward large-scale farming and urbanization in sub-Saharan Africa in order to reduce food insecurity. Opposing experts, including Derek Byerlee, maintain that there are benefits in focusing on smallholder production systems. For further insight into this debate, see Falcon and Naylor (2014) and http://foodsecurity.stanford.edu/events/africas_food_systems_in_2030/. Accessed August 21, 2013.

the importance of understanding history and legal institutions. It also emphasizes the point that laws on paper are not necessarily the laws in practice, and that any major transition from smallholder to large-scale agriculture in sub-Saharan Africa is likely to lead to greater income disparities and more severe food insecurity. The struggle over land is specific to each country, but a broader generalization holds: when disparities in income and assets, such as land, are substantial, relieving poverty through broad-based economic growth is a very tough proposition (FAO/WFP/IFAD 2012).

AGRICULTURE'S DEPENDENCE ON RESOURCES AND THE ENVIRONMENT

Virtually all countries that have transitioned from very low-income to middle-income status during the past half-century have experienced agricultural intensification and sustained rural growth for a significant period of time (World Bank 2008). Many of these countries have also developed large-scale dependencies on fertilizer and water inputs to support their staple crop production. The combination of population- and income-driven demand for food and feed has left few options but to intensify agriculture further, and patterns of fertilizer use and irrigation, once adopted, are not easily broken. Intensification also requires high inputs of energy, especially for agricultural inputs (e.g., the manufacturing of nitrogen fertilizers via the Haber-Bosch process) and the pumping of irrigation water. Given its dependence on energy and sensitivity to climate conditions, intensive agriculture is both a contributor to global climate change and a victim of it.

The exercise of tracing the energy-nutrient-land-water-climate linkages in agriculture is anything but a linear process. The chapters in this section therefore present a variety of analytical frameworks for assessing the interactions among resource and environmental variables, and their implications for food security. In Chapter 9, David Lobell and colleagues discuss the roles of energy and climate in agricultural systems in both rich and poor countries. Although energy has always been an important agricultural input, the connections between agriculture and energy markets have become even tighter since the mid-2000s with the expansion of crop-based biofuels. Investments in energy technologies—ranging from biofuels to hydraulic fracturing to renewable wind and solar systems—strongly influence agricultural costs and returns. They also fundamentally determine the future course of global climate change, which in turn affects crop yields worldwide.

Crop yields and food security are examined from a different angle in Chapter 10, where Peter Vitousek and Pamela Matson explore fertilizer use and its environmental consequences, drawing on their work in two of the world's most intensive and highly productive agricultural regions: Sonora, Mexico, and the North China Plain. Higher rates of nutrient inputs and improved fertilizer efficiency have been essential for improving global food security for the past half-century, and also represent a pathway of climate adaptation—if yield growth can stay ahead of climate-induced yield declines, then the potential negative impact of climate change on food security may be reduced. Nutrient uptake by crops has limits, however, and the environmental impacts of nutrient loss from agriculture create important spillovers for food security over space and time. For example, emissions of fertilizer-derived nitrous oxide (N_2O) from intensive agricultural systems contribute to climate change, placing increased stress on crop yields worldwide—from the U.S. Corn Belt to rice paddies in Thailand to wheat fields in Argentina and beyond—for decades to come.

Environmental impacts are not the only factor threatening world food supplies. Resource constraints on food production, especially water, raise the critical question of "food at what price and for whom?" As Barton Thompson discusses in Chapter 11, legal and administrative institutions governing water rights for agriculture are complex in all countries. They often evolve as economies expand, from a communal rights structure at the early stages of development toward a more centralized property rights structure in more advanced economies. An important conclusion of Thompson's work is that poor, smallholder farmers are rarely the beneficiaries of institutional change in water allocations over the course of economic development. Agricultural supplies and food availability may continue to grow as irrigated area expands or as institutions evolve, but income growth among the poor—and hence economic access to food—will worsen if smallholders are not able to intensify their crop production.

Despite efforts to intensify agriculture throughout the world, the expansion of agriculture into previously unfarmed areas still occurs at large scales in areas where land is available and where rainfall or irrigation can support agricultural production.[15] This extensification, as opposed to intensification, is characterized by lower input use per hectare, but it has major implications for ecosystem services that support agricultural production regionally and globally. For example, clearing land in tropical rainforests, where much of agriculture's land-use change occurs, adds to greenhouse gas emissions, affects regional rainfall patterns, and reduces biodiversity, including crop genetic diversity. In Chapter 12, Ximena Rueda and Eric Lambin show that rising incomes and urban demand in one part of the world often leads to large land-use changes in other regions that supply exports of food and feed. The authors reach the sobering conclusion that current patterns of land conversion for agriculture may contribute more to obesity in emerging economies and rich countries than

[15] The global cultivated land base is currently 1.5 billion hectares (ha), and between 1990 and 2005 it grew by 2.7 million ha per year (World Bank 2010). With rising demand for agricultural products, particularly from emerging economies, conservative estimates suggest that 6 million ha of additional land will come into production each year through 2030 (ibid).

to reduced malnutrition in low-income countries—a "lose-lose" proposition for global food security and the environment.

This result does not always hold, however. In Indonesia, the world's fourth most populous country, converting forestland into oil palm plantations—while destructive to native ecosystems and to climate stability—has provided direct employment and incomes for millions of people, many of whom live in near-poverty.[16] "Mouths to feed" and "hands to hire" are especially daunting challenges for countries with large populations or high population growth rates, and raise the contentious issue of policies intended to support decreased family size and lower population growth rates.. If the environmental, resource, and demographic dimensions of economic development are overlooked, however, there is little hope of achieving, or even approaching, global food security over the long term.

FOOD IN A NATIONAL AND INTERNATIONAL SECURITY CONTEXT

Food insecurity is a vast humanitarian issue. One might ask under what circumstances is it more than that—when should food insecurity also be viewed as a national or international security threat? Competition over agricultural land and water, climate impacts on crop production, and food price spikes leading to urban riots, all represent potential security hazards at national or regional scales (Arezki and Bruckner 2011; Hsiang et al. 2013). The availability of phosphorous, critical for intensified agricultural production but limited in supply, can also be considered a security concern, since phosphorous reserves are more concentrated geographically than oil.[17] In addition, income and educational disparities arising from chronic malnutrition can potentially lead to prolonged civil unrest and conflict. All of these worries are valid, but should they be categorized as national or international security issues?

In 2003, the then-Secretary General of the United Nations, Kofi Annan, convened a High Level Panel on "Threats, Challenges, and Change" to examine what the real security threats were to people in both developing and

---

[16] Indonesia's oil palm sector provides income to roughly two million people through direct employment or small shareholder arrangements. It also creates an estimated $3 of gross national income (GNI) for every $1 of oil palm profits, which can potentially support education, health, and other social programs (see Hunt, 2010).

[17] There is no substitute for phosphorous in agriculture, and some experts project that peak phosphorous extraction may be reached by 2030–2035, leaving rising fertilizer demands dependent on a fixed supply (Cordell et al. 2009, 2011; Elser and White 2010). Of the estimated 67 million tons of mineral phosphate rock reserves on the planet in 2012, 75 percent were in Morocco and the Western Sahara, followed distantly by China (5.5 percent), and Algeria, Syria, Jordan, South Africa, Russia, and the United States (each with 3 percent or less) (USGS 2013). See http://minerals.usgs.gov/minerals/pubs/commodity/phosphate_rock/. Accessed September 15, 2013.

developed countries. In Chapter 13, Stephen Stedman, the Panel's Research Director, reports on the diplomatic process and describes how the meaning of the term "security" has evolved since the Cold War era. Defining security is much more than an intellectual exercise. Understanding how food insecurity fits into broader security definitions is important for identifying which government agency or agencies (e.g., U.S. Department of Defense or U.S. Department of State) should be responsible for solving the problem. The definition and jurisdiction of food security in the broader security realm can have serious consequences for a country's budget allocations, diplomatic relationships, social control, or just plain ethics.

Why This Volume, and Why Now?

The rationale for analyzing food security through such a multifaceted lens is clear: given the changeable nature of events in the twenty-first century, no other approach can hope to capture a meaningful view of the problems and potential solutions. If significant gains are to be made in improving food security at global or regional scales, the best minds from many disciplines must be engaged.

From an academic viewpoint, this book is targeted at the numerous interdisciplinary programs on food security and the environment that have sprung up all over the world. Professors and students with a wide range of backgrounds—economics, engineering, business, law, biology, physics, earth and atmospheric sciences, medicine, nutrition, history, anthropology, sociology, philosophy, ethics, and more—are drawn to the topic. They want to learn more and to contribute to solutions, and thus there need to be multiple entry points to the study of food security.

The main reason for writing this book now, however, is more personal. I have never been a traditional academic scholar with a narrow disciplinary focus. I was trained in economics from my undergraduate days through my PhD, but I have always found the most interesting questions at the intersection between economics and other disciplines in the food security and environmental realm. These are not abstract questions. They are questions about people—their struggles and successes—in countries in Asia, Latin America, and Africa where I have conducted field research. They are questions that arise in conversations with colleagues over a beer after a long day of interviews or field measurements. Wanting the real answers to these questions is what inspires this interdisciplinary approach. Attention to the analytical details provides the foundation. Many students have come to my office wanting to know the formula for how to ask important questions or conduct interdisciplinary research on issues related to food security and the environment. My hope is that this book will be their guide.

There are many excellent scholars around the world who could write well on various topics covered in this volume. I selected the authors based on our collaborations in teaching and research over years, if not decades. The authors consist of 19 professors and research scholars closely affiliated with the Center on Food Security and the Environment (FSE) at Stanford University, a center that I have directed since 2006. I also selected a professional editor from Stanford, who teaches science communication in the School of Earth Sciences. Although I have many talented colleagues outside of Stanford, an important objective was to use the process of writing this book as an illustration of what is possible by developing an engaged, interdisciplinary approach within a single institution.

The book is designed to draw out different voices, alternative perspectives, and a wide range of field experiences that speak to the challenges and opportunities for feeding the world adequately in the twenty-first century. Our hope is that through these various disciplinary and interdisciplinary explorations, the field of food security will come alive for students, practitioners, and anyone interested in the future of humanity.

References

Arezki, R. and M. Bruckner, 2011. Food prices and political instability. IMF Working Paper 11/62 (March 1). Accessed August 19, 2013, http://www.imf.org/external/pubs/cat/longres.aspx?sk=24716.0.

Banerjee, A., and E. Duflo. 2011. *Poor Economics: A Radical Rethinking of the Way to Fight Global Poverty*. New York: Public Affairs (a member of the Perseus Books Group).

Banerjee, A. V., and E. Duflo. 2008. What is middle class about the middle classes around the world? *The Journal of Economic Perspectives* 22(2): 3.

Barrett, C. B., and M. R. Carter. 2013. The economics of poverty traps and persistent poverty: policy and empirical implications. *Journal of Development Studies* 49(7): 976–990.

Battisti, D. S., and R. L. Naylor. 2009. Historical warnings of future food insecurity with unprecedented seasonal heat. *Science* 323(5911): 240–244.

Black, R. E., C. G. Victora, S. P Walker, Z. A. Bhutta, P. Christian, M. de Onis, M. Ezzati, S. Grantham-McGregor, J. Katz, R. Martorell, R. Uauy, and the Maternal and Child Nutrition Study Group. 2013. Maternal and child undernutrition and overweight in low-income and middle-income countries. *The Lancet* 382(9890): 427–451.

Burney, J. A., and R. L. Naylor. 2011. Smallholder irrigation as a poverty alleviation tool in Sub-Saharan Africa. *World Development* 40: 110–123.

Burney, J. A., R. L. Naylor, and S. L. Postel. 2013. The case for distributed irrigation as a development priority in sub-Saharan Africa. *Proceedings of the National Academy of Sciences* 110(31): 12513–12517.

Burney, J. A., L. Woltering, M. Burke, R. L. Naylor and D. Pasternak. 2010. Solar-powered drip irrigation enhances food security in the Sudano-Sahel. *Proceedings of the National Academy of Sciences* 107: 1848–1853.

Cordell, D., J. O. Drangert, and S. White. 2009. The story of phosphorus: Global food security and food for thought. *Global Environmental Change* 19(2): 292–305.

Cordell, D., A. Rosemarin, J. J. Schröder, and A. L. Smit. 2011. Towards global phosphorus security: A systems framework for phosphorus recovery and reuse options. *Chemosphere* 84(6): 747–758.

Cordell, D., T. Schmid-Neseta, S. Whiteb, and J. O. Drangerta. 2009. Preferred future phosphorus scenarios: A framework for meeting long-term phosphorus needs for global food demand. In *International Conference on Nutrient Recovery from Wastewater Streams*. http://www.susana.org/lang-en/library?view=ccbktypeitem&type=2&id=1030.

Easterly, W. 2009. Can the West Save Africa? *Journal of Economic Literature* 47(2):373–447.

Elser, J., and S. White. 2010. Peak phosphorus. *Foreign Policy*. Accessed August 30, 2013, http://www.foreignpolicy.com/articles/2010/04/20/peak_phosphorus?page=0,0.

Evenson, R. E., and D. Gollin. 2003. Assessing the impact of the Green Revolution, 1960 to 2000. *Science* 300(5620): 758–762.

———. 2003. Crop genetic improvement in developing countries: Overview and summary in *Crop Variety Improvement and Its Effect on Productivity: The Impact of International Agricultural Research*. Cambridge, MA: CABI.

Falcon, W. P. and R. L. Naylor (eds.). 2014. *Frontiers in Food Policy: Perspectives on sub-Saharan Africa*. Stanford, CA: Printed by CreateSpace.

Food and Agriculture Organization (FAO). 2013a. "FAOSTAT". Accessed August 29, 2013, http://faostat3.fao.org/home/index.html.

———. 2013b. "Food Balance Sheets." Accessed August 12, 2013, http://faostat.fao.org/site/354/default.aspx.

FAO/WFP/IFAD. 2012. The State of Food Insecurity in the World 2012. Economic growth is necessary but not sufficient to accelerate reduction of hunger and malnutrition. Rome: FAO.

Friedman, T. L. 2012. "The Other Arab Spring." *The New York Times*. Accessed August 9, 2013, http://www.nytimes.com/2012/04/08/opinion/sunday/friedman-the-other-arab-spring.html?_r=1&.

Gugerty M. K., and M. Kremer. 2004. The Rockefeller effect: Looking at organization of the disadvantaged in Kenya. Working paper. Accessed August 30, 2013, http://ipl.econ.duke.edu/bread/papers/041604_Conference/RockefellerEffect_04_12_04_BREAD.pdf.

———. (2008) Outside funding and the dynamics of participation in community associations. *American Journal of Political Science* 52: 585–602.

Headey, D. D., and O. Ecker. 2013. Rethinking the measurement of food security: from first principles to best practice. *Food Security* 5(3): 327–343.

Hsiang, S. M., M. Burke, and E. Miguel. 2013. Quantifying the influence of climate on human conflict. *Science* 341(6151): 1235367.

Hunt, C. 2010. The costs of reducing deforestation in Indonesia. *Bulletin of Indonesian Economic Studies* 46(2): 187–192.

International Monetary Fund. 2010. "World Economic Outlook Database". Accessed August 7, 2013, http://www.imf.org/external/pubs/ft/weo/2010/02/weodata/index.aspx.

Jayne, T. S., D. Mather, and E. W. Mghenyi. 2006. Smallholder farming under increasingly difficult circumstances: Policy and public investment priorities for Africa. Food Security International Development Working Paper 54567, Department of Agricultural, Food, and Resource Economics, Michigan State University.

Jayne, T. S., J. Chamberlin, and M. Muyanga. 2014. Emerging land issues in African agriculture: implications for food security and poverty reduction strategies. In Falcon, W. P. and R. L. Naylor (eds.), *Frontiers in Food Policy: Perspectives on sub-Saharan Africa*, 265–308. Stanford, CA: Printed by CreateSpace.

Johnston, B. F., and J. W. Mellor. 1961. The role of agriculture in economic development. *The American Economic Review* 51(4): 566–593.

The Lancet. 2013. Maternal and Child Nutrition. Accessed August 4, 2013, http://www.thelancet.com/series/maternal-and-child-nutrition.

Leathers, H. D., and P. Foster. 2009. *The World Food Problem: Toward Ending Undernutrition in the Third World*. Ann Arbor, MI: Lynne Rienner Publishers.

Mosher, A. T. 1966. *Getting Agriculture Moving: Essentials for Development and Modernization*. Published for the Agricultural Development Council: Praeger.

Naylor, R. L. 2014. Biofuels, Rural Development, and the Changing Nature of Agricultural Demand. In Falcon, W. P. and R. L. Naylor (eds.), *Frontiers in Food Policy: Perspectives on sub-Saharan Africa*. Stanford, CA: Printed by CreateSpace.

Naylor, R. 2011. Expanding the boundaries of agricultural development. *Food Security* 3(2): 233–251.

Naylor, R. L., and W.P. Falcon. 2010. Food security in an era of economic volatility. *Population and Development Review* 36(4): 693–723.

Naylor, R. L., and W. P. Falcon. 2011. The global costs of American ethanol. *The American Interest* 7(2): 66–76.

Naylor, R. L., A. J. Liska, M. B. Burke, W. P. Falcon, J. C. Gaskell, S. D. Rozelle, and K. G. Cassman. 2007. The ripple effect: Biofuels, food security, and the environment. *Environment: Science and Policy for Sustainable Development* 49(9): 30–43.

Pingali, Prabhu. 2014. "The Green Revolution Forty Years Later: Lessons Learned and Unfinished Business." In Falcon, W.P. and R.L. Naylor (eds.), *Frontiers in Food Policy: Perspectives on sub-Saharan Africa*. Stanford, CA: Printed by CreateSpace.

Timmer, C. P, Falcon, W. P. and S. R. Pearson. 1983. *Food Policy Analysis*. Washington, D.C.: The World Bank and Baltimore, MD: The Johns Hopkins University Press.

United Nations Development Programme (UNDP). 2012. Africa Human Development Report 2012: Towards a Food Secure Future. Accessed May 9, 2014, http://www.undp.org/content/undp/en/home/librarypage/hdr/africa-human-development-report-2012/.

United Nations Population Division. 2012. World Population Prospects: The 2012 Revision. Accessed August 7, 2013, http://esa.un.org/wpp/unpp/panel_population.htm.

United States Geological Survey (USGS). 2013. "Phosphate Rock Statistics and Information." Accessed May 9, 2014, http://minerals.usgs.gov/minerals/pubs/commodity/phosphate_rock/.

World Bank. 2013. "Poverty". Accessed August 7, 2013, http://web.worldbank.org/WBSITE/EXTERNAL/TOPICS/EXTPOVERTY/EXTPA/0,,contentMDK:2004

0961~menuPK:435040~pagePK:148956~piPK:216618~theSitePK:430367~isCURL:Y,00.html.

———. 2010. Rising global interest in farmland: Can it yield sustainable and equitable benefits? Washington, D.C.: World Bank.

———. 2007. "World Development Report 2008: Agriculture for Development." Accessed August 29, 2013, http://go.worldbank.org/2DNNMCBGI0.

World Health Organization (WHO). 2012. "World Health Statistics 2012." Accessed August 29, 2013, http://www.who.int/gho/publications/world_health_statistics/2012/en/.

PART TWO

The Political Economy of Food and Agriculture

2

Food Security for the Poorest Billion

POLICY LESSONS FROM INDONESIA

Walter P. Falcon[1]

This chapter is bracketed by two important years in Indonesian history: 1967 and 2008. The early year was the formal beginning of General Soeharto's presidency; the latter date was the year of his death. By sheer coincidence, my first visit to Indonesia, facilitated by the Ford Foundation, was also in 1967. My memories are still vivid: seeing the Jakarta skyline with only four buildings greater than four stories; maneuvering in traffic jams that consisted of a few old wood-paneled cars ensnarled in a sea of rickshaws on what passed for roads; viewing a totally impoverished countryside rampant with hunger; and struggling to get through rural military checkpoints that were disrupting the limited flows of rice and everything else at each county (*kabupaten*) line. My remembrances of 2008, on my last professional visit, are equally clear: seeing a Jakarta skyline that now resembled Manhattan's; being caught in traffic jams, but of the Los Angeles variety; viewing all kinds of construction and small industries in the countryside; and zipping across the rice plain of northern Java, dodging 18-wheelers on an eight-lane highway. The economy-wide change was immense, and this chapter is mostly about the improvements in Indonesia's food security during those 41 years as seen by someone who was actively involved in the process.

---

[1] "Freeman Spogli Institute for International Studies and Woods Institute for the Environment, Stanford University". This chapter is one-third about development theory, one-third Indonesian economic history, and one-third memoir. It owes much to many people; however, they are not responsible for any remaining errors of omission or commission. I am particularly grateful for discussions with C. Peter Timmer and for access to his personal library on Indonesia. The many years of working in Indonesia with him and Scott R. Pearson, and more recently with Rosamond Naylor, have helped to shape my thinking on food security, as did the mentoring I received from three great Indonesian practitioners, all of whom have recently died: Professor Widjojo Nitisastro, Professor Saleh Afiff, and General Bustanil Arifin. I appreciate, as well, the specific comments of Thomas Hayden, William Burke, Tom Fingar, William Fuller, Joanne Gaskell, J. Tomas Hexner, Donald Kennedy, Eric Lambin, David Lobell, Edwin Oyer, Ximena Rueda, and Johan Swinnen.

Hunger in Global and Indonesian Settings

Food security is not assured with the acquisition of adequate calories and protein; however, escaping malnutrition surely starts with these two dietary components.[2] Yet even these bare nutritional necessities remain painfully out of reach for far too many. In 2009, more than one billion people were listed as being food insecure in this most fundamental protein-calorie sense of that term (FAO 2010). At 27 percent, sub-Saharan Africa had the highest percentage of its population living—sometimes barely—with hunger. South and Southeast Asia together had the largest absolute number of individuals (435 million) who are unable to meet their baseline dietary needs (FAO 2011). That one-seventh of the world still goes to bed hungry each night must be one of the most dismal global failures of the twenty-first century.

The past 60 years have seen some of the greatest advances in food production, storage, and distribution technologies in history. How has calorie-protein malnutrition remained so widespread? Low incomes that limit access to food top the list of reasons. Twenty-seven countries in the world had average per capita incomes of less than $700 in 2010 (World Bank 2012). At less than $2 per person per day, it is extraordinarily difficult to maintain adequate diets, even when two-thirds or more of the income goes for food. And average incomes can hide the darker consequences of extreme income inequality in many nations. For example, in Bolivia and Southern Sudan, the poorest 20 percent of the population earn but 2.1 percent and 2.7 percent, respectively, of the country's total income. Even in middle-income countries, such as Brazil and South Africa, people in the poorest income quintile are typically food insecure (FAO 2011; World Bank 2012). The problems of low and unequally distributed incomes interact with inadequate governance, corruption, depleted or non-existent natural resource bases, gender and urban biases, armed conflict, and natural disasters to exacerbate hunger. Haiti, with nearly 60 percent of its population undernourished, is subject to all of these factors, making it an unfortunate poster child for the poorest billion.

Ironically, three-fourths of these poorest billion people live in rural areas and participate directly in agricultural production. Agriculture is obviously the source of food, but only a progressive agriculture can also be the source of additional jobs and rising incomes. This latter point is of critical importance, since the bulk of the poor are net purchasers of food, even if they are farmers. Their plots of land are too small, their yields too low, and their labor productivity too constrained to produce enough food for their typically large families.

Food security is inherently a consumption concept. But food consumption is necessarily intertwined with agricultural production during the early

---

[2] See Naylor (Chapter 1) for various food security definitions. See also Leathers and Foster (2009) for a rigorous, yet readable account of undernutrition in the developing world.

stages of economic development when most of the labor force is in agriculture. In later stages of development, this need not be so.[3] The nature of the food-security problem and its solution are therefore fundamentally different in poor countries than in rich societies, and the task of "getting agriculture moving"[4] is at the heart of improved food security within low-income societies.

The escape from hunger, whenever it has been achieved on a national level, has involved new technology, altered institutions, and greater public and private investments in agriculture. It has also involved sensible policies both for the agricultural sector and for the economy as whole. Indeed, macro policy—interest rates, trade arrangements and tax provisions—often turn out to be more important than food policies per se. Agricultural production is highly decentralized in virtually all countries, since command and control approaches for millions of small producers have never proved workable over the long run. That is why creating a reasonable constellation of macro-economic and agricultural policies turns out to be key for overcoming food insecurity at the early stages of development. These policies must provide the incentives for the productivity increases needed to get agriculture moving, and do so in ways that do not create policy legacies that come back to haunt the country at later stages of development (see Rozelle et al., Chapter 3, and Cuéllar et al., Chapter 4).

This chapter is thus about policy and about resolving the series of food policy questions that always seem to arise during the early stages of development. It is also about Indonesia, now the fourth largest country in the world in terms of population. The choice of country is not accidental. In 1967 Indonesia was among the poorest of all nations. Sixty percent of its population was below the $1 per day poverty line, making it roughly comparable to modern-day Haiti (World Bank 1990; Naylor and Falcon 1995). Indonesia had just emerged from a civil war and a hyperinflation of greater than 1,000 percent annually that had left the government virtually bankrupt. Indeed, conditions were so dire that new theories were developed that used economic dualism (Boeke 1953) and agricultural involution (Geertz 1963) to explain why Indonesia would *always* remain a poor rural society doomed to "shared poverty."[5] With 17,000 islands (Figure 2.1), 700

---

[3] In the United States, for example, the SNAP (food stamp) legislation resides in the "Farm Bill" as a matter of legislative convenience and political coalitions, not because it is dependent on the level of grain production. Declaring that, "you can't feed the hungry by starving the farmer [in the U.S.]", may be great Southern politics; however, the statement fails as a matter of logic. Representative Austin Scott (2012) as quoted in *Politico*. See also Cuéllar et al. Chapter 4.

[4] "Getting agricultural moving" is a phrase (and book) made famous by Arthur Mosher (1966), an important pioneer in the field of agricultural development. It has become a shorthand expression for overcoming all of the constraints associated with increasing productivity within agriculture as part of the early phase of an economy's structural transformation.

[5] Geertz (1963) was especially pessimistic about markets and economic growth as potential solutions to Indonesia's poverty problems. He instead foresaw a system of institutional arrangements whereby income and wealth would be shared so as to maintain very minimal living standards.

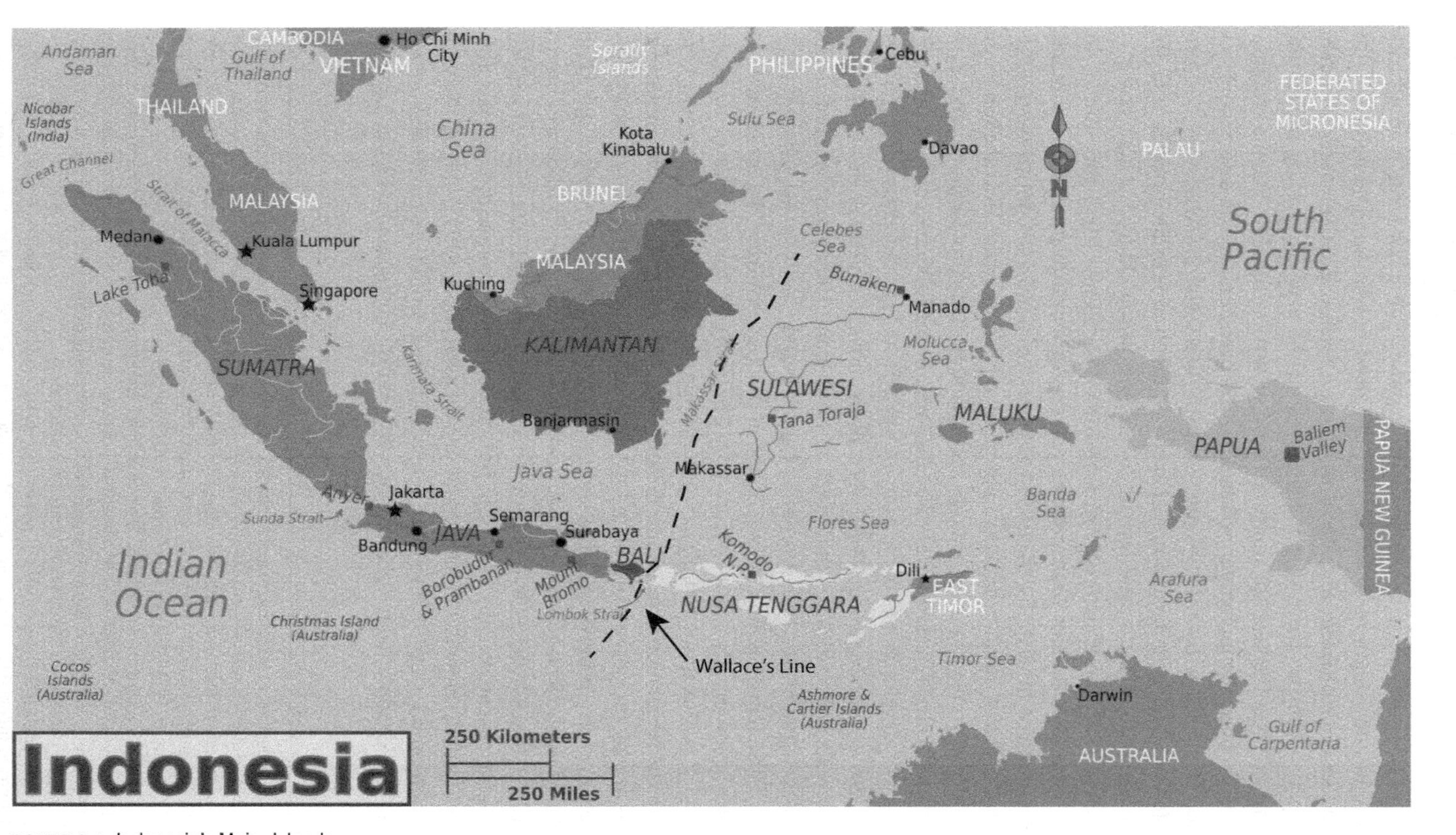

FIGURE 2.1 Indonesia's Major Islands

Map Credit: Fitzgerald 2009, http://commons.wikimedia.org/.

language groups, and a government in total disarray, the outlook for Indonesia in 1967 was incredibly gloomy.

Over the next 40 years, however, Indonesia confounded the experts. Its economy grew rapidly, and by 2009, the country had lowered the percentage of food-insecure persons to only 13 percent (FAO 2011). During those 40 years, Indonesia moved more people out of poverty than any other country level except China and India, and the reduction in Indonesia's percentage of food-insecure population has been substantially more impressive than in India.

Indonesia's progress in food security has been substantial by any set of metrics. However, the record there is far from perfect, and at various stages Indonesia stumbled in its efforts. Examining what worked and what did not and what role policy played in this process are the objectives of this analysis. After setting brief historical, cultural and geographic contexts, I focus on five policy questions that affect food and agriculture; how these questions were resolved in Indonesia, primarily during the Soeharto era; and what lessons Indonesia's answers might provide to other countries.

On a more personal level, I should note that I was an active participant-observer in this process. Between 1967 and 2009, I visited Indonesia more than 100 times, typically spending two months each year in residence. At various times, two colleagues, C. Peter Timmer and Scott R. Pearson, and I worked within the Ministries of Planning, Finance, Trade, and BULOG (*Badan Urusan Logistik*, the national logistics command, which often served as a *de facto* Food Ministry). It is my hope that sharing these experiences will provide insights to policy analysts working in other countries.[6]

Context

History, culture, and geography always matter when it comes to food security. In Indonesia, these three forces came together around rice, which traces its wet-cultivation in Java to the eighth century. Although Indonesia annually produces about 20 million metric tons (mmt) of cassava (Falcon et al. 1984) and 18 mmt of maize (Timmer 1987), its staple food system centers on rice (Pearson et al. 1991; FAO 2012). In the early 1970s alone, rice supplied, on average, 55 percent of the calories and 20 percent of the protein to a predominantly rural population. Much of Indonesia's success story in food security thus centers on how farmers, consumers, the private sector, and the government managed that single commodity.[7]

---

[6] Research, teaching, policy advice, and mentoring of graduate students were key parts of our activities. During that 40-year period, Stanford awarded 10 PhD degrees to students working on various aspects of Indonesia's food system, one of whom is the editor of this volume.

[7] Indonesia, the world's third largest producer, now produces about 60 million metric tons (mmt) of paddy (unmilled rice still containing the husk) annually, equivalent to 40 mmt of milled (white)

HISTORY

Much of economic development is path dependent—where countries end up is importantly related to where, when, and how they get started.[8] Indonesia's history, as with many other developing nations, had a profound effect on how modern-day efforts to reduce hunger actually began. Two wars—World War II and Indonesia's War for Independence—overlapped to create very difficult initial conditions for the new country. The Netherlands had been in Southeast Asia since the seventeenth century, having established Batavia (now Jakarta) in 1619, but it was only in the early twentieth century that the Dutch dominated and controlled the entire Indonesian archipelago. The Japanese occupation of Indonesia during World War II effectively ended Dutch rule with even more intensified levels of exploitation. The occupation was very harsh, and more than 4 million people are estimated to have died in Indonesia during the war as a result of famine and forced labor (Frederick and Worden 1993).

Indonesia declared independence two days after Japan surrendered in August 1945, and a fiery young architect, Soekarno, was named president. The Netherlands attempted to reassert control, however, and it was not until December 1949 that independence was truly achieved. Independence created a peak in nationalism, but it provided little lasting cohesion among the many islands and language groups. Soekarno's "guided democracy"—doctrinaire socialist policies by proclamation—proved disastrous, especially when coupled with a growing animosity between the military and the Communist Party (Hughes 1967). In 1965, an attempted coup was countered by the army, leading to a violent anti-communist purge in which 500,000 people were estimated to have been killed—some for political reasons, some for ethnic reasons, but many for the settlement of long-standing personal animosities in the countryside (Cribb 2002).[9] A Javanese general, Soeharto, maneuvered his way through the disruptions in 1965–67 to become president, a role he maintained until 1998.

Much has been written about Soeharto—his rise to power, his group of "crony capitalist" friends, and his rapacious family. Much less has been written about other characteristics, which were arguably more important for Indonesia's successful development. Two dimensions of his development outlook seem especially important for food security. First, he understood the enormous instability that arose from the country's geographic and ethnic diversity, and that decentralized growth in rural areas was vital for a stable society. But

rice. Transplanted rice in standing water is the dominant cultivation system, but there is considerable regional variation in how rice is grown. Various systems are described in Heytens (1991).

[8] Paul Krugman (1994) laments that the path-dependent and disequilibrium nature of economic development cause it to be an "unsuitable" topic for modern mathematical approaches in economics. He argues further that these difficulties have been major factors in the decline of development economics as a field.

[9] Estimates of the number killed range from 300,000 to 2 million people.

he also recognized that a strong military presence would be needed, and that a centralized regime was necessary to deal quickly with widespread protests and insurgencies. His early management of the tension between centralization and decentralization was impressive. However, during the last 10 years of his presidency, tension mounted as Soeharto and his family consolidated increasing amounts of military and economic activity under their immediate control.

Second, Soeharto has been accused of widespread corruption, and his rule was certainly authoritarian. But at the personal level, he also genuinely cared for peasants and their welfare. His rural orientation set the tone from the very beginning of his tenure as president. He visited the countryside frequently, and he sometimes made spot checks of rural programs to see what was actually happening. This concern about farmers carried over into a concern for their families. He was a vigorous supporter of rural health clinics and especially of family planning. To be sure, he made certain that his friends and family received plenty of payoffs as a consequence of trade, resource extraction, and industrial policies, but his rural instincts served the cause of improved food security extremely well. To have had a rich (especially in his later years), male, Javanese, Muslim general as president was perhaps not so surprising for Indonesia. Having that same person be the champion for small farmers, women, rural health clinics, and family planning was quite remarkable.

Throughout the Soeharto presidency, there were competing groups within government. The role of the military has already been noted, and Soeharto's family and a relatively small set of industrialists formed other centers of power. Perhaps the most interesting group, however, was the "Berkeley Mafia." What set Indonesia apart from most other countries in the 1967–98 period was the quality of its macro-economic policy. Soeharto relied upon a group of young economists to run large parts of the economy, particularly the Ministries of Planning, Finance, and Trade, as well as the Central Bank.[10] This group was led by the late Professor Widjojo Nitisastro,[11] who had been trained (along with several others) at the University of California, Berkeley. They were a world-class group of "technocrats"—another name given to them, although they never used it—who were crucial in keeping Indonesia's economy directed toward pro-poor growth for more than 30 years.

[10] There are few groups comparable to the "Berkeley Mafia" in other countries. The "Chicago Boys" in Chile during the 1970s and 1980s had some similarities, but the Chicago group was much more doctrinaire in their approach to economic policy (Valdis 1996).

[11] The U.S. training of this group was underwritten primarily by the Ford Foundation—a high-payout investment in human capital that is also part of the Indonesia story. Interestingly, the group was referred to as the Berkeley Mafia, both inside Indonesia and internationally, even though only about half of them had their PhDs from UC Berkeley. Other key members included Professor Ali Wardhana, Professor Mohammed Sadli, Professor Emil Salim, Professor Subroto, and Professor Saleh Afiff (though he was a half-generation younger). For many of these men, their professorships at the University of Indonesia came rather late in their careers. Given the choice between being called "minister" or "professor," they almost invariably chose the latter.

It was the technocratic group with whom the Harvard (and later Stanford) advisory groups predominantly worked. As the Berkeley Mafia moved among ministries so too did our advisory activities in a kind of symbiotic relationship. We knew full well that they were in charge, and that at the end of the day, the decisions were theirs to make or not make. In turn, they knew that quiet, off-the-record conversations were possible with us, and that we often had the time they lacked for local travel and analyses. Those of us working on agriculture were referred to by them in many ways: "our eyes in the countryside," "the rice doctors," and "my troubleshooters." It may also seem surprising to some that from about 1985 onward, in-country stays by the Stanford group were done without formal contracts—just on the basis of handshakes from one visit to the next.

Although none of the technocrats ever headed the agricultural ministry, the food and agricultural impacts of their decisions on trade, exchange rates, interest rates, and investment allocations dominated most decisions made in the Ministry of Agriculture. Indeed, this is one of the most important development lessons of the entire Indonesian experience: macro-economic policy really matters for agriculture, especially during the early phases of economic development.

This selective sketch of Indonesian history sets the scene for describing early efforts to improve food security. It is a backdrop not unlike what many poorer developing countries are facing today. Wars had left Indonesia devastated in both economic and political terms. Food stocks were almost non-existent; the army "ran" much of rice marketing—meaning that field units commandeered much of the rice trade to feed the troops. Perhaps worse, fear and suspicion were pervasive in 1967. All of the officials with whom I first worked asserted that they were on someone's kill list. In such a situation, the first question was how even to get a policy process started (see Box 2.1).

CULTURE

Achieving food security is difficult enough if a country's landscape and people are homogeneous. But "cultural diversity" hardly begins to capture the complexity that is Indonesia. Although this island country has the largest number of Muslims of any nation in the world, it is not an Islamic state; in fact, six religions are recognized formally. The island province of Bali is largely Hindu, but with strong Buddhist influences. Christianity, in many different forms, is well represented, especially in provinces such as North Sulawesi. In addition, animism remains at the religious surface on a number of islands. The densely populated island of Java—approximately the size of Greece—contains nearly 60 percent of Indonesia's entire population. It is subdivided into several cultural groupings, of which Javanese and Sundanese are the largest. For the country as a whole there are some 15 ethnic groups that individually contain greater

BOX 2.1

Starting the Agriculture Policy Process

Much of what constituted agricultural policy in post-independence Indonesia was simply dealing ad hoc with the particular food crisis of the moment. By 1970, however, two principal needs had emerged: the urgent need for improved technology, and the need for stronger incentives to farmers. These two principles then became embedded in policy beliefs that farmers would respond to new opportunities, and that improvements had to be done sequentially because of the country's limited human and financial resources.

Those of us working with Indonesian policymakers found an early framework by Arthur Mosher helpful in talking about key issues. Mosher's approach focused on three concepts: an achievement distribution (A), which described the yields/hectare (ha) farmers were actually achieving; a technical ceiling (T), which indicated the highest yields/ha possible under known experimental conditions; and an economic constraint (E), which showed what yields/ha were profitable for farmers to produce. This framework is depicted in these panels, where the horizontal axis is (conceptually) a distribution of all fields of rice in the country ranked from highest yields/ha to lowest, and where the vertical axis is yield/ha. The shaded portion of the graph depicts total production (i.e., area times yield).

Panel A shows the rice situation for Indonesia in 1967: yields were low, constrained both by technology and incentives, and the achievement distribution tailed off sharply as a result of inadequate information and ill-functioning irrigation systems. Panel B shows the lifting of the technical ceiling from T to T' as a consequence of the partnership with the International Rice Research Institute; Panel C shows that result of changed rice policy from E to E'; for example, more rural credit and subsidized fertilizer; and Panel D shows the technical and economic constraints having been lifted, and improvements in the level and slope of the achievement distribution as a consequence of investments such as irrigation. Through a series of not-always-linear steps, Indonesia went from Panel A in 1967 to Panel D in 2008. These panels are stylized versions of reality but they proved effective in discussions about what was occurring, and which set of decisions should next be made.

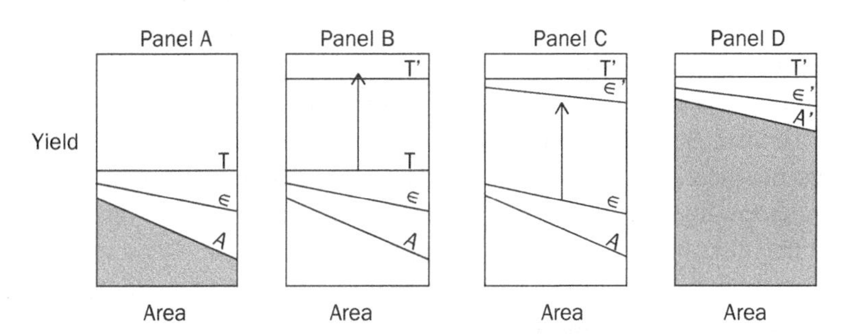

Source: Timmer, Falcon, and Pearson (1983), Chapter 3.

than 1 percent of Indonesia's population—and scores more than that comprise far less.

The ethnic Chinese community in Indonesia has been particularly important to food security. The Chinese represented less than 5 percent of the total population in 1967, but this group had long dominated the rice processing and rice marketing sectors; they were often also the suppliers of rural credit. Since the Chinese tended to be wealthier and non-Islamic, ethnic tensions were (and still are) ever-present, especially during periods of food scarcity. Storing rice, for example, provoked radically different responses from different groups—what was antisocial hoarding to some was prudent management of stocks through time for others! Indeed, ethnic issues raised critically important policy questions about how hard the new Soeharto government should strive in creating alternative marketing and milling channels for rice.[12] Like history, culture also matters for food security.

GEOGRAPHY

Three features of Indonesia's geography placed additional burdens on creating the initial environment for getting agriculture moving. With more than 7,000 inhabited islands, communication in the 1970s was exceedingly difficult. Integrating markets required seagoing vessels and the connection of literally hundreds of ports where corruption was known to be pervasive. To avoid taxes and some (but not all) forms of payoffs, much of the tonnage moved more informally in 100-ton prows operated by a sailing group (the "Bugis") from South Sulawesi. They were able to dock in many of the sheltered coves that dotted the many thousands of miles of Indonesian shorelines. This unregulated trade was sometimes a blessing and sometimes a curse for food policy. A standard quip was that God intended Indonesia for free trade. Domestic price policy for rice was necessarily linked to the international markets. When Indonesian rice prices were more than about $25 per ton above or below world rice prices—at least for very long—the Bugi sailors suddenly became regional importers or exporters who made markets work, or simply smugglers, depending on one's point of view.

During one debate about the level of domestic rice prices in the mid-1980s, Peter Timmer, Scott Pearson, and I argued that smuggling was undermining current rice price policy. To counteract the argument that "smuggling just could not be happening," we went to the local market, bought sacks of rice from several different countries and simply placed the bags on our desks at BULOG. Two days later the bags of rice had disappeared. We were informed, fortunately smilingly, by General Arifin that such "jokes" were in bad taste, and that he did not wish his troubleshooters to become troublemakers. But policy did get changed without any more words having been written or spoken—at least by us.

Resource distribution was, and still is, a second key feature of Indonesia's geography. The central government, the high religious culture, the heartland for labor-intensive rice agriculture, and the bulk of the population all resided

[12] Development of the rice milling industry is covered well in the now-classic paper by Timmer (1973).

on Java. But much of the natural resource base, including oil, copper, and timber, resided in the "outer" islands. So too did Indonesia's important tree crop sector, especially oil palm, which has become an important part of the Asian vegetable oil revolution since 1990 (Rueda and Lambin, Chapter 12). These resource-based activities typically took place in less densely settled communities that often had different traditions and languages than did Java. This geographic configuration had obvious political-economy components that were centrifugal in character: Where were government revenues to come from? Where were revenues to be spent? How much autonomy should be given to local governments? And, in the case of food, was policy to be the same for all islands, given especially that Wallace's line (Figure 2.1) divided the country's ecosystem origins and the types of agriculture these systems could best support?[13] For example, if the price of rice or fertilizer was to be stabilized or subsidized, should prices be the same for remote locations as they were in Java? And, how was the goal of creating a unified nation to be reconciled with "sensible" economics when the two objectives clashed?

The country's location at the western edge of the Pacific Ocean was a third feature of Indonesia's geography that greatly influenced food policy. Like Peru at the eastern end of the Pacific, Indonesia is slammed by irregularly regular El Niño events that happen every four to seven years. The slackening or reversal of winds during El Niño episodes causes rainfall over most of Indonesia to be delayed, and reduced in amount and duration (Naylor et al. 2007). The month-long "hungry" season (*paceklik*) before the main harvest is then extended, rice becomes more scarce, and rice prices typically rise, sometimes drastically so. As a consequence, food *policy* necessarily had to deal with climate variability if food *security* was to be a meaningful concept.[14]

Soeharto and the technocrats were tested early and severely in 1973 by one of the most extreme El Niño events of the twentieth century. It devastated the Indonesian rice crop, and in combination with other weather and market forces, this particular El Niño also totally disrupted international trade in rice. During most El Niño events, rice is available internationally, albeit at high prices. But on this specific occasion, there was literally no rice for sale (Timmer 2010). Coming as it did early in Soeharto's presidency, the climate-created chaos of 1973 caught everyone's attention, and then held it for a critical period of two

[13] A. R. Wallace, the famous evolutionary biologist, noted in 1859 that Indonesia's ecosystems were divided, with fauna on those island formations east of the Bali/Lombok divide related to Australia and those west related to Asia. Agricultural systems in Indonesia are dependent on many factors other than fauna; however, Wallace's line remains a useful indicator of the great agricultural differences between "eastern" and "western" Indonesia.

[14] Developing a reliable forecasting model that linked El Niño data (Pacific Ocean sea-surface temperatures) to the size of rice harvests six to nine months later, and then embedding that model within the Ministry of Agriculture, was a research and policy accomplishment that we note with pride (Falcon et al. 2004).

years. Just when the government thought that it was beginning to see progress with new rice technologies and policies, climate variability created unequivocal misery in the countryside. For months the senior leadership in government worried about rice production, consumption, and trade, and very little else. Looking back, it is clear that this particular El Niño episode stiffened the leadership's resolve to make agricultural improvement a cornerstone of national policy. Everyone concluded that more had to be done to help farmers increase rice output and to assure that poor consumers were protected from climate-induced price volatility. In fact, much good came out of a very bad early event.

Resolving Policy Dilemmas

The foregoing history is far too brief for such a large and complex country as Indonesia. However, it helps to set the stage for the real-time policy tensions that arose among various sectors and principal players as they grappled with food security. The focus in this section is primarily on the 1967–98 period, when I believe the lessons are both clearer and more applicable to current developing countries at the lower end of the income scale. They are described mostly in the form of vignettes, rather than in terms of formal development theory. My hope is that these stories will convey more meaningfully than equations how the answers evolved, and why.

INDUSTRY VERSUS AGRICULTURE

Planning for Indonesia's first five-year plan (*Repelita I*) began in late 1967, mainly under the leadership of the technocrats. In early 1968, Professor Widjojo concluded that a "bull session" (his term) was needed to make operational President Soeharto's vision for the country. A series of those informal discussions took place over the decade to follow, but it was the first one that set the entire course for Indonesia's development strategy.[15] It was attended by about twenty Indonesian policy economists, who would eventually play key leadership roles for the country over the next 35 years, and by a half-dozen foreigners from the donor and advisory communities. I was among the latter group, and in retrospect, it was the most interesting and important professional gathering that I have attended in 50 years as a development economist.

Many developing countries at that time were taking their cues on planning from either India, which was then entering its fourth five-year plan (1969–74), or from Argentina and Colombia, which had begun serious economy-wide

[15] This section draws heavily on personal notes from the bull session held March 16–17, 1968 in Jakarta.

planning in the late 1950s. The Indian plans had typically focused on industry, particularly heavy industry; they were domestic in their orientation; and they seemed somewhat complacent about agriculture. Most Latin America countries were embarked on high-cost industrialization plans designed mainly to replace industrial imports. What came out of Indonesia's first "bull session" was different from both approaches by nearly 180 degrees. The first emphasis was on stabilizing the macro economy: gaining control of inflation; replacing doctrinaire pronouncements with more market-oriented policies; reestablishing trade and other relationships with the rest of the world; and limiting the role of government. Another set of conclusions emphasized the need for increasing agricultural productivity with improved seeds, fertilizer, and improved infrastructure, especially irrigation and ports. Perhaps most importantly, the strategy knowingly concentrated on only a few activities because of the government's financial and human-resource constraints.

Cynics might argue that the focus on agriculture was quite consistent with a concomitant strategy of leaving manufacturing and the extractive sectors to "generals and friends." This interpretation might have had more credence during later stages of Soeharto's presidency, but initially, the rural focus was driven by Soeharto's vision for a more livable countryside, and by Professor Widjojo's intellectual training in development economics. That early direction proved to be very important. Between 1967 and 1996, Indonesian GDP per capita grew at the very respectable annual rate of 5.1 percent. Perhaps more amazingly, by emphasizing the rural sector, it also improved the incomes for the poorest quintile by 5.1 percent annually during the 1967–96 period (Timmer 2004). This development performance helped give rise to a new term: pro-poor growth.

INTEGRATED RURAL DEVELOPMENT PROJECTS VERSUS SEQUENCED CENTRAL POLICIES

During the post–World War II era, two broad approaches have characterized strategies to get agriculture moving in developing countries. The first of these has been integrated rural development. This concept focuses on geographically demarcated projects, and attempts to bring all ministries together at the local level under a regional planning authority. Using this approach, the dual benefits of coordination and interaction could, in principle, be obtained; each group would know what the other agencies were doing, and there would be the more-than-additive effects of dealing, in particular places, with seeds, credit, and pesticides, and also irrigation, roads and health. Over time, all sub-regions of the country would be covered and rural development would have taken place.

The alternative approach to agricultural development embraces a more centralized approach that relies on line ministries, not regional authorities, and places much more emphasis on policies rather than on specific project areas. In the 1970s and 1980s, integrated rural development was very much in vogue.

The World Bank had numerous such projects in Africa, and Pakistan was using a similar approach with respect to irrigation development. The seductive appeal of more-than-additive effects seemed almost irresistible. But resist the Indonesians did, with one important exception (transmigration) that will be discussed.

With 40 years of history now as a teaching aid, the Indonesians were wise to have resisted the regionally integrated approach. As noted by the World Bank itself, that approach to rural development largely failed (World Bank 1988). The local communication that was supposed to occur among agencies often did not—it was still easier for ministry people to talk vertically within their own siloed agency, rather than horizontally with other organizations. In addition, local groups often felt that they were not integral to the projects and did not take full ownership of them. When donors left the region upon "completion" of the infrastructure, regional project governance tended to be undercut seriously by the line ministries and often fell apart completely.

Indonesia chose instead a more centralized policy approach—perhaps because of a combination of fear and faith. The fear had two components: the technocrats worried that there was insufficient local human capital for a decentralized approach, and also that there were too few resources in Jakarta for much-needed monitoring. They were also worried about the divisive approach of such a regionally explicit strategy. They were genuinely concerned about the unity portion of Indonesia's motto: "unity in diversity." On the faith side, they held the strong belief that peasants would respond to sensible policies if given half a chance. In 2013, that belief would not be remarkable; in 1968, it was very enlightened thinking.[16]

Having chosen the more centralized policy-driven approach, policymakers were still left with serious dilemmas: the problems were huge; the country's geography was sprawling; and government resources were few. A central feature of the Indonesian approach was to focus on a sequence of policies for dealing with food problems. Such an approach is in sharp contrast to one that seeks to implement a catalog of everything that might be useful for agricultural development. While it would have been nice to have the interaction effects of doing everything at once, the technocrats realized that the government was capable of only doing two or three things at a time.

Early efforts at food policy were directed at improving the capabilities of the poor through the provision of inputs and advice, and by lowering the transaction costs of operating in the rural economy via investments, especially in roads and irrigation facilities (Timmer 2004). Getting the army out of the rice trade in the countryside was also important, if only quietly discussed. The

[16] By way of intellectual benchmarking, T. W. Schultz's Nobel winning volume, *Transforming Traditional Agriculture*, was published only in 1964.

technocrats knew that only if the army could be provisioned adequately with food would there be much hope for making rural markets work.

The sequence of policy actions from 1967 to 1998 is rich in important details (Booth 1988; Pearson et al. 1991a; Timmer 2004; Resosudarmo and Yamazaki 2011).[17] The first phase was dominated by various forms of the BIMAS (*Bimbingan Massal*) program, which involved small packages of improved rice seeds and nitrogen fertilizer, combined with subsidized credit. (The subsidized credit component was largely eliminated in 1986.) An expanded irrigation program, mostly funded by the World Bank, also was a key component of the early phase. National programs for rice production—combining improved seeds, nitrogen, and irrigation water with incentives—brought forth a surge in rice output. Most of all, farmers responded to the incentives and new technology in ways that few predicted at the time.

The second phase of policies came a decade or more later and focused on rural credit and financial intermediation. Although Grameen Bank and many other microfinance institutions have captured recent microfinance headlines, the much larger KUPEDES Program in Indonesia was arguably more important in terms of scale and organization for lending to the poor. As background, it was said (perhaps even believed) that small farmers would or could not pay interest rates comparable to the actual transaction costs associated with small loans. This had led Indonesia, and many other countries, to create special subsidized credit programs that necessarily then had to be rationed. The Indonesian experience indicated that farmers would pay "high" interest rates to the banking system—which actually did not seem so high to them as compared to local moneylender rates—if credit were available locally and if they had access to high payout assets (e.g., cows) and expenditures (e.g., fertilizer). It also showed that farmers would put savings into accounts if interest rates were competitive (Patten and Snodgrass 1987; Robinson 2002).

Rural health clinics also became a key feature of the second phase. Since the instincts of the health ministries of almost all countries tend to favor richer clients in cities, Indonesia's child and family planning focus on poor families in the countryside was notable. A 1973 nutritional survey had shown that two-thirds of Indonesia's children under 5 years were undernourished—either wasted, stunted, or both. In response, and with prodding from both the president and the technocrats, the Ministry of Health and the National Family Planning Board developed large field-based programs. The activities centered on women and children: neonatal care, weighing programs, nutritional supplements for young children, family planning advice and contraceptives, and limited forms of medical advice. By 1990, Indonesia had over 20 million children

[17] A number of policy innovations on pricing and trade are presented in subsequent sections.

enrolled in 250,000 clinics nationwide, at an annual cost of about $0.75 per child (UNICEF 1996).

Numbers tell the tale of what happened to rural communities in Indonesia. For the 20-year period between 1970 and 1990, rice (paddy) production more than doubled, growing from 19.3 million metric tons (mmt) to 45.2 mmt (International Rice Research Institute 2012). Average consumption of calories per person per day from rice rose from 1,030 to 1,300. Real rural wages rose by about 20 percent between 1976 and 1988 (Naylor 1991). Infant mortality fell from 100 deaths per 1,000 live births in 1970 to 56 deaths in 1990 (and halved again to 27 deaths by 2010). And total fertility dropped from 5.5 to 3.1 between 1970 and 1990 (World Bank 2012). The centralized policy approach, which focused on a limited set of sequenced policies, had achieved results even beyond the expectations of those involved with early planning and implementation.

The success story outlined here deserves two important addenda. First, a substantial gap existed in 1967 for rice yields in Indonesia relative to yields being achieved in other comparable countries. As described in a subsequent section, the country was thus able to realize large gains from imported forms of agricultural technology, especially improved seeds. Moreover, the long history of intensive rice cultivation, especially on Java, meant that farmers reacted quickly to the incentives placed before them. Second, the success that occurred in food and agriculture could not have happened without sound and innovative macro policy. In particular, sensible management of the exchange rate kept agriculture competitive, even during periods of substantial foreign exchange earnings from petroleum. "Dutch disease" (the undesirable distortions and impacts of large resource revenues on a country's long-run development) did not occur, and the comparison between what happened in Indonesia and Nigeria in this regard could not be more stark.[18]

But not everything worked. One of Indonesia's few attempts at project-based integrated development was mostly a large and costly failure. As early as the nineteenth century, the Dutch had sought to relieve crowding and tiny landholdings on Java, Madura, and Bali, and simultaneously to provide workforces for the outer islands through resettlement programs. The process was known as transmigration (*transmigrasi*), and both the concept and investments in resettlement were expanded during the Soekarno and Soeharto regimes. The program peaked in the 1979–84 period when the World Bank funded seven such projects at an aggregate cost of about $600 million. Although two projects (based on rubber) proved economically viable, most did not.[19]

Settlers found out that soils were different from those they were accustomed to and that the labor-intensive rice cultivation techniques that they had

---

[18] A cogent summary of macro economic issues, especially as they relate to Indonesian food policy, can be found in Pearson (1995).

[19] Much of this section is drawn from the Independent Evaluation Group of the World Bank (2012a).

known on Java were inappropriate for the new locations. Off-farm employment proved difficult to find, and in several projects the environmental consequences, especially deforestation, created substantial negative externalities. Perhaps most importantly, almost no one was happy with the program. Settlers often felt that they had been duped into moving and that they then had no way out once they became disillusioned. The traditional communities in the project regions were also upset. Land use and conflicting land rights were a part of the problem, but even more serious was the clash of cultures and religions. For example, in 2001 the local Dayaks on the island of Borneo clashed with transmigrants from Madura, killing hundreds and displacing thousands of the Madurese. More generally, there was fierce resistance by locals who, rightly or wrongly, perceived "Javaization" and "Islamization" to have been primary project objectives.

In total, about 4 million transmigrants were resettled at a probable cost exceeding $1 billion. One suspects that there would have been much greater benefits if that $1 billion had been spent simply on facilitating more general forms of voluntary migration, or better still, if invested in agricultural R&D. Perhaps the good news is that the transmigration program was scaled back drastically in 2000 following the Asian financial crisis. There is, I believe, one overriding Indonesian lesson on transmigration for other countries, especially those now involved with large land sales (Smith and Naylor, Chapter 8). Many things can and do go wrong when outside settlers become involved with what locals believe to be their land. Clashes, often violent, seem almost inevitable in these circumstances.

PRODUCERS VERSUS CONSUMERS

One of the classic dilemmas surrounding food security lies in balancing the simultaneous needs for low food prices to assist consumers and for high agricultural prices to provide incentives to farmers. That difficult tradeoff formed the core of our 1983 volume on *Food Policy Analysis*, and the issue is as alive today as it was earlier (Timmer, Falcon, and Pearson 1983; Swinnen 2012). As we wrote then, "the incomes of the poor depend on their employment opportunities, many of which are created by a healthy and dynamic rural sector. Incentive prices for farmers are, in the long run, important in generating such dynamism and the jobs that flow from it. But poor people do not live in the long run. They must eat in the short run, or the prospect of long-run job creation becomes a useless promise."

The Indonesian strategy for solving the foregoing food-price problem emphasized approaches that could help producers and consumers simultaneously. They were surprisingly successful in finding win-win solutions, primarily: (1) by increasing yields and productivity more generally so as to increase farmer incomes via increased production rather than via higher prices; (2) by investments and interventions designed to avoid simultaneously the deep fall in rice prices at the time of harvest (helping farmers) as well as the sharp rise in rice prices in the hungry season (helping consumers); (3) by using international trade in rice to offset year-to-year variability in rice supplies caused mainly by climate and pests; and (4) by distributing rice physically to special groups and the poor. A fundamental assumption for most of these activities was that helping to stabilize real rice prices, both intra- and inter-annually, would be welfare increasing. That assumption was, and still is, a debated point among development economists, which makes the Indonesia case of particular interest.[20]

At the heart of Indonesia's food-security success was the Green Revolution technology that combined improved varieties, increased amounts of fertilizer, and improved and expanded irrigation systems. Between 1970 and 1990 the area with improved rice varieties went from 7 percent to 77 percent; rice (paddy) yields grew from 2.3 mt per hectare (ha) to 4.3 mt per ha; fertilizer (urea) use on wetland rice went from about 15 kg per ha to 185 kg per ha; and rice under irrigation went from 5.7 to 8.0 million hectares (Heytens 1991; World Bank 1992). Technology, investments, and policy all played key roles.

Inadequate infrastructure, particularly irrigation systems and rural roads, was a serious constraint in 1967. For farmers, these deficiencies limited the amount of rice being grown and added substantially to marketing costs. The constraints manifested themselves in many ways, perhaps most importantly, by "excessive" seasonal variation in rice prices. With rice production heavily concentrated in the November to March rainy season, rice prices tended to crash in April at the end of the large "wet season" harvest and to spike the following February just before the new crop came in.

Most of the early investments in irrigation were funded by the World Bank in a series of 15 projects, each of which averaged about $200 million.[21] These investments, aimed both at the rehabilitation of old systems as well as development of new areas, caused rice production to increase; moreover, the production was more equally distributed throughout the year. Evening out intra-year production dampened seasonal movements in rice prices. In addition, many farmers were able to grow two crops per year when before they had only grown one. And because the new rice varieties took fewer days to grow, farmers sometimes could even grow three crops per year instead of two.

Investments in warehousing were also crucial in Indonesia's approach to rice policy. With the seasonality in rice production, and with imperfectly connected island economies, there was a glaring need for seasonal and emergency rice reserves across Indonesia. BULOG, the national logistics command,

---

[20] The most informed assessment of Indonesia's food stabilization effort can be found in Timmer (1996).

[21] See World Bank (1989) and Rosegrant et al. (1987). There is no doubt that those irrigation projects increased rice output; however, there is professional debate about the economic rates of return to those projects.

erected an integrated set of warehouses across Indonesia (Falcon et al. 1985). These physical investments were important in assuring regular rice supplies; moreover, regional warehouses were key instruments in the policy approach Indonesia used to stabilize rice prices. The basic policy strategy was straightforward. BULOG, through its regional organization, would implement a floor price for paddy (set by economic and agricultural ministers, and the president). The agency would be a buyer of last resort after harvest. Such purchases would then be stored in secure warehouses, to be released later in the year to help curtail price increases prior to the next harvest.[22]

What appeared conceptually simple, however, was difficult to implement. Tight controls were needed to curtail corruption, and to keep local BULOG offices from going rogue. While BULOG's record was far from perfect, regional communication with headquarters was on a daily basis, many inspectors were used, and both physical and economic losses were surprisingly small.[23] To achieve that outcome, however, two things had to be done with consummate skill. The first was setting the floor price correctly. If set too high, the government would destroy the private sector and would need to purchase "too much" of the crop—both of which would have been undesired outcomes.

The unstated goal was to set prices such that BULOG would procure about 5 percent of the crop on average. Between 1970 and 1990, the actual percentage varied from 1 percent in 1973 to 9 percent in 1995 (World Bank 1992). But finding the correct price was easier said than done.[24] The right floor price depended on international prices and on the size of the crop, while crop size was not yet known and was frequently being buffeted by El Niño events. The difference between procuring 1 percent versus 9 percent also had cash flow and other implications for the national budget. To operate such a procurement policy, so-called price triggers were required; for example, BULOG could not be given the task simply of buying a given number of tons, since that would likely have added to price instability rather than stability. To defend a support price through purchases, however, BULOG needed a second instrument: an open letter of credit from the Central Bank to buy an uncertain tonnage of rice. Suddenly food policy was right in the middle of monetary policy.

The International Monetary Fund (IMF), which frequently kept a watchful eye on Indonesia through various standby agreements, was vocal in its opposition to the policy approach. This led to several fierce and memorable policy

---

[22] The basic formulation of the approach was first described in Mears and Afiff (1969).

[23] During the early phases of BULOG operations, we estimated that about 3% of value of its rice operations was "unaccounted for." By the late 1990s, under different leadership, we estimated that number had risen to 30%, at which point BULOG the institutional solution had become BULOG the institutional problem.

[24] At the end of the day, the paddy support price was negotiated. Key players included the ministries of finance, planning and agriculture, the Central Bank, and BULOG, with the president having the final word.

battles. In the end, the Indonesian leadership prevailed, and 50 years later, perhaps even the IMF concludes that the right decision was made. Leveling out production, introducing an integrated warehousing network, lowering interest rates, and running a price-band system for rice led to two important results. First, the seasonal price movements in rice were considerably leveled. The average (pre-harvest) peak in rice prices relative to the (post-harvest) trough in prices fell from 18 percent during the 1969–78 era, to 12 percent during 1979–89, to 5 percent from 1990–97. And second, and in contrast to international rice prices that remained highly variable, Indonesia was able to maintain domestic rice prices that were remarkably stable in real terms, yet at levels that provided strong production incentives for farmers. After the government gained control of domestic prices in the mid-1970s following the El Niño crisis in 1973, domestic real rice prices were essentially unchanged between 1974 and the mid-1990s (Figure 2.2). In 1997–98, however, another rice price spike occurred. The spike coincided with a very severe El Niño event and an economy-wide financial disruption. These proved to be a lethal combination that caused Indonesia's foreign exchange rate to go from about 2,500 rupiah (Rp) per US$ to more than Rp16,000 per US$, before settling down to about Rp11,000 per US$ (Timmer 2004a).

The stabilization program was not costless. The mythical idea that the government could somehow buy low and sell high, and make money in the process turned out indeed to be a myth. Pearson (1993) estimated that the cost of rice price stabilization ranged from about $50 million to $100 million annually

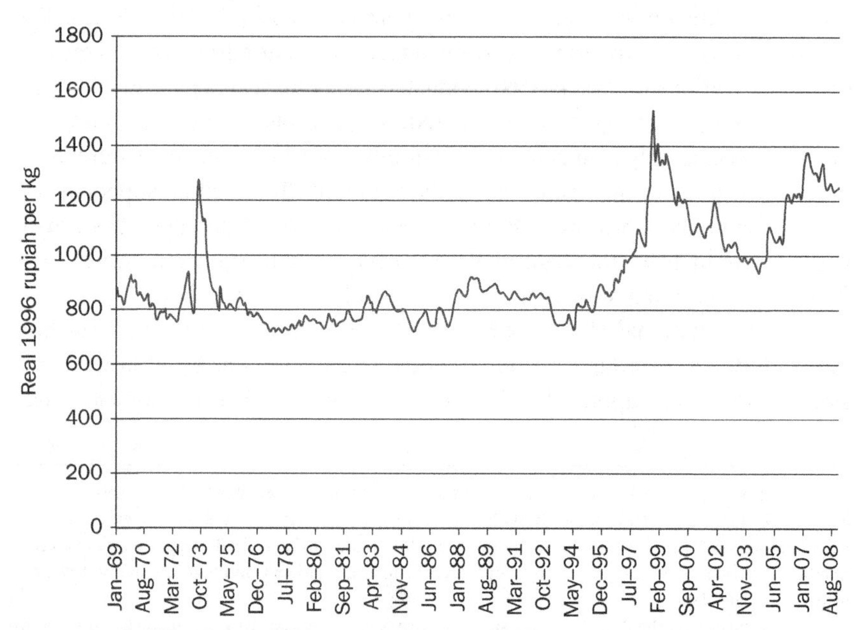

FIGURE 2.2 Real Rice Prices (1996 base) in Indonesia, 1969-2002, Rp/kg

Source: L. Peter Rosner and C. Peter Timmer [Timmer 2004a].

(depending largely on the amount of rice procured domestically relative to the amount of rice imported). Was this cost of about $0.50 per person per year worth it? I believe the answer is unequivocally yes, both for the planning and investment stability it provided to farmers, and for the sheltering it also provided to consumers. Should other countries undertake similar price band operations? Perhaps, but if they do, they need to understand the difficulty of the operation, the analytical and human capital required for doing it right, and the potential impacts of such an approach on macro-economic policy.

One final component of Indonesian consumer policy is worthy of mention. Throughout the Soeharto era, the government distributed physical rations to some regions and to several specific groups. On average, about 0.1 mmt of rice annually was sent to distressed areas affected by floods, drought, and volcanic explosions. On an aggregate rice consumption base of about 30 million mmt, this number appears quite modest, and was arguably too low to correct the poverty and distress that prevailed in the early period.

By contrast, from 0.8 mmt to 1.5 mmt tons of rice were distributed each year, in monthly allotments, during the 1970s and 1980s to members of the civil service and to the army (World Bank 1992). These groups were clearly not the poorest of the poor, and critics have sometimes described these distributions as being both unnecessary and inefficiently delivered. But to the technocrats, these distributions seemed wise (Timmer and Falcon 2005). They knew and understood corruption, and saw its economic origins in an underpaid and demoralized civil service that was asked to manage a host of regulations. They recognized that the military, often with less than a third of its operational costs financed by the official budget, would turn to other sources, legal or illegal—especially when it came to provisioning the troops. They also knew that the agricultural modernization program could not permit the military to raise havoc with rice markets via the commandeering of rice in the countryside. The special distribution programs may have continued for too long and at greater tonnages than were necessary. But few who understood the broader background wanted to take on the special distributions as a primary policy battle.

DOMESTIC VERSUS INTERNATIONAL STRATEGIES FOR FOOD SECURITY

In the mid-1960s, few foreigners or foreign operations were left in Indonesia. Most foreign firms had been nationalized under Soekarno, and many foreign personnel had been asked to leave the country. By 1970, however, Indonesia had dramatically changed its stance towards the world community. These sweeping changes included altered attitudes toward commodity trade, foreign private investment, and the donor community. They affected many sectors, but had especially important consequences for food security. Three sets of examples indicate how the new attitudes affected Indonesian food policy, mostly for the better.

The rhetoric in virtually every developing nation calls for "self-sufficiency" in food production, especially for the primary staple. Indonesians also used such language in many speeches and documents; however, a closer inspection of actual expenditures and actions reveals a different story. The technocrats realized that with the inter-annual variations in rice due to climate and pests, the country would require massive stocks across years if price volatility were to be managed without trade. They realized as well that the carrying charges (interest, storage costs, and the physical deterioration that occur when paddy is stored in the humid tropics) would be very large. On the other hand, they realized that the size of Indonesia and the thinness in the international market for rice dictated that most of Indonesia's rice be grown domestically. They thus evolved the idea of "self-sufficiency on trend," meaning that in bumper years the country would export small amounts, and that in lean years it would import. This slight change in emphasis was crucial for assuring adequate rice supplies, and for underpinning the domestic price stabilization effort.

International rice prices are notoriously volatile, with typically triple the variability seen in maize and wheat prices. This variability arises because of the extensive government-to-government involvement in rice trade, the lack of a broadly based rice futures market that can be used to hedge risk, and the small ratio (less than 5 percent in the 1970s and 1980s) of global rice trade to production (Dawe and Slayton 2010). Indonesia was averse to importing international rice-price instability. The government therefore developed arrangements that shielded domestic rice prices from sharp peaks and troughs in international prices, although still following trends in international rice prices. To accomplish this feat, BULOG was given monopoly rice import rights for the country, buying rice as cheaply at it could and then storing the imports in its warehouse system. These imports were then used, along with the rice procured domestically, to defend rice prices to consumers (i.e., to protect the ceiling price in the stabilization scheme). Had there not been changed attitudes about imports, domestic price stabilization simply would not have been possible.

These arrangements prompted international concern about the state monopoly in rice trading, and there were also fears that Indonesia would attempt to *support* rice prices, rather than merely to *stabilize* them. As noted previously, the potential for smuggling actually helped to keep domestic and international prices broadly aligned. Moreover, the technocrats had watched the disasters that had happened in the Philippines and elsewhere when government grain agencies went bankrupt as they moved from price stabilization to price supports. For that reason, they monitored the rice price-stabilization effort carefully from the very beginning. There were occasional irregularities, but on the whole, BULOG's performance as Indonesia's sole rice importer was quite good.

Overall, the stabilization effort, which included substantial trade in rice, was enormously successful. Early international rice trade varied annually between 0.4 mmt of exports (1985) to 2.0 mmt of imports (1981), and was a key pillar in the stabilization strategy. Numbers tell the tale: the coefficient of variation (CV)[25] of monthly international rice prices for the 1960s and 1970s was 47 percent; whereas the CV for monthly Jakarta wholesale prices was only 13 percent (Timmer 1989).

Direct foreign involvement in Indonesian food production also provides interesting case material. Most of the investments in Indonesia's food sector, and especially in food staples, were domestic in origin.[26] One early (1968) foreign effort in the rice sector involved a number of foreign firms (Hoechst, Ciba-Gigey, COOPA, AHT, and Mitsubishi); it was an unmitigated disaster. That disappointing effort was an early variation of the BIMAS program, and carried the name BIMAS *Gotong Royong* (mutual aid). Under this program the government contracted with foreign firms to supply fertilizer, pesticide, and equipment at highly subsidized rates to farmers who were forced to be a part of the program. Those companies were paid a fixed rate per hectare, yet lacked any responsibility for ensuring production successes or loan repayments. By 1970, the program covered about 15 percent of Indonesia's rice area, but it was stopped after only four seasons (Resosudarmo and Yamazaki 2011). The input costs were high (more than $50 per ha), the companies had no financial responsibility for program success, and many of the production techniques were grossly inappropriate. Practices such as aerial spraying of strong pesticides on tiny fields often less than 0.1 ha within densely populated areas were clearly not what Indonesia needed. How this program got started has always been a murky question, involving frequent allegations of kickbacks and corruption; how it got stopped is clearer. Presidential visits into the countryside showed that the program was not working; it was summarily replaced by the BIMAS program described earlier. Although foreign private investment in tree-crop plantations (rubber and oil palm, in particular) continued to be a part of Indonesian agricultural development, there was almost no direct foreign involvement in staple food production after the BIMAS *Gotong Royong* fiasco.

The opening of Indonesia to international trade and thinking also foreshadowed new international institutional arrangements of immense significance. None was more important than the partnership forged with the International Rice Research Institute (IRRI) in 1972. International transfers of technology are often taken for granted; in reality, nationalistic attempts to prevent the import of improved germplasm are all too common (Nestel 1985; IRRI 2012). By working closely with IRRI, Indonesia was soon actively involved with the Green Revolution for rice. A long series of improved varieties—IR 5, 8, 24,

[25] Defined as the standard deviation divided by the mean.

[26] A counterexample was the ill-fated Mitsogoro attempt at a corn plantation in Sumatra. It failed for a number of reasons, most especially because the seed varieties being used at the time were unable to withstand outbreaks of downy mildew (Timmer 1987).

36, and 64—evolved over a 20-year period. The varieties were so dominant in terms of yield and disease resistance that both scientists and policymakers worried about too much of the country being planted to a single variety of rice. For example, IR36 was planted on about 40 percent of all Indonesian rice land in 1981. Fortunately—and perhaps more by good luck than by good design—IR36 resistance to pests did not break down and there was no ecological disaster. New seeds resistant to brown leaf hoppers, when combined with large quantities of relatively low-priced fertilizer, formed the core of the successful rice production strategy.

A great deal of training and human-capital development also came with the varieties, and there were many IRRI spillover effects on Indonesian agriculture. The Agricultural Development Council, USAID, and the Ford and Rockefeller foundations made key contributions. Nestel (1985) describes in some detail the series of linked activities through which IRRI helped to develop and strengthen Indonesia's Agency for Agricultural Research and Development (AARD). As a consequence of these efforts, the number of MS holders went from 21 to 310, and the number of PhDs from 0 to 36 in AARD during the period 1975–82 (E. Oyer, personal communication 2012). Major efforts were also made during this period to facilitate English-language training as a precursor for many of the foreign degree programs.

One more dimension of Indonesia's changed international attitudes concerned foreign-aid relationships. Despite arguments on particular topics, the quality of leadership of both the resident IMF and World Bank missions, particularly during the early phases of the economic transformation, was outstanding.[27] The Bank's role in coordinating the annual aid-pledging sessions was admirable in two respects: these sessions provided for a serious annual review of policy, and they also offered a chance for arm's length assistance by the Bank in helping to coordinate activities of all donors and to reduce the number of strings and special provisions tied to that aid.[28] The Indonesians also gained a great deal from their close 30-year association with Harvard's Development Advisory Service (later called the Institute for International Development).[29] Not many countries have shown such a willingness to have continuing in-house advice on delicate reform issues that involved food, exchange rates, fiscal policy, foreign private investment, and monetary policy.

---

[27] There were clearly stormy periods with the World Bank, the IMF, USAID, and other donors. For example, the IMF standby arrangements were always a source of tension, and changing U.S. policies on military assistance, food aid, and family planning frequently proved frustrating for the Indonesians.

[28] Early in the Soeharto era, the Netherlands convened countries for the aid-pledging meeting. Frictions developed between the Netherlands and Indonesia over this role, however, and the World Bank wisely took on the task of convener.

[29] By way of disclosure, I was a member of the Harvard Group until 1972. At that time, I (and much of the Indonesian agricultural portfolio) shifted to Stanford University. The Harvard-Stanford working arrangements remained close after 1972, and to most Indonesian and international observers, the distinction between the two was inconsequential.

Among the most important contributions of the Harvard and Stanford advisory groups were the help they provided the technocrats in conversations across ministerial silos, especially the coordination between the central economic agencies and sectoral organizations such as BULOG. The technocrats, with help from the international community, proved especially skillful at understanding the links between good food policy and sound macro policy. The power of those connections when done successfully is another key lesson that Indonesia provided to the rest of the developing world.

GOVERNMENT INTERVENTION VERSUS MARKETS

Among the most difficult strategic choices governments face is between intervening to achieve policy objectives and letting markets provide solutions that are typically more efficient, but often more socially painful. Indonesia's technocrats were great believers in markets, but they were not doctrinaire. One of the boldest of their policy actions involved the aforementioned price stabilization effort for rice. They also believed in the learning-by-doing aspects of subsidies, especially at the early stages of agricultural development. One of the greatest policy successes occurred via fertilizer subsidies; on the other hand, the worst food-policy mistake in the entire 1970–90 period involved a subsidy on pesticides.

As for the first of these, Professors Widjojo and Afiff saw from the beginning that providing fertilizer to farmers clearly lay in the realm of a public good (Falcon 2007). They spent endless hours ensuring that urea was available in the countryside, and also that it was priced properly. They found the former task, involving as it did both the public and private sectors, to be almost as difficult as the latter. The outer islands posed a particular problem in that the fertilizer markets were both distant and thin. The decision to intervene in fertilizer markets also had large political-economy significance at the regional level. Despite efficiency arguments to the contrary, the technocrats argued for and won a pan-Indonesian pricing strategy (i.e., fertilizer costs were to be the same for all Indonesian farmers wherever they were located). Those of us who argued for more efficient regional pricing strategies were summarily told that this was one way the Government could and would demonstrate it was building one nation.

The next step lay in setting the level of the fertilizer subsidy. Professor Afiff had earlier developed a "farmers formula" (*rumus tani*), in which he argued that there was to be a more-or-less fixed relationship between rice and urea prices. That indeed became the policy objective.[30] There were several elements in the

---

[30] The technocrats saw new seeds and fertilizer as key to increased yields. As a matter of policy, they wanted to assure that paddy prices and urea prices were linked in ways that assured high rates of return for increased fertilizer applications.

subsidy on fertilizer, though exactly how much each component cost kept many economists busy with calculations. For example, how much of the subsidy was to be allocated to the domestic fertilizer industry, how much to transport costs within Indonesia, and how much to farmers? Moreover, subsidies varied considerably by island and by type of fertilizer. On average, and using the difference between international and domestic prices as the basis for calculation, the economic subsidy for nitrogen fertilizer (urea) averaged about 40 percent between 1970 and 1985, and for triple super phosphate (TSP) the average was about 55 percent (Rosegrant et al. 1987).[31] Though there was debate on the exact subsidy levels for fertilizer and on where in the budget those subsidies belonged, everyone agreed that they were substantial.

Did these subsidies make any difference? It is impossible to say what the outcomes would have been without government assistance. With government assistance, however, fertilizer use per hectare increased by a factor of 10 between 1969 and 1986 (Heytens 1991). In a very detailed study, Timmer (1986) estimated econometrically that about half of the growth in rice production from 1970 to 1982 could be attributed to fertilizer pricing and stable rice prices. Decomposing the sources of growth, especially between the new varieties and their fertilizer complements, has always been difficult analytically. Overall, however, it appears that the budgetary cost of the fertilizer program, which peaked at about $700 million per year in 1985 (roughly 7 percent of total development expenditures), was government funding well invested.

The 1997–98 financial crisis mostly solved the last of Indonesia's fertilizer subsidy problems—namely, how to end them. The IMF, in particular, was strong in its arguments against ongoing subsidies as a part of its standby agreement with Indonesia. The Indonesian leadership agreed: the fertilizer market was well-established; agricultural credit had become less of a bottleneck; and the learning-by-doing phase of the subsidy had been achieved. In some sense, the IMF's insistence provided cover for a necessary but difficult decision.[32]

At the same time the fertilizer subsidy success was occurring, a great subsidy failure took place with pesticides. Indonesia faced severe pest problems, largely from the brown leafhopper, whose infestations greatly reduced rice yields. Farmers sprayed for the pests, but in doing so destroyed insect species helpful to rice production. This result led to more spraying, and insect damage spiraled upward. The longer-run solution was found in resistant seed varieties;

---

[31] The domestic/international comparisons, besides being island and fertilizer specific, are also sensitive to the exchange rate used to convert international to local prices.

[32] The 1997 standby agreement was mostly a disaster. It started with the notorious picture of IMF Managing Director Camdessus standing with folded arms over Soeharto at the signing ceremony. Instead of focusing on a half-dozen policy actions on which the Indonesians needed to focus, such as the fertilizer subsidy, the agreement went on for pages and pages with dozens of actions—great and tiny—that the government might, should, or must undertake. In the end, the horribly cumbersome document created major frustrations for both sides.

however, the short-run policy decisions made in the mid-1980s were misguided at almost all levels. First, Indonesia took up the local manufacture of a pesticide (Diazinon) that had been banned in many countries because of its negative health and environmental side effects. Second, the Minister of Agriculture, who had a direct economic interest in the plant, was instrumental in pushing through a 100 percent subsidy on pesticides. The ministry also instructed the extension service to recommend repeated sprayings—resulting in growing financial and ecological problems.

Even as the pesticide problems became clear, a solution was not apparent. Finally, the situation changed. It involved several factors, including the independent work of several brave Indonesian entomologists; a determined foreign advisor who lived in the countryside for two years gathering empirical data on plant hopper damage; and several unannounced helicopter visits to affected sites by two economic ministers. Evidence from the countryside was brought to bear in a cabinet meeting, the agriculture minister was disgraced, the pesticide subsidy was removed, and the plant was shut down. It was not the practice of Soeharto to fire ministers, nor did he fire this one. But the agricultural minister was rendered neutral for the rest of his incumbency.[33]

A final footnote to this story concerns field schools whose objective was to teach farmers about integrated pest management. These schools, limited in number, were ongoing during the era of pesticide subsidies, yielding claims that farmers taught in these schools were accountable for the big drop in pesticide use. A closer look at the evidence indicates that the cut back in spraying was due primarily to the removal of the subsidy and the closing of the pesticide plant. Surveys indicate that famer responses to spraying were similar in areas with and without schooling (Feder et al. 2003). However important and useful these institutions may have been in other ways, they should not be thought of as having solved the problems of pesticide spraying within Indonesia.

The general success of the rice stabilization program also created unexpected challenges. Virtually every commodity group came forward, urging government involvement in pricing and marketing. The technocrats made clear that their main interest was in rice, the major staple crop. They were able largely to stop an ill-fated price stabilization scheme for cloves being promoted by Soeharto's son. They were also able to convince chili producers and consumers that pricing problems with chiles, created primarily by increased demand from the manufacture of instant noodles, were more efficiently addressed by several planeloads of chilies from Thailand than by a government program. We

[33] One additional dimension of the pesticide story is worth reporting. Professor Afiff was able to work with USAID to develop a multi-million dollar grant for pest management and extension, conditional on the pesticide subsidy being stopped (William Fuller, personal communication 2012). This example is but one of several instances when the technocrats were able to align donor interests with their own in getting important decisions made.

frequently mused that our most important policy advice could often be delivered in one word: "don't."

The technocrats fared less well in their battles involving the marketing of several other commodities, especially sugar, vegetable oil, and wheat. One technique for allocating favors to "friends and generals" was to provide them with monopoly distribution rights for a given commodity such as vegetable oil or sugar, for a particular province. The typical results were excessive margins and high profits for the distributors, and inflated prices borne by consumers. Good arguments were brought to bear against these distribution systems, but were usually defeated. On occasion, however, desirable consequences occurred from these policies. Gaskell (2012) shows, for example, that the control over vegetable oil distribution had a major impact on the phenomenal switch from coconut to palm oil for cooking purposes.[34]

No progress was made on pricing wheat in a manner that would have assisted food security objectives. Indonesia produced no wheat domestically and thus had no wheat-farmer lobby to contend with. Moreover, it had some of the largest and most technically efficient wheat flour mills in the world. The question was how wheat flour from imported wheat was to be priced (Timmer 1971). Internationally, the ratio of rice to wheat flour prices was about 2 to 1. By controlling flour imports, and by using favored distributors, wheat flour in Indonesia was treated as a luxury good and sold at prices equal to rice.[35] Many of us saw great opportunities for a different pricing strategy whereby consumers, especially urban consumers, could shift at the margin in times of rice scarcities. But several attempts to change wheat price policy were to no avail; indeed, they prompted quiet advice that further discussion of this topic might result in the loss of visas! Although the pricing logic was appealing, it flew in the face of Soeharto family interests. Use of milling profits went to the family foundation and for helping to underwrite off-budget military expenses.

Conclusions

Indonesia's approach to food security was quite remarkable in terms of design, implementation, and outcome. A significant number of policies and institutional arrangements were involved, seven of which seem to be especially

---

[34] The Indonesian oil palm story is interesting, important, and complex; however, it is largely outside the scope of this chapter. Between 1965 and 2010, palm oil's share of domestic cooking oil use went from 2 percent to 94 percent. This change occurred as part of plantation and smallholder initiatives for oil palm on the supply side, and aggressive use of marketing rights to influence consumer choice on the demand side.

[35] It is also ironic that virtually all of the wheat imported as food aid passed through these mills and was priced similarly.

important for Indonesia's next generation of policymakers and for policy analysts more generally.

1. Indonesia's leadership was acutely aware that food policy was different in character from agricultural policy. From the outset, leaders were interested both in poor consumers and smallholder farmers.
2. The technocrats understood the key role of macroeconomic policy—exchange rates, interest rates, trade policy—on food security. The jurisdictional bridging of the central economic ministries with agriculture, BULOG, and other sectoral agencies was among their most notable accomplishments.
3. Stabilizing domestic rice prices in Indonesia was crucial for providing a safety net for consumers and incentives for farmers. This approach required significant analytical, financial, and implementation capacity—requirements that may be lacking in many developing countries.
4. Input subsides were seen not as good nor bad per se, but conditional on particular circumstances. Early fertilizer subsidies appear to have been a very wise use of public funds. Pesticide subsidies, by contrast, were one of the worst food-policy errors in the Soeharto period.
5. Improved rice technology, much of it imported, was fundamental to improved food security. Equally important, Indonesian farmers, when provided the opportunity, responded rapidly to improved technology and to incentives more generally.
6. Sequenced policy-reform packages, limited in scope at any given time, progressively improved farmer incentives, raised efficiency in agricultural marketing, and provided better access to staple foods by poor households.
7. The human dimensions of Indonesia's success story were crucially important: Soeharto, for all of his faults, stressed the quality of rural life, family planning, and rural health; and the technocrats, world-class development economists, used an inclusive, one step-at-a-time approach to get agriculture moving.

The foregoing points could easily be expanded into a list of 10 or 20. For example, Indonesia had a substantial resource base that provided the country financial flexibility (particularly after 1975) with respect to its food policy—a flexibility that the country then used wisely. The development of a rural financial sector provided important funds to smallholder farmers and destroyed many myths about rural banking in the process. Similarly, a long-standing tradition of labor-intensive, rice-based agriculture provided the background know-how within agriculture that does not always exist in poor countries.

Many facilitating activities worked together to provide a remarkable story, and this tale of success raises a final question: What can other countries and practitioners learn specifically from the Indonesian experience? There can never be a direct transplanting of strategies and policies from one country to another; the differences in history, culture, and geography are simply too significant. Yet there are important lessons from Indonesia's experience. Operationally, the vital components involved productivity-raising technology for agriculture; consistent and sequenced sets of economic policies; inclusion of groups (consumers) and sectors (health) in addition to agriculture (smallholder farmers); and active ministerial consultation across spatial and organizational jurisdictions.

At a broader level, Indonesia demonstrates that governments and policies need not be perfect to improve food security. Corruption was, and still is, a very significant problem. The technocrats were able partially to play a watchdog role, but even they lost many battles. No one likes corruption except the corrupt; nevertheless, great progress on poverty reduction was still possible.

Indonesia also raises interesting questions about governance and the timing of economic and political reforms and their effects on food security (Stedman, Chapter 12). There was general political stability and consistency of economic policy during the Soeharto period. There were also elections during his time in office, but at that time no one ever called the country a well-functioning democracy. The political situation in Indonesia is changing after the turn of the century, and the swing is toward more democratic processes. But Indonesia suggests the likelihood that, for many countries, democracy may well follow successful development rather than precede it.

References

Boeke, J. H. 1953. *Economics and economic policy of dual societies*. New York: Institute of Pacific Relations.

Booth, A. 1988. *Agricultural development in Indonesia*. Sydney: Asian Studies Association of Australia in association with Allen & Unwin.

Cribb, R. 2002. Unresolved problems in the Indonesian killings of 1965–1966. *Asian Survey* 42(4): 550–563.

Dawe, D., and T. Slayton. 2010. The world rice market crisis of 2007–2008. In *The rice crisis: Markets, policies, and food security*, ed. D. Dawe. Earthscan: London, 15–28.

Falcon, W. 2007. His [Widjojo's] contributions to food security and poverty alleviation in Indonesia. In *Tributes for Widjojo Nitisastro by friends from 27 countries*, eds. M. A. Anwar, A. Ananta, and A. Kuncoro, 25–31. Jakarta: Kompas Book Publishing.

Falcon, W., W. O. Jones, S. Pearson, J. Dixon, G. Nelson, F. Roche, and L. Unnevehr. 1984. *The Cassava economy of Java*. Stanford: Stanford University Press.

Falcon, W., S. Pearson, C. P. Timmer, L. Mears, and M. Hastings. 1985. *Rice policy in Indonesia, 1985–1990: The problems of success*. Jakarta: BULOG.

Falcon, W., R. Naylor, W. L. Smith, M. B. Burke, and E. McCullough. 2004. Using climate models to improve Indonesian food security. *Bulletin of Indonesian Economic Studies* 40(3): 355–377.

FAO. 2011. *The state of food insecurity in the world*. Rome: FAO.

FAO. 2010. *The state of food insecurity in the world*. Rome: FAO.

FAO. 2012. "FAOSTAT." Retrieved December 15, 2012, from http://faostat3.fao.org/home/index.html.

Feder, G., R. Murgai, and J. Quizon. 2003. Sending farmers back to school—the impact of farmer field schools in Indonesia. Research Working Paper WPS3022, World Bank, Washington, DC.

Fitzgerald, P. 2009. "Indonesia Regions Map." Retrieved July 8, 2013, from http://commons.wikimedia.org/.

Frederick, W., and R. Worden, eds. 1993. *Indonesia: A country study*. Washington, D.C: GPO for the Library of Congress.

Gaskell, J. 2012. The vegetable oil revolution in Asia. PhD diss., Stanford Univ.

Geertz, C. 1963. *Agricultural involution*. Berkeley: University of California Press.

Heytens, P. 1991. Technical change in wetland rice agriculture. In *Rice Policy in Indonesia*, ed. S. Pearson, W. Falcon, P. Heytens, E. Monke, and R. Naylor, 99–113. Ithaca: Cornell University Press.

Hughes, J. 1967. *Indonesian upheaval*. New York: D. McKay Co.

International Rice Research Institute. 2012. World rice statistics. Los Banos: IRRI. Accessed August 15, 2012, from http://www.irri.org/index.php?option=com_k2&view=itemlist&layout=category&task=category&id=744&Itemid=100346&lang=en.

IRRI. 2012. "IRRI and Indonesia." Retrieved August 15, 2012, from http://irri.org/our-work/locations/indonesia.

Krugman, P. 1994. The rise and fall of development economics. In *Rethinking the development experience*, ed. L. Rodwin and D. Shoen, 39–58. Washington DC: The Brookings Institution.

Leathers, H., and P. Foster. 2009. *The world food problem*, 4th ed. Boulder, CO: Lynne Reinner Publishers.

Mears, L., and S. Afiff. 1969. An operational rice price policy for Indonesia. *Ekonomi dan Keuangan Indonesia*. Reprinted in B. Arifin. 1995. *Beras, Koperasi Dan Politik Orde Baru*. Jakarta: Pustaka Soinar Harapan, 367–381.

Mosher, A. 1966. *Getting agriculture moving*. New York: Praeger, for the Agricultural Development Council.

Naylor, R. 1991. The rural labor market in Indonesia. In *Rice policy in Indonesia*, ed. S. Pearson, W. Falcon, P. Heytens, E. Monke, and R. Naylor, 58–99. Ithaca, NY: Cornell University Press.

Naylor, R., D. Battisti, W. Falcon, M. Burke, and D. Vimont. 2007. Assessing risks of climate variability and climate change for Indonesian rice agriculture. *PNAS* 14(19): 7752–7757.

Naylor, R., and W. Falcon. 1995. Is the locus of poverty changing? *Food Policy* 20(6): 501–519.

Nestel, B. 1985. *Indonesia and the CGIAR centers*. Washington, DC: World Bank.

Patten, R. H., and D. Snodgrass. 1987. *Monitoring and evaluating KUPEDES (general rural credit) in Indonesia.* Cambridge: Harvard Institute for International Development.

Pearson, S. 1993. Financing rice price stabilization in Indonesia. *Indonesian Food Journal* 7(4): 83–96.

Pearson, S. 1995. Exchange rate policy in Indonesia, 1968–94. In *Beras, Koperasi, dan Politik Orde Baru*, ed. B. Arifin. Jakarta: Pustaka Sinar Harapan, 386–405.

Pearson, S., W. Falcon, P. Heytens, E. Monke, and R. Naylor. 1991. *Rice policy in Indonesia*. Ithaca, NY: Cornell University Press.

Pearson, S., R. Naylor, and W. Falcon. 1991a. Recent policy influences on rice production. In *Rice policy in Indonesia*, ed. S. Pearson, W. Falcon, P. Heytens, E. Monke, and R. Naylor, 8–21. Ithaca, NY: Cornell University Press.

Resosudarmo, B., and S. Yamazaki. 2011. "Training and Visit (T&V) Extension vs. Farm Field School: The Indonesian Experience." Working Paper 2011/01, Australia National University, Department of Economics: Canberra.

Robinson, M. 2002. *Microfinance finance: Lessons from Indonesia*, Vol. 2. Washington, DC: World Bank.

Rosegrant, M., F. Kasryno, L. Gonzales, C. Rasahan, and Y. Saefudin. 1987. *Price and Investment Policies in the Indonesian Food Crop Sector.* Final report submitted to the Asian Development Bank. Washington DC: International Food Policy Research Institute.

Schultz, T. W. 1964. *Transforming traditional agriculture*. New Haven, CT: Yale University Press.

Swinnen, J. 2012. Global food supply: Mixed messages on prices and food security. *Science* 335(6067): 405–406.

Timmer, C. P. 1971. Wheat flour consumption in Indonesia. *Bulletin of Indonesian Economic Studies* 7(1): 78–95.

———. 1973. Choice of technique in rice milling on Java. *Bulletin of Indonesian Economic Studies* 9(2): 57–76.

———. 1986. The role of price policy in increasing rice production in Indonesia, 1968–82." In *Research in domestic and international agribusiness management*, Vol. 6, ed. Ray Goldberg, 55–106. Greenwich, CT: JAI Press.

———. 1987. *The corn economy of Indonesia*. Ithaca, NY: Cornell University Press.

———. 1989. Food price policy: The rationale for government intervention. *Food Policy* 14(1): 17–27.

———. 1996. Does BULOG stabilize rice prices in Indonesia? Should it try? *Bulletin of Indonesian Economic Studies* 32(2): 45–74.

———. 2004. The road to pro-poor growth: The Indonesian experience in regional perspective. *Bulletin of Indonesian Economic Studies* 40(2): 177–207.

———. 2004a. Food security in Indonesia: Current changes and the long-run outlook. Working Paper 48, Center for Global Development: Washington, DC.

Timmer, C. P., and D. Dawe. 2010. Food crises past, present (and future?): Will we ever learn? In *The rice crisis: Markets, policies and food security*, ed. D. Dawe, 3–11. London: Earthscan.

Timmer, C. P., and W. P. Falcon. 2005. Professor Dr. Saleh Afiff: An appreciation. *Bulletin of Indonesian Economic Studies* 41(3): 303–305.

Timmer, C. P., W. P. Falcon, and S. R. Pearson. 1983. *Food policy analysis*. Baltimore: Johns Hopkins University Press.

UNICEF. 1996. "Community-Based Health Care: Indonesia Sets the Pace." New York: UNICEF. Accessed August 15, 2012, from http://www.unicef.org/sowc96/indonesi.htm.

Valdis, J. 1995. *Pinochet's economists: The Chicago school of economics.* Cambridge: Cambridge University Press.

World Bank. 1989. *Project Performance Audit Report, Indonesia, Irrigation Projects X, XIV, and XV*. Washington DC: World Bank.

———. 1990. *Poverty Assessment and Strategy Report*. Washington, DC: World Bank.

———. 1992. *Indonesia: Agricultural transformation challenges and opportunities* (two volumes). Washington, DC: World Bank.

———. 2012. Data. Washington, DC: World Bank. Acccessed August 15, 2012, http://data.worldbank.org/indicator/NY.GDP.PCAP.CD.

———. 2012a. Independent Evaluation Group. Transmigration in Indonesia. Washington, DC: World Bank. Acccessed August 15, 2012 http://lnweb90.worldbank.org/oed/oeddoclib.nsf/DocUNIDViewForJavaSearch/777331DDD0B6239C852567F5005CE5E2.

World Bank Operations Evaluation Department. 1988. *Rural development: World Bank experience*. Washington, DC: World Bank.

3

The Food Security Roots of the Middle-Income Trap

Scott Rozelle, Jikun Huang, and Xiaobing Wang

Argentina was one of the world's wealthiest countries at the start of the twentieth century, with per capita income well above that of Japan, many Western European nations, and neighboring Brazil. Over the course of the century, its economy weakened and become highly volatile, and Argentina failed to maintain high-income status. Despite efforts by the popular Perón leadership following World War II to support labor unions and nationalize several industries, income disparities mounted, real incomes fell for the majority of workers, and the rule of law deteriorated. By 1950, Argentina's per capita income was half that of the United States, and by the end of the century it had fallen well below that of the EU and the Asian Tigers (Taiwan, South Korea, Hong Kong, and Singapore) (Aiyar et al. 2013).

Argentina's experience is not unique—many developing countries in Latin America and elsewhere have become stuck in what economists call the "middle-income trap." It is not surprising to see growth rates taper off after intense periods of escalation, especially in countries that have experienced rapid progress during the poor- to middle-income phase of development. But when countries fall into the middle-income trap, economic growth stagnates and sometimes even declines. It is then difficult to escape a self-reinforcing, downward economic spiral and instead move onto a high-income trajectory (Kharas and Kohli 2011; Agenor and Canuto 2012).

What causes the broader phenomenon of the middle-income trap? Many development specialists blame misguided industrialization and investment policies (Ben-David and Papell 1998; María, Papageorgiou, and Perez-Sebastian 2011; Lin and Treichel 2012), while others attribute the problems to underdeveloped macro-economic institutions such as banking and fiscal agencies (Grilli and Millesi-Feretti 1995; Quinn, 1997; Edwards 2001; Bussiere and Fratscher 2008). Experts also highlight the importance of robust political institutions that are needed to manage different growth patterns and shifting expectations (Mauro 1995; Knack and Keefer 1997; Berg, Ostry, and Zettelmeyer 2012).

There is another, quite distinct explanation for the middle-income trap, however, which was visible in Argentina's case during the twentieth century and is increasingly relevant today for rapidly growing countries like China. It is based on the observation that countries become more vulnerable to economic stagnation or collapse when their internal inequality is high—a situation that is typically accompanied by insufficient investments in health, nutrition and education for those at the lower end of the income distribution (e.g., Maddison 2003; Kohli and Mukherjee 2011; Zhang et al. 2013). The lack of investment in basic human capital leaves large segments of the labor force ill-equipped to join the high-wage, high-technology, service sectors, which play a role for any country aiming to transition from middle- to high-income status. As labor costs rise during the middle phase of development, workers without adequate education and skills are left behind, unable to participate in high productivity jobs and increasingly alienated and polarized from society. In short, these countries fail to establish a productive labor force at a time when a larger domestic pool of high-quality labor is needed to stimulate economic growth, and at the same time, to curb rising wage rates.

Increasing income disparities and alienation also give rise to crime, public riots, violence, and other anti-social behavior, forcing governments to allocate resources toward social control and away from productive investments and the development of the complex and expensive institutions that are needed in fully developed economies. Facing an increasingly unstable environment, investors (foreign and domestic alike) often search for alternate investment opportunities outside the country. The shortage of productive human resources, the reallocation of government resources, and the deteriorating foreign investment climate conspire to reduce the already slowing rates of economic growth.

The predicament of the middle-income trap raises another set of critical questions with regard to food security. Is it possible that the observed poor health and education outcomes among the poor in some middle-income countries are due to systematic and chronic deficiencies in their diets—that the foods they eat fail to provide the nutrients needed for advancing education and cognitive skills at this stage of development? Put more simply: is there a *second food security challenge* that some countries face even as they meet basic caloric needs and enter the ranks of middle income? If so, what is the evidence that there is a nutrition problem helping to set the middle-income trap? How is the second food security challenge related to traditional concepts of food security? What are the causes of nutritional deficiencies, and what are some possible solutions? And is there a role for the state in trying to address the challenge?

Our goal in this chapter is to answer some of these emerging questions. We begin by describing the traditional food security challenge in order to highlight its similarities to and differences from this second food security challenge. (Further discussion of the traditional food security challenge can be found in the first two chapters of this volume, as well as in other sources including Rosegrant and Cline [2003] and Babu and Sanyal [2009].) We then assess the second food security challenge in greater detail. Specifically, we examine the economic and social context of countries vulnerable to the middle-income trap, the evidence that there are nutritional factors contributing to this trap, and potential solutions for overcoming the second food security challenge.

Within this broader discussion, we present a case study of China, in which we provide primary research results on the links between rural poverty, nutrition, health, and cognitive development, focusing specifically on iron-deficiency anemia. Our work underscores the point that poorly diversified diets that are deficient in micronutrients can have serious impacts on health and education, particularly during childhood. Without addressing the second food security challenge, China and other middle-income countries are likely to be constrained in developing economically and socially, in becoming affluent societies, and in achieving security more broadly defined.

Traditional Food Security

The traditional food security problem is defined at both the individual and nation-state levels. At the individual level, food security is about having adequate supplies of affordable food for each household throughout the year to ensure a healthy and productive life. At the national level, food security is about having adequate supplies of affordable food inside the country throughout the year to ensure a stable economic growth path.

When defined in this way, solutions to the traditional food security challenge are almost always focused on supplying individuals or nations with sufficient macronutrients (calories and protein). There is a clear link between traditional food security and economic growth, in that without enough calories and protein, individuals fall into poor health, experience increased morbidity, become stunted or wasted, and are subject to chronic diseases. These conditions negatively affect incomes and end up driving individuals into a poverty trap (as discussed further in the sub-Saharan African context in Chapters 6 and 7). At the national level, high rates of morbidity and mortality restrict the potential of the labor force, causing poverty to become entrenched and creating a series of long-run development challenges.

Food insecurity in the traditional sense is a problem faced by nations that are in the early stages of economic development. Traditional food security is a concern both for nations that are poor and not growing and for nations that are poor and have recently begun to develop. For that reason we associate traditional food security concerns with *phase 1 of development* or the stage in which a nation is poor or between poor and middle income (as reviewed by Naylor in Chapter 1).

The root cause of traditional food insecurity at the household level is well known (Sen 1981; Ivanic and Martin 2008; Von Braun et al. 2008; FAO 2012). The fundamental problem is one of economic access, or the ability of households to purchase and/or produce enough food to meet the needs of all family members. Economic access for the rural poor, in turn, depends on two factors: incomes and food prices. When prices are low, even those with low levels of income are typically able to access food in quantities sufficient to escape the most severe malnutrition from the lack of calories. When prices are low and incomes are rising, food is even more affordable and, hence, economic accessibility is enhanced.

Because of this relationship between food access and security, solutions to traditional food insecurity have been founded on efforts to increase agricultural productivity (Huffman and Evenson 2006). Policies designed to promote productivity include investments in land tenure reform, agricultural research and development (e.g., Green Revolution programs), investments in natural resources (e.g., in water and soils) and agricultural extension services, and subsidy programs that encourage the use of improved seeds, chemical fertilizers, and other modern inputs (as discussed in detail by Falcon in Chapter 2). The success of many traditional food security programs makes the linkages clear: higher agricultural productivity enhances the incomes of farmers who adopt the new innovations and investments. Higher agricultural productivity can also lead to lower and more stable food prices (Datt and Ravallion 1998; de Janvry and Sadoulet 2002; Ivanic and Martin 2010). When prices are lower, even those without land (e.g., poor, landless rural laborers and the urban poor) can become more food secure.

It is no surprise that the price of food is one of the main metrics used when gauging the success of traditional food security policies. Sufficient, low-priced calories and protein (macronutrients) provide economic access—that is, affordable food—for most of the population. Those working in the low-wage, labor-intensive factories and construction sites can be made healthier, stronger, and more productive. Those left working on farms and in the newly emerging factories will also become healthy, strong, and productive.

Achieving traditional food security is thus a key component of national development strategies in many poor countries—one that can ignite growth in phase one of development—although other policy efforts are also needed to supplement the efforts of agricultural officials. When rural households (which make up a large share of the nation's population in phase one of development) become food secure, morbidity falls. If populations are healthier, programs that promote basic education and literacy are often more successful (Newman et al. 2002; Miguel and Kremer 2004). Hence, if industrial and trade policies can attract investment in low-wage, labor-intensive manufacturing, policies promoting traditional food security can help sustain or accelerate the growth of industry and the overall economy (Johnston and

TABLE 3.1

Calorie Supply per Capita (Kcal/capita) in 2009, Crops Equivalent

Insufficient calories		Sufficient calories	
Country	Kcal/capita	Country	Kcal/capita
Haiti	1,979	Brazil	3,173
Chad	2,074	China	3,036
DR Congo	2,056	Mexico	3,146
Ethiopia	2,097	Thailand	2,862
Kenya	2,092	Turkey	3,666

Source: FAOSTAT (2010).

Mellor 1961). The positive results of improving calorie and protein availability and access on a national scale can help nudge economies into virtuous cycles. Although wages are low, hours of employment can be increased, boosting aggregate income. Wages from these diversified sources of income, as well as from household earnings from agriculture, can raise consumption and reinforce positive health, nutrition and education outcomes. A healthier and better-nourished labor force can, in turn, stimulate more industrialization and urbanization, generating further employment opportunities and higher wage rates.[1]

The difference between nations that are food insecure and those that have achieved traditional food security is reflected in average calorie availabilities, as shown in Table 3.1. Columns 1 and 2 reveal the average calorie supply in some of the world's poorest countries—ones that are just starting or that have not yet begun on the path to economic development. When average per capita calorie supply is less than 2,200 kcal per day, a large share of the population lacks access to sufficient macronutrients and remains food insecure, particularly in countries where income inequalities are large and the depth of poverty is great. Columns 3 and 4, in contrast, reflect the situation in some of the successful middle-income countries. Average calorie availability of these five middle-income countries is 3,177 kcal per day. For the average individual in these nations, there is sufficient food energy available for a healthy and productive life. Most of these countries are considered to have solved their most basic food security challenges—calorie deficits are minimal and there are few reports of starvation.

[1] The connections between nutrition, morbidity, labor productivity, and economic growth were a focus of economist Robert Fogel's (2004) work and the topic of his address for the Nobel Prize in Economic Sciences, which he shared with Douglass North in 1993. See http://www.nobelprize.org/nobel_prizes/economic-sciences/laureates/1993/fogel-lecture.html. Accessed September 26, 2013.

Nations Facing the Second Food Security Challenge

Although affordable supplies of calories and protein exist for the majority of citizens in countries that have advanced into the middle-income stage of development (stage two), micronutrient deficiencies often remain widespread. What are the economic and social conditions within these countries that give rise to this second food security challenge?

Table 3.2 presents a partial list of countries in stage two of development, all of which share several key characteristics beyond their middle-income status (defined by the World Bank [2012] as per capita purchasing power parity income of between US$4,000 and US$11,000. Rising real wages, accelerating levels of permanent urbanization, and a move away from low-wage manufacturing and subsistence agriculture are evident in all of these countries. They have all experienced significant industrialization during phase one of development, and by the time they have entered phase two of development, they have begun to re-industrialize into higher technology and service activities that yield greater financial returns to labor inputs. This process occurs because as wages rise, firms need to move into high-value, innovation-based industries and more sophisticated services in order to retain profit margins.

Under these conditions there are new demands placed on the skills and cognitive abilities of individuals in the labor force. When firms were engaged mainly in low-wage manufacturing in stage one of development, labor demand centered on strong, healthy, and disciplined workers. But in phase two, a greater premium is placed on higher education, vocational skills, and the ability to learn (Barro 1991). In order to be worth the high wages that employers must pay (due to market conditions), workers often need to have skills in math, science, English,

TABLE 3.2

Middle Income Countries with Very High Levels of Inequality

Countries	Average per capita ppp-adjusted income in 2011	Gini ratio in 2009
Brazil	11,410	54.7
Chile	19,820	52.1
China	8,390	42.1
Costa Rica	11,910	50.7
Malaysia	15,720	46.2
Mexico	15,930	47.7
Russia	21,700	40.1
Thailand	8,710	40.0
Tunisia	8,860	41
Uruguay	14,600	46.3
Venezuela	12,380	44

Source: World Bank, see note 3.

and information and computer technology (ICT). The workplace becomes more demanding and is constantly changing. Even in jobs that are relatively labor-intensive, workers are asked to take on more responsibilities, interface more with consumers, and perform more varied tasks. In short, workers receive higher wages but must perform at higher competencies as a result.

The good news for a growing economy in phase two of development is that incomes are higher and families have more resources at their disposal. The bad news is that most middle-income countries are also plagued by underdeveloped economic and social institutions (Mankiw et al. 1992; Hall and Jones 1999; Durlauf et al. 2009), including distorted or dysfunctional credit markets (Cerra and Saxena 2008). In many of these nations, individuals and families must use their personal savings to pay for all consumption items, even high-valued assets such as housing and education. Health insurance schemes are often partial or nonexistent (Acharya et al. 2013), and health plans rarely cover catastrophic illnesses or injuries. Similarly, unemployment and disability insurance, as well as social security and welfare programs, are often absent (Hall and Jones 1999; Mont 2007; Durlauf et al. 2009). Most middle-income countries strive to develop these economic and social institutions, but their safety nets can best be described as low and permeable.

Perhaps the most striking characteristic shared by middle-income economies is their high levels of inequality—a point well developed by Simon Kuznets (1955) and reflected in the traditional Kuznets curve.[2] Table 3.2 (column 3) shows the Gini ratio (relative measure of inequality) for a suite of middle-income countries.[3] In all cases, the Gini ratio for middle-income countries is greater than 40, and some have a Gini ratio in the high 50s (e.g., the wealthiest group controls 40–60 percent of the nation's wealth). According to the World Bank (2009), Gini ratios in this range signal substantial inequality within a nation's population, with great wealth and great poverty often existing side-by-side.

In countries where average per capita incomes have risen rapidly—moving them into middle-income status—but where large income inequalities are also present, many citizens fall into the lower tail of the distribution and thus remain poor. In the case of China, for example, average gross national income (on an international purchasing power parity basis) in 2009 was $6800 per capita, and

---

[2] The Kuznets curve depicts a relationship between economic growth and income inequality, with inequality growing as countries move from low to middle income, and declining at higher levels of income. Since the early 1990s, an environmental Kuznets curve has been presented (not designed by Kuznets himself) as a parallel framework to demonstrate the relationship between environmental quality and economic development.

[3] The Gini ratio (also known as the Gini index) measures the extent to which the distribution of income or consumption expenditure among individuals or households within an economy deviates from a perfectly equal distribution. A Gini ratio of 0 represents perfect equality, while a Gini ratio of 100 implies perfect inequality. For further details, see The World Bank: http://data.worldbank.org/indicator/SI.POV.GINI. Accessed September 26, 2013.

only 11.8 percent of the population fell below the extreme international poverty line of US$1.25/day.[4] However, nearly 30 percent lived on less than US$2 per day in 2009, and an even higher share of the population remains under the US$3 per day line. The combination of modest per capita income levels and high inequality implies similar patterns in almost all other middle-income countries.

Evidence of Nutrition Problems in Middle-Income Countries

Households living in poverty are often restricted to simple, starchy diets, and as their incomes rise, their consumption diversifies into a broader mixture of grains, vegetables, fruits, and animal products. This pattern, commonly referred to as "Bennett's Law" (Timmer, Falcon, and Pearson 1983), suggests that households in middle-income countries have more diversified diets than do households in low-income countries. However, high inequality in most middle-income countries means that a large share of residents continues to live in extreme or near poverty, and their diets are often limited to basic starches that are low in micronutrients. Micronutrients include basic vitamins and minerals, such as vitamin A, iron, and iodine, which are essential for human health and cognitive development (Leathers and Foster 2009). Many poor households, even in middle-income countries, lack leafy green vegetables, fruits and meat in their diets. As a result, they commonly suffer from iron deficiency anemia—a debilitating health condition that affects hundreds of millions of people worldwide, mostly in developing countries (Yip, 2001).

IRON DEFICIENCY ANEMIA

Prolonged iron deficiency impairs hemoglobin production, limiting the amount of oxygen that red blood cells can carry to the body and brain. As a consequence, anemia develops, causing lethargy, fatigue, poor attention, and prolonged physical impairment. A large body of research links anemia (as well as iron deficiency not serious enough to impair hemoglobin synthesis) with cognitive impairment and altered brain function (Yip 2001). That means anemia is doubly burdensome, with serious implications for the educational performance of those with the condition; indeed, iron deficiency and anemia have been shown to be negatively correlated with educational outcomes, such as attendance, grades, and attainment. It is potentially responsible for part of the education performance gap between rural and urban students experienced in many developing countries, including rapidly growing economies like China.

[4] The World Bank provides income and poverty data. See http://data.worldbank.org/country/china. Accessed September 26, 2013.

Even in more wealthy societies, anemia has been linked to poor concentration and low educational outcomes, and treating iron deficiency has been shown to have positive results.

Numerous examples exist in the literature. For example, controlled studies of pre-school age children in East Africa indicate improvements in language and motor development following increased levels of iron intake (Stoltzfus et al. 2001). Programs to overcome iron deficiency anemia also have been shown to increase school learning in parts of India (Bobonis et al. 2006). Even in the United States, pockets of lower standardized math test scores among school-age children and adolescents have been attributed to iron deficiency, even in non-severe cases (Halterman et al. 2001). School-age children and adolescents deficient in iron register lower scores on various mental performance and educational achievement tests (Nokes et al. 1998). Treating the iron deficiency of school-age children and adolescents can improve and may even reverse the diminished cognitive and educational performance (Nokes et al. 1998). Treatment of iron deficiency later in life may also be effective for improving health and human capital, as shown in a study of Indonesian adults participating in the Indonesia Family Life Survey; with treatment, adults were less likely to be absent from work due to illness, more likely to be energetic, have better psycho-social health, and have higher earnings.[5]

In virtually all countries, the prevalence of iron deficiency anemia falls when incomes rise (Gwatkin et al. 2007; de Benoist et al. 2008).[6] Nonetheless, anemia rates in many middle-income countries remain high, particularly where income disparities are large. Table 3.3 shows that anemia rates among preschool-aged children in Brazil, Mexico, Turkey, Thailand, and China range from 20 to 55 percent. Rates for pregnant and reproductive-aged women are

TABLE 3.3

Percent of Children and Women with Iron-Deficiency Anemia, Selected Countries (1993–2005)

Country	Preschool	Pregnant women	Reproductive-aged women
Brazil	55	29	23
China	20	29	20
Mexico	30	26	21
Thailand	25	22	18
Turkey	33	40	26
US	3	5	7

Source: United Nations World Food Programme.

[5] The Indonesian Family Life Survey (IFLS) is an ongoing longitudinal survey sponsored by the RAND Labor and Population Program (http://www.rand.org/labor/FLS/IFLS.html). The research on iron deficiencies within the IFLS was conducted by Duncan Thomas and colleagues in 2003, see http://emlab.berkeley.edu/users/webfac/emiguel/e271_s04/friedman.pdf. Accessed September 26, 2013.

[6] For national data on anemia, see the World Health Organization's *Global Database on Anemia*: http://www.who.int/vmnis/anaemia/en/. Accessed September 26, 2013.

likewise high, with a wide range. According to these data, large shares of the populations in these middle-income countries are malnourished, even though they are consuming sufficient calories.

A Case Study of Anemia in Rural China

In recent decades, China has seen dramatic improvements in income, health, and education as its economy has boomed. The government has made significant investments in agricultural productivity, and both the rural and urban populations have achieved traditional food security as measured by calorie and protein availabilities and access. The average person's daily calorie supply exceeds 3,000 (as shown in Table 3.1), and only a very small share of the population consumes less than 2,200 calories per day.

Nonetheless, a number of studies conducted in the late 1990s and early 2000s suggested that a significant share of children across rural China were so severely iron deficient as to be classified as anemic. For example, a study in Shaanxi Province run by the provincial Center for Disease Control found anemia in as many as 40 percent of freshmen in a rural junior high school (Xue et al. 2007). A study in Guizhou Province found anemia rates to be as high as 50 to 60 percent (Chen et al. 2005). These studies were small and non-representative, but they still indicate that anemia may be a serious problem in rural China, at least for a segment of the population. Because of China's increasing income gap between rural and urban areas, it is likely that only a small share of the urban population suffers from anemia (Fu et al. 2003). But if a large share of China's rural population is still experiencing widespread nutritional deficiencies, including anemia, one might conclude that recent government efforts to reduce inequality and improve rural welfare have not been sufficient.

The problem is that high-quality data on anemia and other micronutrient deficiencies are difficult to come by for China. The earlier local studies were mainly based on anecdotal evidence, and the larger-scale studies reported figures for the country as a whole without reporting to the regional level. Worldwide, few countries have conducted rigorous, empirically based analyses of micronutrient problems in areas where large numbers of poor people live. If such analyses could be done, it would be easier to systematically identify the segments of the population that suffer from micronutrient deficiencies, and to address the second food security challenge regionally and globally.

CONDUCTING FIELD TRIALS

Recognizing the dearth of reliable data in this area, our research team began a series of field trials on micronutrient deficiencies and treatment in rural China in 2008, some of which are still ongoing (Luo et al. 2012a,b; Kleiman-Weiner et al., 2013; Sylvia et al. 2013). Two sets of studies are particularly relevant for the discussion here. (Our methodology for these studies is described in greater detail in Appendix A.) The first set examined new mothers and babies between 6 and 12 months old from over 1,800 households in 351 villages. All participants were tested for anemia, and babies were assessed for cognitive and motor skill development. In total we tested 1,808 babies. The sample covered 21 nationally designated poverty counties in Han-populated areas in Southern Shaanxi province, in the Qingling Mountain range. This mother-baby study is among the largest infant anemia studies ever conducted in China.

The second set of studies was based on student- and school-level tests of third-, fourth- and fifth-grade students. We collected data from 41 nationally designated poverty counties in four western provinces: Ningxia, Qinghai, Shaanxi, and Sichuan. These four provinces are among the poorest in China based on per capita income (Luo et al. 2011).[7] Over 737 million people live in rural regions of China, accounting for 56 percent of the population. Even if we consider only the rural populations of the poor counties in our four sample provinces, the results from this set of studies are relevant for more than 50 million children.

FIELD TRIAL RESULTS

The results of our field trials confirmed that the incidence of anemia in poor areas of China is extremely high. In the baby study, nearly half of the infants (49 percent) had hemoglobin levels below the anemic threshold (indicating insufficient iron, i.e., anemia), and over a quarter (28 percent) were in the near-anemic range (see Appendix A). In total, 77 percent of babies in our study were anemic or near anemic. When scaled up, these results suggest that if an estimated 40 percent of China's children are being born and raised in poor rural counties, then just under one-third (77 percent of the 40 percent) of China's infants are vulnerable to iron deficiency anemia.

The immediate consequences of such high rates of anemia are reflected in low developmental scores as measured by the Bayley Scales of Infant Development (BSID) (see Appendix A). Our results show that 20 percent of babies were significantly delayed in cognitive development (>2 standard deviations below the international mean), and 32 percent were significantly delayed in their motor development. In total, 40 percent of the babies in our study demonstrated one or both types of impairment. Interestingly, a relatively small fraction of the babies were stunted or wasted; the anthropometrics section of the study revealed that fewer than 2 percent were wasted and fewer than 5 percent

[7] Average annual per capita income in Ningxia, Qinghai, Shaanxi, and Sichuan provinces ranges from about US$345 to US$465 and falls below the mean national income by 23%, 35%, 14%, and 36%, respectively.

were stunted. One might infer from these results that the sample population of infants was not suffering from traditional food insecurity (calorie and protein deficiencies) but instead from micronutrient deficiencies.

Anemia rates were also very high in our studies of school-aged populations (Appendix B). One-third of the students in our sample had anemia or were at risk of becoming anemic. The variation across counties and even across schools within counties was large; for example, more than 90 percent of the students in each of the four sample schools in Qinghai Province were anemic, while fewer than 10 percent of the students in the four schools in Ningxia province were anemic. According to the World Health Organization, anemia should be considered a serious problem in populations with a 5 percent or greater prevalence. Of the 283 schools we sampled, only four had anemia levels that would *not* be considered serious. All 41 counties in our study contained schools with anemia levels above this cutoff, suggesting widespread malnutrition and micronutrient deficiencies. It is no surprise, therefore, that we commonly saw students taking naps at recess or sitting listlessly in class during the course of our research.

IMPLICATIONS OF OUR RURAL FIELD STUDIES

Our field results have serious implications for the health and education of a significant share of China's future labor force. Although our studies have been focused on just four provinces in western China, they target poverty designated counties and are relevant for a school-aged population of between 10 and 15 million. The total number of malnourished youth in China is obviously much larger. If nutrition is not improved in rural areas—and specifically if micronutrient intake is not enhanced among China's rural population—these children will have difficulties in learning and will fall behind in the school system. Studies have shown that many students drop out of school as early as grade seven when their grades drop and they become uncompetitive in school (Yi et al. 2011). As China's low-wage jobs disappear with economic development, the problem of school dropouts is exacerbated. In this way, China is a victim of its own success—economic growth leads to higher wages, but economic disparity leads to greater polarization in the labor force and a greater rural-urban income gap. The poorly educated segment of the labor supply will increasingly move into the informal sector with no long-term benefits. This scenario portends greater social conflict in a country that is widely known for economic development under conditions of strong social control.

Even in the absence of social strife, it is still likely that there will be significant segments of the future labor force in China that will have difficulties acquiring the competency in the math, science, literacy, foreign language, and ICT skills needed to support industrial growth in the country's changing economy. In short, China may be setting itself up to be in danger of falling into the middle-income trap. This possibility is not absent from the minds of key leaders in China. In June 2013, China's Vice Premier, Zhang Gaoli, gave a speech in Chengdu (the capital of Sichuan province in Southwest China) claiming specifically that the country *"has to stand up to the test of striding over the middle-income trap."*[8]

Explanations of the Second Food Security Challenge

What is really behind the second food security challenge? Why is it that large segments, or at least non-trivial shares, of populations of middle-income countries suffer from malnutrition when their economies are expanding rapidly and absolute poverty is shrinking? Why don't families invest more into their babies, children, mothers-to-be and others?

REASON ONE: MARKET SIGNALS AND POOR INSTITUTIONS

The first reason is that countries in the early stages of phase two of development may be victims of their own success during phase one of economic development. Investments in agricultural technology and institutions, along with policies supporting agricultural production and trade, have been an important engine of growth for many of these countries, leading them toward middle-income status (see Chapter 2 by Falcon). As a result, large supplies of low-priced staples have become available to their populations. As prices fall, basic economics suggests that demand rises for normal goods (as seen with downward-sloping demand curves); the growth in demand for staple crops is often especially strong in low-income economies because the share of budgets spent on food by poor households is large (Timmer, Falcon, and Pearson 1983).[9] Moreover, the income elasticity of demand for staple foods tends to be positive and relatively high for low-income societies, meaning that when their incomes rise, people demand more of these foods. Thus as countries invested heavily in staple crops to solve traditional food insecurity problems early in the development process, they set up a situation of rising demand for crops with high calories—but broadly deficient in micronutrients—particularly among the lower end of the income distribution. Households generally do not substitute into more expensive calories (e.g., meat, milk, eggs, vegetables, fruits) until their incomes rise sufficiently to support a more diversified diet.

Even when their incomes rise, there may be other barriers that keep households from consuming more expensive, micronutrient-rich calories. For

---

[8] *Bloomberg Businessweek*, June 13, 2013: http://www.businessweek.com/articles/2013-06-13/correlations-the-middle-income-trap. Accessed September 27, 2013.

[9] This result comes from the fact that the price elasticity of demand is a function of both substitution and income effects. See Timmer, Falcon, and Pearson (1983).

example, the real costs of these products might be higher than the market price. While staple grains and pickled vegetables can be produced at home and stored for long lengths of time relatively cheaply, this is not necessarily true for many meats, fruits, vegetables, and dairy products that require refrigeration. Refrigeration, in turn, requires reliable electricity or diesel. And, even when there is reliable power, refrigeration requires expenditures on monthly usage. Moreover, micronutrient-rich foods need to be accessible, and markets are not always nearby. Finally, meats, fish, and other foods are often more expensive to prepare in terms of both fuel and time.

There are other reasons, as well, that explain why households may be reticent to spend their income on more expensive calories. As Banerjee and Duflo describe in their popular book on "*Poor Economics*" (2011), there are many competing uses for the income of families in developing countries—a point that is reinforced, as discussed earlier, by the fact that many middle-income countries have underdeveloped credit markets, health care systems, and other economic and social institutions. Without these institutions, families must place high priorities on the use of their liquid assets. Without access to reasonable mortgage loans, families need to save for housing. The absence of consumer loans means that families also must save for the marriage of their sons and daughters. Poorly developed social security and pension systems mean that households also need to save for retirement. Imperfect health insurance systems mean that households must save for catastrophic illnesses/injuries. There are also new demands in prospering communities for temples, festivals, celebrations, and other cultural activities. In short, the fundamental logic is: "Why spend additional money on a higher-priced, more varied diet, when there are so many other important demands on the family budget?"

REASON TWO: ABSENCE OF KNOWLEDGE ABOUT NUTRITION

The reluctance to allocate family budgets to higher quality, more expensive diets is also linked to the absence of knowledge about the importance of good nutrition. Micronutrient-based diseases, such as anemia, are often called "hidden hunger." This term is used because there are no overt symptoms of the disease. In addition, there is only slow and imperfect correlation between nutritional interventions (e.g., the use of supplementation or the adoption of new diets) and anemia status. For example, the short-term correlation between anemia status and physical stature, motor development, and cognitive skills is often weak and difficult to observe.

In many settings there is another important reason why knowledge may be lacking. With high rates of migration, caregiving is often the responsibility of grandparents or other older relatives. In most developing, middle-income countries, grandparents are illiterate (or only semi-literate) and have never had an opportunity to read or learn about the importance of nutrition. In addition, in a number of countries economic development and social change have been so rapid that it is hard for older people to imagine why nutrition would be needed for math and language, and not just for good farming. In their younger days (and even today), basic calories and protein have supported their productivity in the fields and in the factories. Switching from inexpensive staples to more expensive foods rich in micronutrients is counterintuitive for many of these people who have survived on basic needs and frugality.

Finally, there tends to be a complete absence of formal nutrition education and training in most developing country school systems. In an earlier generation when most of the population was deficient in macronutrients, there was little impetus to teach students (in school) or adults (in extension classes) about the importance of micronutrients. The absence of nutrition education was most stark when the grandparents, and even parents, of today's youth where school-aged children, which was during stage one of development—a time when traditional food security concerns were at the forefront.

An illustration from our school survey in Northwest China tells the story clearly. In our study, we wanted to test the premise that one of the most educated persons in a rural community is the principal of the school. The results were striking. In a sample of more than 200 school principals (from 200 randomly selected elementary schools in Shaanxi province), only about 1 out of 20 principals even knew what "anemia" was. Moreover, data from our baseline survey showed that principals believed that less than 5 percent of their students were nutritionally deprived—when subsequent analysis showed that the number was more than 30 percent. These results pose a challenging question: If the school principals know so little about nutrition, is it reasonable to expect the parents or grandparents of their students to know much more?

Sadly, our study demonstrated that the knowledge of parents of students in these schools was even lower than that of the principals. Only 2 out of 100 caregivers reported ever having had any formal education or training in nutrition. Virtually none of the caregivers of students who tested positive for anemia had any indication that his or her child was sick. Even fewer knew how to treat the condition.

REASON THREE: TIME INCONSISTENCY

A third and final explanation for the emergence of the second food security challenge is time inconsistency—the inconsistency between the demand for labor skills in the present period, and the need to invest in labor skills for a future that relatively few people can accurately perceive. Current health and cognitive skills may be fine for now (e.g., for farming or for work in a labor-intensive, low-skill, low-wage environment). However, these health and cognitive abilities will not be sufficient for a time period 10 to 20 years from now when employers will require workers to have computer skills, language capabilities, science knowledge, or math aptitude. Without such skills, employers are likely to decide that workers are not productive enough to be paid the going (and relatively high) wage rate.

Conditions for the second food security challenge appear to be set in motion by the dynamics of overcoming traditional food insecurity. But they are maintained, it seems, by the three-way interaction of the absence of nutrition knowledge, the competition for family budgets, and poor long-run vision on what sorts of investments in human capital are needed and how nutrition interventions are likely to play out. If development for these middle-income countries had taken many generations (as it did in many Western European countries and their colonies including the United States, Canada, Australia, and New Zealand), there may have been less of a time inconsistency and more time for adjustment and learning. But in the fastest-developing countries today, especially those with high levels of inequality, the second food insecurity challenge is pervasive and could well become a powerful and dangerous barrier to further development.

Need for Policy Action

Should policymakers thus try to produce better nutrition outcomes in order to combat the second food security challenge? The case for government intervention is justified if the social return is greater than the private return, and/or if there are negative externalities caused by market failure. In addressing the second food security challenge, both of these conditions appear to hold. Both the social and private returns to intervention are high for the countries in question when one contemplates the benefits, which range from human health to education to economic development to peaceful societies. In addition, the "hidden hunger" nature of anemia and other micronutrient deficiencies, as well as their treatments, seem to work against the potential solutions being offered by the market or adopted and financed by individuals. Specifically, it is hard for individuals who are poor to invest in something with a return that is 10 to 20 years away, even if that potential return is high. This point is especially true if credit markets are imperfect, information is imperfect and other institutions that supply insurance against health problems and retirement are imperfect.

Finally, the case for government action is strengthened by the low discount rates for individuals and households in developing countries. A healthy and cognitively adept labor force is essential for future growth and development, and to avoid stagnation and collapse, which would affect all of society. As a result, there appears to be a strong role for the state in trying to address micronutrient-deficiency based food insecurity.

So what can states do? A recent review of papers reporting on interventions that involve nutrition with the goal of improving learning outcomes (McEwan

2013), presents a number of different approaches. These include nutritional education offered in schools; nutritional training in villages through the public health system; efforts to diversify agriculture; programs to promote nutrient fortification of salt, flour, and other products; and programs that favor the direct use of micronutrient supplements. According to McEwan's analysis, these programs have had an overall positive impact on learning. But this meta-analysis also documents considerable heterogeneity. Although some interventions are shown to have a positive and statistically significant impact on learning, many of the programs have failed to produce measureable, positive impacts.

According to the experiences of other countries as well as our own experience in China, there are a number of factors that help some programs succeed while others end up in failure. First, in poor rural areas and urban slums, it is difficult to use information treatments to get large positive impacts. The human capital (and ability to learn) in these areas is often low. The subtle and difficult-to-observe correlations between nutrition and learning make it hard for individuals to accept and act on information provided in a short-term training project. In addition, even if households wanted to act on new nutrition knowledge, other barriers including the absence of refrigeration or poor access to markets make that difficult. Encouraging the diversification of crops in the field, meanwhile, often means that policymakers must reverse the actions that led to liberalized, productive agricultural economies in the first place. This strategy might end up creating serious market distortions. Fortified foods can be a powerful tool. But in many areas the foods that are to be fortified, such as wheat flour, are produced by the households themselves rather than purchased. As a result they never make it through the fortification process. And because households are reluctant to pay for nutrition, supplementation programs need to be free to users, making them expensive for governments to run. Raising benefits further by offering money for participation in the nutrition program may enhance the uptake but further increases costs, as well as complicating programs with the need to verify nutritional intake. It already takes massive efforts by the state on many fronts to make progress in solving the traditional food security challenge. In all likelihood it will take a multi-pronged effort that will include training, education, fortification, supplementation and more to combat the second food security challenge, too.

One of the most difficult barriers to addressing the second food security challenge, in fact, may have to do with political economy. Policies and interventions to counter the second food security challenge will work best if policymakers are thinking far ahead. The best time to address the second food security challenge is immediately after progress has been made in solving the traditional food security challenge, while nations are still in the first phase of development. Even when wages are still low (but about ready to begin rising), the education system needs to be ensuring student progress through high school, with education in math, science, language, English, and ICT skills. The education system must in effect train students for jobs that they will be taking 20 to 30 years in the future—jobs that in many cases do not yet exist in the developing country. For example, a large majority of students finished high school in South Korea and Taiwan even during the era of low wages in those countries, which is one reason that these countries became known as Asian Tigers. To advance education early and effectively, micronutrient deficiencies need to be overcome before a country reaches middle income. This means that there needs to be an early and concerted effort that almost certainly needs to be led by the state—and leaders who have a vision that stretches decades into the future.

Appendix A: Methodology of Rural Field Trials in China

In choosing our student sample observations, we followed a uniform selection procedure. First, we obtained a list of all nationally designated poverty counties in each of the study regions. In China, "poverty" is a designation given by the National Statistics Bureau as a way of identifying counties that contain significant concentrations of people who live under the national poverty line. There are 592 nationally designated poverty counties in China; they comprise about one third of the total counties and account for 20 percent of China's total population (PALGO 2012). There are 109 poor counties in our four study provinces. From these poor counties we took a random sample of 41 counties, based on our statistical power calculations of necessary sample size.

Inside each sample county, our survey team obtained a list of all townships, and in each township we then obtained a list of all *wanxiao* (rural elementary schools with six full grades, grades 1–6). Sampled schools had over 400 students and at least 50 boarding students. In total, we identified 368 schools that met these criteria and randomly chose 283 schools for inclusion in our study. The location, size, date, and other information about the survey are summarized and grouped by province and study year in Appendix Table 3.A1.

Data were collected by eight enumerator teams. In each team one person collected data on the school from principals and third-, fourth-, and fifth-grade homeroom teachers, while others collected individual and household socio-economic information from students. Trained nurses from the Xi'an Jiaotong University's School of Medicine measured hemoglobin levels on-site using HemoCue Hb 201+ finger-prick technology.

We used the international standard threshold for determining anemia in both studies. Babies are considered anemic at hemoglobin levels at or below 110 g/L, and near-anemic (in danger of falling into anemia) at hemoglobin levels above 110 g/L and below 120 g/L. For school-aged children, the threshold for anemia is 120 g/L. Cognitive skills of babies were assessed using the Bayley Scale of Infant Development (BSID).[10]

[10] BSID is used to measure the cognitive and motor development of infants 1 to 42 months of age.

APPENDIX TABLE A1

Description of the Sample Populations Used for Our Rural Studies of Anemia

	Sample province	Number of sampled counties	Per capita income of sample area (PPP-adjusted, in USD) [a]	Number of sampled schools	Number of sampled students	Survey date
Dataset 1	Shaanxi	9	683.48	70	4151	October 2008
Dataset 2	Shaanxi	8	660.20	24	1,476	June 2009
Dataset 3	Shaanxi	10	769.14	66	2,066	October 2009
Dataset 4	Qinghai	5	813.91	37	1,474	October 2009
Dataset 5	Ningxia	5	794.21	37	2,658	October 2009
Dataset 6	Sichuan	3	1085.81	21	516	April 2010
Dataset 7	Shaanxi	1	579.02	28	427	April 2010
Total/Avg	—	41	769.44	283	12,768	—

[a] All values are reported in US dollars in real PPP terms by dividing all figures that were initially reported in yuan (Chinese currency) by the official exchange rate (7.62 yuan: 1 dollar in 2007) and multiplying by the purchasing power parity multiplier (1: 2.27543).

Data sources: Authors' surveys.

Appendix B: Results of Anemia Field Trials in Rural Schools

Across all of the schools surveyed (41 counties), we found the overall mean hemoglobin average was 124.6 g/L. Hemoglobin levels were normally distributed across all seven datasets, with a standard deviation of 12.5 (Table 3.B1, row 1, column 3). In our sample, 4,303 of the 12,768 students had hemoglobin levels lower than 120 g/L, yielding a population anemia prevalence of 33.7 percent (row 9, column 3). If we were to instead use an anemia cutoff of 115 g/L, anemia prevalence would be lower but still significantly high at 21 percent.

There was considerable variation in anemia prevalence (< 120 g/L) across the sample, ranging from 25.4 percent in Ningxia to 51.1 percent in Qinghai. From a multiple regression of county dummy variables on anemia levels (results not shown for brevity), the p-value of the test (an F-test of the joint significance

APPENDIX TABLE B1

Hemoglobin Counts and Anemia (Hb < 120g/L) Prevalence in Sample Students

	12 Years old and below	Above 12 years old	Total
Hemoglobin (g/L)			
Total[a]	124.5 (12.3)	125.4 (14.3)	124.6 (12.5)
Shaanxi—2008 (Dataset 1)	122.8	124.6	122.9
Shaanxi—2009a (Dataset 2)	124.7	125.1	124.8
Shaanxi—2009b (Dataset 3)	126.7	131.0	126.9

	12 Years old and below	Above 12 years old	Total
Qinghai—2009 (Dataset 4)	119.2	118.0	118.9
Ningxia—2009 (Dataset 5)	128.2	131.7	128.7
Sichuan—2010 (Dataset 6)	126.1	N.A.	126.1
Shaanxi—2010 (Dataset 7)	125.2	124.6	125.2
Anemia (%)			
Total	33.8	33.2	33.7
Shaanxi—2008 (Dataset 1)	37.7	33.0	37.5
Shaanxi—2009a (Dataset 2)	31.6	31.3	31.6
Shaanxi—2009b (Dataset 3)	26.8	15.5	26.2
Qinghai—2009 (Dataset 4)	50.3	53.1	51.1
Ningxia—2009 (Dataset 5)	26.3	19.8	25.4
Sichuan—2010 (Dataset 6)	24.8	N.A.	24.8
Shaanxi—2010 (Dataset 7)	33.2	32.1	33.1

[a] Numbers in parentheses indicate the standard deviation of hemoglobin count distribution.

Data source: Authors' surveys. See Table A1 for more information about the datasets.

of the dummy variables) indicated that there was a significant county effect ($p < 0.001$) within provinces.

References

Acharya, A., S. Vellakkal, F. Taylor, E. Masset, A. Satija, M. Burke, and S. Ebrahim. 2013. The impact of health insurance schemes for the informal sector in low-and middle-income countries: A systematic review. World Bank Policy Research Working Paper 6324.

Aiyar, S., R. Duval, D. Puy, Y. Wu, and L. Zhang. 2013. Growth slowdowns and the middle-income trap. Working Paper, International Monetary Fund.

Babu, S., and P. Sanyal. 2009. *Food security, poverty and nutrition policy analysis: Statistical methods and applications*. Washington, DC: Academic Press.

Banerjee, A. V., and Esther Duflo. 2011. Poor economics: A radical rethinking of the way to fight global poverty. PublicAffairs Store.

Barro, R. J. 1991. Economic growth in a cross section of countries. *Quarterly Journal of Economics* 106(2): 407–443.

Ben-David, D., and D. H. Papell. 1998. Slowdowns and meltdowns: Postwar growth evidence from 74 countries. *Review of Economics and Statistics* 80(4): 561–571.

de Benoist, B., E. McLean, I. Egli, and M. Cogswell, eds. 2008. Worldwide prevalence of anaemia 1993–2005: WHO global database on anaemia. World Health Organization.

Berg, A., Jonathan D. O., and J. Zettelmeyer. 2012. What makes growth sustained? *Journal of Development Economics* 98(2): 149–166.

Bobonis, G. J., E. Miguel, and C. Puri-Sharma. 2006. Anemia and school participation. *Journal of Human Resources* 41(4): 692–721.

Bussière, M., and M. Fratzscher. 2008. Financial openness and growth: short-run gain, long-run pain? *Review of International Economics* 16(1): 69–95.

Cerra, V., and S. Chaman Saxena, 2008. Growth dynamics: The myth of economic recovery. *American Economic Review* 98(1): 439–457.

Chen, J., X. Zhao, X. Zhang, et al. 2005. Studies on the effectiveness of NaFeEDTA-fortified soy sauce in controlling iron deficiency: A population-based intervention trial. *Food & Nutrition Bulletin* 26(2): 177–186.

Datt, G., and M. Ravallion. 1998. Farm productivity and rural poverty in India. *The Journal of Development Studies* 34(4): 62–85.

de Janvry, A., and E. Sadoulet. 2002. World poverty and the role of agricultural technology: direct and indirect effects. Journal of Development Studies 38 (4): 1–26.

Durlauf, S. N., A. Kourtellos, and C. M. Tan. 2008. Are any growth theories robust? *The Economic Journal* 118(527): 329–346.

Edwards, S. 2001. Capital mobility and economic performance: Are emerging economies different? No. W8076, National Bureau of Economic Research.

FAO (Food and Agriculture Organization of the United Nations). 2012. *State of food insecurity in the world 2012*. Rome: FAO. Accessed September 26, 2013, http://www.fao.org/docrep/016/i3027e/i3027e00.htm.

Fu, Z. Y., F. M. Jia, W. He, G. Fu, and C. M. Chen. 2003. The status of anemia in children under 5 and mothers in China and its factor analysis. *Acta Nutrimenta Sinica* 25(1): 70–73.

Grilli, V., and G. M. Milesi-Ferretti. 1995. Economic effects and structural determinants of capital controls. *IMF Staff Papers* 42(3): 517–551.

Gwatkin, D. R., S. Rutstein, K. Johnson, E. Suliman, A. Wagstaff, and A. Amouzou. 2007. Socio-economic differences in health, nutrition, and population. Country Reports on HNP and Poverty, The World Bank.

Hall, R. E., and C. I. Jones. 1999. Why do some countries produce so much more output per worker than others? *The Quarterly Journal of Economics* 114(1): 83–116.

Halterman, J. S., J. Kaczorowski, C. Aligne, P. Auinger, and P. Szilagyi. 2001. Iron deficiency and cognitive achievement among school-aged children and adolescents in the United States. *Pediatrics* 107(6): 1381–1386.

Hayami, Y., and V. W. Ruttan, 1971. *Agricultural development: An international perspective.* Baltimore/London: Johns Hopkins Press.

Huffman, W. E., and R. E. Evenson, 2006. *Science for agriculture: A long-term perspective.* Oxford: Wiley-Blackwell Publishing.

Ivanic, M., and W. Martin, 2008. Implications of higher global food prices for poverty in low-income countries. *Agricultural Economics* 39(s1): 405–416.

———. 2010. Poverty impacts of improved agricultural productivity: Opportunities for genetically modified crops. *AgBio Forum* 13(4): 308–313.

Johnston, B. F., and J. W. Mellor. 1961. The role of agriculture in economic development. *The American Economic Review* 51(4): 566–593.

Kharas, H., and H. Kohli. 2011. What is the middle income trap, why do countries fall into it, and how can it be avoided? *Global Journal of Emerging Market Economies* 3(3): 281–289.

Kleiman-Weiner, M., R. Luo, L. Zhang, Y. Shi, A. Medina, S. Rozelle. 2013. Eggs versus chewable vitamins: Which intervention can increase nutrition and test scores in rural China? *China Economic Review* 24:165–176.

Knack, S., and P. Keefer. 1997. Does social capital have an economic payoff? A cross-country investigation. *The Quarterly Journal of Economics* 112(4): 1251–1288.

Kohli, H. A., and N. Mukherjee. 2011. Potential costs to Asia of the middle income trap. *Global Journal of Emerging Market Economies* 3(3): 291–311.

Kuznets, S. 1955. Economic growth and income inequality. *American Economic Review* 45:1–28.

Leathers, H. D., and P. Foster 2009. *The world food problem: Toward ending undernutrition in the Third World*. Ann Arbor, MI: Lynne Rienner Publishers.

Lin, J. Y., and V. Treichel. 2012. Learning from China's rise to escape the middle-income trap: A new structural economics approach to Latin America. World Bank Policy Research Working Paper 6165, World Bank.

Luo, R., Y. Shi, L. Zhang, H. Zhang, G. Miller, A. Medina, and S. Rozelle. 2012. The limits of health and nutrition education: Evidence from three randomized controlled trials in rural China. *CESifo Economic Studies* 58(2): 385–404.

Luo, R., Y. Shi, L. Zhang, C. Liu, S. Rozelle, B. Sharbono, A. Yue, Q. Zhao, and R. Martorell. (2012). Nutrition and educational performance in rural China's elementary schools: Results of a randomized control trial in Shaanxi province. *Economic Development and Cultural Change* 60(4): 735–772.

Maddison, A. 2003. Development Centre Studies. The World Economy: Historical Statistics. Paris, France: OECD Publishing.

Mankiw, G. N., D. Romer, and D. N. Weil. 1992. A contribution to the empirics of economic growth. *The Quarterly Journal of Economics* 107(2): 407–437.

Mauro, P. 1995. Corruption and growth. *The Quarterly Journal of Economics* 110(3): 681–712.

McEwan, P. J. 2013. Improving learning in primary schools of developing countries: A meta-analysis of randomized experiments. Working paper, Wellesley College.

Miguel, E., and M. Kremer. 2004. Worms: Identifying impacts on education and health in the presence of treatment externalities. *Econometrica* 72(1): 159–217.

Mont, D. 2007. Measuring disability prevalence. World Bank Social Protection Discussion Paper 0706, World Bank.

Newman, J., M. Pradhan, L. B. Rawlings, G. Ridder, R. Cao, and J. L. Evia. 2002. An impact evaluation of education, health, and water supply investments by the Bolivian Social Investment Fund. The World Bank Economic Review 16(2): 241–274.

Nokes, C., C. van den Bosch, and D. A. P. Bundy. 1998. The effects of iron deficiency and anemia on mental and motor performance, educational achievement, and behavior in children. A report of the INACG. Washington, DC: International Life Sciences Institute.

Quinn, D. 1997. The correlates of change in international financial regulation. *American Political Science Review* 91(3): 531–551.

Rosegrant, M. W., and S. A. Cline. 2003. Global food security: Challenges and policies. *Science* 302: 1917–1919.

Sen, A. 1981. *Poverty and famines: An essay on entitlement and deprivation*. Oxford England: Oxford University Press.

Stoltzfus, R. J. 2001. Defining iron-deficiency anemia in public health terms: A time for reflection. *The Journal of Nutrition* 131(2): 565S–567S.

Sylvia, S., R. Luo, L. Zhang, Y. Shi, A. Medina, and S. Rozelle. 2013. Do you get what you pay for with school-based health programs? Evidence from a child nutrition experiment in rural China. *Economics of Education Review* 37:1–12.

Timmer, C. P., W. P. Falcon, and S. R. Pearson. 1983. *Food policy analysis*. Baltimore: Johns Hopkins University Press.

Von Braun, J., A. Ahmed, K. Asenso-Okyere, S. Fan, A. Gulati, J. Hoddinott, R. Pandya-Lorch, M. W. Rosegrant, M. Ruel, M. Torero, T. van Rheenen, K. von Grebmer. 2008. High food prices: The what, who, and how of proposed policy actions. Policy brief 1A. Washington, DC: International Food Policy Research Institute.

World Bank. 2007. Poverty headcount ratio at $1.25 a day. Accessed September 4, 2013, http://data.worldbank.org/indicator/SI.POV.DDAY.

———. 2009. "GINI Index." Accessed September 4, 2013, http://data.worldbank.org/indicator/SI.POV.GINI.

———. 2012. Gross national income per capita 2012, Atlas method and PPP. World Development Indicators database.

———. 2013. World Development Indicators: Distribution of income or consumption.

Xue, J., Z. Wang, J. Zhang, Y. Han, C. Huang, Y. Zhang, and Z. He. 2007. Study on current situation of anemia and intelligence of country adolescents in Shaanxi. *Modern Preventive Medicine* 34(9): 1800–1801.

Yip, R. 2001. Iron deficiency and anemia. Nutrition and health in developing countries. In *Nutrition and health in developing countries*, ed. R. D. Sembad and M. W. Bloem, 327–342. Totowa, NJ: Humana Press.

Zhang, L., H. Yi, R. Luo, C. Liu, and S. Rozelle. 2013. The human capital roots of the middle income trap: the case of China. *Agricultural Economics*, 44(s1): 151–162.

4

Institutions, Interests, and Incentives in American Food and Agriculture Policy

Mariano-Florentino Cuéllar, David Lazarus,
Walter P. Falcon, and Rosamond L. Naylor

Anyone who has had the good fortune to fly across the United States on a clear summer day cannot help but be impressed by the vastness of American agriculture. From the salad bowls of the California valleys, to the western cattle ranches, to the Wheat Belt of the Great Plains, to the pivot-irrigated circles of corn in Nebraska, to the rainfed corn and soybean systems of the Midwest, to the small dairy farms in New England, agriculture is literally everywhere. And if the airplane has a southerly routing, peanuts, rice, and cotton are also visible. Together these regions underscore the scale of food production activities that contribute to America's international reach as a global food power; they also demarcate divisions that are important to domestic policies and politics.

The vastness of U.S. agriculture as seen visually from 30,000 feet, however, juxtaposes with its small role on the ground economically. In 2012, agriculture contributed only 1.2 percent of GDP (World Bank 2013). Although the U.S. Department of Agriculture (USDA) asserts that there are 2.2 million farmers, in fact, fewer than 300,000 of them supply 82 percent of the value added in U.S. agriculture (Wilde 2013).[1] To complete the circle of contradictions, this small sector of the economy exerts a disproportionately large influence on American politics and policy. To make matters still more confusing, three-quarters of the total expenditures authorized under the current Farm Bill are spent on consumers, not on farmers. Put simply, U.S. agricultural policy is messy.

[1] In 1975, the USDA and the Census Bureau defined a farm as a place from which there are sales of agricultural products of more than $1,000. This definition gives rise to a great many part-time farmers. Defining a full-time farmer is arbitrary, though gross sales of greater than $250,000 is one frequently used metric. A sale's definition also underscores the important difference between "farm income" and "income of farmers," the latter including income from off-farm employment. These concepts are frequently confused in the literature. For more details, see O'Donoghue and Hoppe (2009).

There are reasons why U.S. agricultural policy is in the shape it is, however, and they help to explain why it is extremely difficult to reform. The goals of this chapter are to explain and reconcile some of the contradictions in U.S. agricultural policy, to trace their origins historically and institutionally, to note the similarity and differences between the United States and other developed nations, and to assess the impacts, intended and unintended, of food and agricultural policies on domestic and international food security.

U.S. agriculture plays a massive role in the global food economy despite its meager contribution to domestic GDP and employment. That role, along with its many internal contradictions, makes it particularly interesting to study. The United States is one of the world's leading producers of food, yet until 2011, when surpassed by China, it was also the world's largest importer of food and food products. It exports by far the largest amounts of agricultural commodities, provides the largest share of global food aid donations, and in 2013 produced more crop-based biofuels than any other country, including Brazil. By sheer size in international markets, the United States has a strong influence on the level and stability of world food prices, and hence on global food security. Meanwhile, the United States also maintains a very large domestic nutrition safety net, thereby shielding many of its low-income consumers from more serious threats of food insecurity that might arise from its own agricultural policies, the policies of other countries, or sheer economic misfortune.

Herein lies another important juxtaposition: America's domestic policy agenda versus its foreign policy agenda. U.S. food and agricultural policies are motivated largely by domestic political pressures arising from farm coalitions, agribusiness, and politicians representing farm states. The focus is rarely on foreign policy concerns, despite the fact that global food insecurity arising from U.S. agricultural policies may foster humanitarian crises and political instability in other countries. The political process underpinning this strong domestic orientation in the United States is entrenched in the historical composition of agriculture-related legislative committees (with committee members dominated by farm states), in the representation of the U.S. Senate (with sparsely populated, agriculture-intensive states equally represented), and in the priorities of executive agencies responsive mainly to domestic food and agricultural demands. Success in passing farm legislation has been made possible mainly by the inclusion of consumer interests. However, American agricultural policy still tends to neglect health and environmental outcomes, at home and abroad, as well as the interests and perspectives of the broader U.S. electorate.

This chapter begins with a brief historical description of American food and agricultural policies, and then turns to the political processes that perpetuate a focus on domestic farm interests. We discuss the roles of the legislative and executive branches in agricultural policymaking, and explore the institutional dynamics affecting policy outcomes such as concentrated interests, coalition politics, and processes that hinder the general public from influencing food and agricultural policies (typically referred to as "collective action problems"). The chapter ends with an assessment of domestic and international food security implications of U.S. policies, and of the tensions, often overlooked, between domestic and foreign policy agendas.

Federal Farm and Food Policy

The federal government's involvement with America's food and agricultural system is as old as the republic itself. Early debates centered on the constitutionality of the federal government intervening on food issues at home and abroad. In 1793, for example, a group of 3,000 French citizens were forced to flee Haiti and arrived destitute on Maryland's shores. Senator James Madison, in response to a Maryland request for food and other assistance for these refugees, stated that he "could not lay my finger on that article in the Federal Constitution which granted a right to Congress of expending on objects of benevolence. . . ." But benevolence won the vote (reported in Riley 2015).

Many volumes have since been written about U.S. agriculture and agricultural policy (for example, Benedict and Stine 1953; Paarlberg 1964; Cochrane and Ryan 1976; Cochrane 1993; Gardner 2002; Imhoff 2011; Wilde 2013). Our purpose in this chapter is not to duplicate these writings, but rather to provide a brief historical context for understanding how modern U.S. agricultural policy took on its present form, and to show why, many decades after earlier legislation, U.S. farm policy continues to affect food security abroad. The founding period of the late eighteenth century obviously had enormous consequences for the apportionment of legislative power. Three more recent eras—the 1860s, the 1930s, and the period of evolutionary change since World War II—also seem especially worthy of comment.

THE 1860s

Three federal actions in the 1860s had lasting consequences for American agriculture. The Civil War, combined with President Abraham Lincoln's abolition of slavery in 1863, fundamentally changed the nature of society, especially in the South. Altering the organizational structure of agriculture changed land and labor markets, as well as the composition of farm output. Many of the plantations were broken up into much smaller farming units, and labor migration from the Southeast had profound social and economic implications for both the agricultural and non-agricultural sectors of the entire nation.

At about the same time, the passage of the Homestead Act in 1862 was critically important to how much of the midwestern portion of the United States was settled. By providing virtually free land to those who wished to farm it, the Act provided the heartland of U.S. agriculture with its deep and abiding egalitarian ethos. The choice of 160 acres as the modal settlement size continues to influence the basic structure of U.S. farm organization. To this day there is dependence on "family farms" that still produce 88 percent of the nation's farm output—widespread misconceptions about the role of corporate agriculture notwithstanding (USDA Chart Gallery 2013). However, not all land was "homesteaded" in this form. Settlers in the Great Plains, faced with the challenge of cultivating more marginal, drought-prone land, were granted 640 acres apiece in Nebraska (under the Kinkaid Act of 1904) and 320 acres elsewhere in the Plains (under the Enlarged Homestead Act of 1909). In addition, the very large land grants given to railroads west of the Mississippi River, and the settlement arrangements in California during the Mexican period (1821–49), created much larger farm units in those regions (Shannon 1977). The resulting managerial style and size distribution of holdings are major reasons why the structure of Western agriculture is so different from the Corn Belt.

A third component was the passage of the Morrill Act in 1862. This legislation helped to establish a land-grant university for each state.[2] A primary focus of these institutions has been on agriculture, and their triple objectives of research, teaching, and extension have produced extraordinary flows of relevant farm and food knowledge and technology.[3] These flows were particularly important until World War II, after which the private sector took on increased dominance in producing and distributing agricultural technology.

THE 1930s

The 1930s ushered in a period of more active—and prolonged—involvement by the federal government in agricultural markets, and thus represented another major landmark in U.S. farm policy. This period is best known for the Great Depression and widespread droughts that contributed to the Dust Bowl. Farm policy at this time was motivated initially by the buildup of large farm surpluses and declining agricultural prices in the 1920s and 1930s. Real prices received by farmers rose dramatically between 1910 and 1920 as a result of World War I and the Russian Revolution, but then fell precipitously in the 1920s as yield growth and land expansion produced a rising stream of farm surpluses under relatively stable economic and political conditions (Gardner 2002). In an effort to support rural incomes, the government passed the Agricultural Marketing Act in 1929. This act established the Federal Farm Board whose mandate was to stabilize prices largely through supply-side management—for example, paying farmers to idle land or slaughter livestock prematurely to keep markets

---

[2] A second Morrill Act of 1890 created land-grant institutions for minority groups under the then-prevailing doctrine of separate-but-equal.

[3] Numerous other countries have attempted the land-grant model, but interestingly, few have achieved great success with the U.S. model. The extension component, in particular, has often proven to be the weak link.

tight and prices high. These institutions set the stage for more pronounced government involvements down the road that remain in play today.

As the Great Depression hit the country in the early 1930s, the demand for crop and livestock products plummeted, creating widespread hardship in rural America that extended from tenant-farming operations (e.g., peanuts, cotton, tobacco) in the post–Civil War South to large livestock operations in the West. The early 1930s also marked the end of an unusually wet period in the Great Plains—a period that had encouraged settlers (with various forms of government support including land allocations) to cultivate the region with unwarranted intensity, believing somehow that "rain followed the plow" (Worster 1979). Having plowed up the native vegetation that historically had held the soil in place, farmers in the Plains experienced massive wind erosion and desiccation of their lands. Persistent droughts over broad areas created dust storms that reached all the way to the East Coast, dramatically reducing yields over 100 million acres and leading to one of the most significant human migration episodes in American history. The fact that the Great Depression and the Dust Bowl occurred simultaneously was probably not a coincidence. The United States had been in an expansionary mode in the 1920s, and the risks of both ecosystem and economic collapses were inadequately assessed (Worster 1979).

Coincidence or not, these events led to a populist movement in the 1930s that backed the interests of farmers throughout the United States and underscored a genuine concern for feeding the country. Agriculture was then a more sizable part of the economy, and President Franklin Roosevelt's New Deal policies reflected these demographic realities and national sentiments. Congress passed a series of bills on soil conservation, crop insurance, farm price supports, and nutrition assistance following his election in 1932. The Agricultural Adjustment Act was put into legislation during the 1930s,[4] giving Congress the authority to stabilize and bolster domestic prices for agricultural products (Gardner 2002). At varying times, policy attempts were made to limit crop and livestock production, to dump surpluses abroad, to restrict imports, and to purchase supplies (for government storage) in efforts to raise prices. Most of the efforts of the 1930s to overhaul farm programs by attempting to fix prices or quantities, or to improve the rural economy more generally, proved unsuccessful. The decade was essentially a dark time for American agriculture, a testing era for various types of agricultural policy instruments, and an active period for setting policy precedents. However, it was World War II, not policy, which

[4] The Agricultural Adjustment Act of 1933 went through a series of iterations in the 1930s, having been deemed unconstitutional by the Supreme Court in 1936 for levying taxes on processors to support farm incomes. The Agricultural Adjustment Act of 1938 shifted the financing burden onto taxpayers but otherwise kept much of the program content intact. The 1938 Act is considered part of "permanent law" for commodity programs and farm income support (along with the Agricultural Act of 1949, described later). If current legislation under the Farm Bill expires and is not renewed by Congress, the law reverts back specifically to these earlier Acts.

fundamentally changed the overall economy, the incomes of farmers, and the input structure for American agriculture.

THE POST–WORLD WAR II ERA

Agricultural policy innovations since World War II have evolved via a series of legislative actions aimed at increasing the income of farm households, providing safety nets for poor consumers, and investing in infrastructure for rural America. Five-year Farm Bills have become the standard operating procedure for Congress, and simply keeping up with the lengthy, cross-referenced, multi-titled legislation has created full-time jobs for many food policy analysts (see Box 4.1).

Between the 1950s and 1970s, agricultural policy was often the focus of intense partisan fights between the major political parties, especially with respect to presidential appointments to the Secretary of Agriculture position. The conservative appointees typically pressed for fewer restrictions on agricultural production and more market determination of prices, i.e., they wanted "freedom to farm" as espoused by Secretary Ezra Benson (1953–61) (Benson 1960). In sharp contrast, the liberals, such as Secretary Orville Freeman (1961–69), sought stricter controls on output but with much higher price guarantees and direct government payments to farmers (Cochrane and Ryan 1976). In retrospect, neither approach worked very well. Varying attempts to cut production via acreage restrictions failed because of legislative loopholes (intended and otherwise), and because of rapid technological change that increased yields per acre at rates faster than acreages were being curtailed by policy.

The only serious attempt in the United States to impose strict "supply management" for a major crop occurred in 1962–63. The central debate was on limiting wheat production, and on whether wheat farmers would vote (via a special referendum) to impose strict production controls on themselves in return for high guaranteed prices. The approach was very controversial, with those in favor of it arguing that it would be the difference between "$2 [per bushel] wheat and $1 wheat." The American Farm Bureau was adamantly against the plan, arguing that it was "the tightest, strictest, most complete control ever considered for a major commodity" (Cochrane and Ryan 1976). In the end, the proposed program failed to win enough votes from wheat producers to implement the plan; moreover, the failure of the wheat referendum effectively killed strict supply management as a broader approach to farm policy (Cochrane and Ryan 1976; Siracusa 2004).

The aggregate result of farm policies from 1950–70 was mostly the creation of massive grain surpluses, which were delivered to the government under the loan provisions of the legislation (Box 4.1). By 1962, the U.S. government held wheat stocks totaling 30 million metric tons (mmt), compared to the total global wheat trade that year of 38 mmt (FAOSTAT 2013). These stocks laid

BOX 4.1

Key Provisions in U.S. Food Legislation

Congress passes new agricultural legislation about every five years. These bills typically exceed 1,000 pages and are often referred to as "Farm Bills", even though they are omnibus in character. They embrace a combination of goals, regulations, and special-interest provisos for both producers and consumers. Five policy concepts have proven durable over the years, and these elements are summarized here. Readers are warned that the devil is in the details, and that specific provisions vary by commodities, years, regions, and bills.

Loan Rates. Farmers have long been able to borrow funds from the government using their harvested commodities as collateral. Loan rates (prices) are specified in the legislation; moreover, if market prices turn out to be below the loan rates, farmers may simply deliver their crop to the government as a means of paying off the loans. From 1950–70, loan rates were often set higher than market prices and the government took delivery of millions of tons of grain ("surpluses") in this manner. In more recent years, loan prices have been reduced in the legislation relative to expected market prices. As a consequence, the program now is primarily a lending service and not an intentional or unintentional mechanism for surplus accumulation.

Commodities. Current legislation provides for several types of payments to farmers. "**Direct Payments**" provide income support to framers who have previously grown the crop for specified periods. The payment is fixed in the legislation (e.g., in terms of dollars per bushel) and is given to farmers whether or not they produce the crop. "**Counter-Cyclical Payments**" are also given to famers if market prices are below the "target" prices specified in the legislation. Costs of the counter-cyclical program are thus heavily dependent on variations in market prices among years.

Conservation Reserve. Farm legislation authorizes that (up to) a certain number of acres of farmland can be diverted (with payment) from crop growing to conservation purposes. Degraded or potentially degradable land, representing about 10 percent of total U.S. cropland, has usually been put into the Conservation Reserve Program. Land is selected from tracts offered by farmers using a bidding process. Land conservation contracts have typically been for 10 years.

Crop Insurance. An increasingly important provision of farm legislation covers crop insurance. The government pays half the premiums for farmers and some underwriting losses to cover risks of weather and prices. Although many farmers suffered severe production losses caused by the severe drought of 2012, most losses greater than 25 percent were covered by insurance for the major crops.

Nutritional Assistance. American families whose income is below the poverty line are eligible for supplemental nutritional assistance ("food stamps" or SNAP). Payments depend on the depth of the family's poverty. About 20 million families (47 million individuals) received payments in 2012, which averaged slightly less than $300 per month per family. Total program costs are counter-cyclical and rise during periods of recession.

Primary Source: USDA, ERS, Chart Gallery, http://www.ers.usda.govidataproducts/chart-gallery.aspx#.UauT50v1X34, accessed June 2, 2013.

the basis for very large food aid shipments. U.S. wheat shipments of food aid totaled 90 mmt during the decade of the 1960s. These aid shipments astoundingly amounted to some 20 percent of total global wheat trade for the decade (Riley 2015). A lasting impression of U.S. surpluses was created during this period, which remains widespread but misguided today. By 1980, surpluses had largely disappeared, partly because of legislation that lowered loan rates relative to expected market prices (USDA Chart Gallery 2013); partly because of food aid; and partly by the unexpected sales to the U.S.S.R. starting in 1973 when that country radically changed its trade policy. Currently, the United States holds almost no government stocks of grain. When the U.S. government now provides food aid in the form of cereals, for example, it must go into the market to purchase the grain.

Payment Schemes

The disappearance of government-held farm surpluses, however, did not mean an end to farm programs designed to curtail downward movements in agricultural prices and to improve farm incomes (Box 4.2). The bulk of this financial support goes to farmers who produce a small number of "program commodities": corn, soybeans, wheat, rice, barley, cotton, oats, peanuts, and sorghum. With the exception of milk production, livestock farmers, who currently produce 44 percent of gross farm receipts, are not major recipients of funding, nor are fruit or vegetable producers. The absence of major support programs for fruits and vegetables may have had negative spillovers on nutrition in the United States, although the evidence is not clear. The fruit and vegetable growers themselves have sometimes contributed to this policy direction by asking for marketing quotas—limits on how much of the commodity can be brought to market—in order to keep prices higher for these largely non-storable products.

BOX 4.2

Types of Farm Policy Interventions

Category	Government Role	Farmer Effect	Consumer Effect
Price support	Buys at commodity support price	Higher price	Higher price
Supply control	Limits how much is grown or imported	Higher price	Higher price
Deficiency payment	Pays farmer difference between target price and market price	Lower market price	Lower market price
Direct payment	Pays farmer based on historical income	Higher income	–
Insurance	Pays farmers for Production losses	Avoids major losses	–

Source: Adapted from Wilde (2013).

Payments are largely based on the acres of land (called "base" acres) that have historically been included in farm programs.[5] In fact, several of the programs provide farmers with payments even if these base acres are left to lie fallow. Payments under the base-acreage programs are highly concentrated—even more so than the concentration of land ownership. In 2012, for example, the largest 10 percent of farms in terms of sales received 67 percent of the $5 billion spent on this program (EWG 2013). The disconnect between political rhetoric about helping small U.S. farmers and the actual distribution of dollar benefits is remarkable, and is a primary cause for widespread public cynicism about agricultural legislation.

Counter-cyclical programs form a second major form of farmer payments. As their name implies, these payments are triggered when crop prices in the market dip below levels established in the text of the Farm Bill. In the post-2005 period, this form of government payment has been low (except occasionally for peanuts and upland cotton) because world commodity prices have been high. Like direct payments, the counter-cyclical payments are based on historic production, and are highly concentrated in terms of crops, regions, and farm size.

Conservation provisions of American agricultural policy are a third form of payments to farmers. Conservation has had a long history in the United States dating back to the Dust Bowl era of the 1930s, but took on new prominence in the 1985 and 1990 legislation; indeed, the latter was entitled the Food, Agriculture, Conservation and Trade Act. Groups seeking to limit agricultural production—thereby raising prices—joined with environmentalists to establish a Conservation Reserve Program (CRP) for the protection of erodible land. Congress establishes a maximum area for the reserve, and farmers bid for the right to place their land in the CRP. Various criteria, such as slope and soil type, go into the selection of tracts to be included; geography and the conscious inclusion of parcels from numerous states also go into the process. Farmers typically sign ten-or fifteen-year contracts, with average contract prices in 2013 at about $55 per acre per year. In 2012, 27 million acres of U.S. cropland, involving nearly 400,000 farms, were in the CRP (USDA Chart Gallery 2013). However, with the higher commodity prices after a dip in 2005, CRP payments have become less competitive relative to profits from crop production, and a sharp falloff in new or re-contracted acres being placed into the program has taken place.

Figure 4.1 summarizes recent government payments by category. These totals have recently run about $12 billion per year, but as the bar for 2005 also shows, total expenditures are very sensitive to world prices and the compensation given to famers via counter-cyclical payments.

---

[5] Providing subsidies based on historical acres rather than on current production has a curious but important logic. As part of the international trade negotiations on agriculture, countries have agreed to "decouple" subsidies from current production lest the subsidies induce ever greater production.

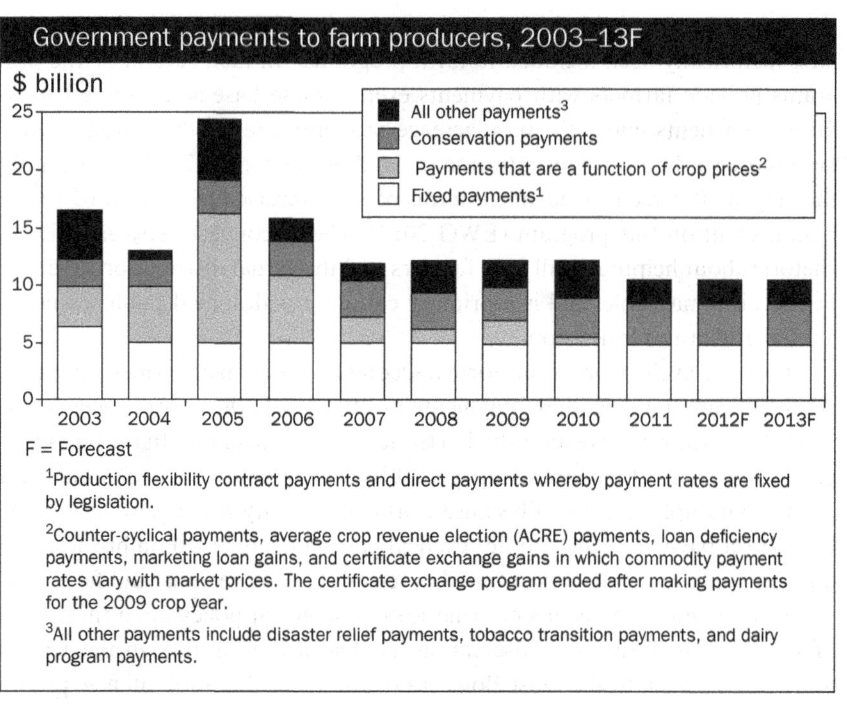

FIGURE 4.1 Farm Payments by Category, 2003–2013

Source: USDA, Farm Service Agency, Natural Resources Conservation Service, and Commodity Credit Corporation. Forecast as of February 11, 2013.

Crop Insurance

Agriculture has always been a risky business as a consequence of uncertain weather, prices, diseases, and pests. Since 2000, however, new insurance instruments have become more widespread to help combat those risks. They are simultaneously becoming a more important part of the federal budget for agriculture. Much of the insurance legislation traces to 1938; however, legislative amendments in 1994 and 2000, plus the expanded role of the Risk Management Agency, greatly spurred the purchase of "multi-peril" insurance by farmers.[6] This program is a public-private partnership, in which the government pays half of the insurance premiums of farmers, and agrees also to underwrite some costs of the private insurance companies. In 2012, for example, the Federal Crop Insurance Program (FCIP) covered 280 million acres and paid out $116 billion in total liabilities (Shields 2012). About 90 percent of U.S. cropland was covered by insurance, typically at a level of 75 percent.[7] In addition to yield

---

[6] The crop insurance provisions are complex and detailed. See Shields (2012) and USDA "Risk Management Policies" http://www.rma.usda.gov/policies/ for various definitions and rules.

[7] For example, a farmer growing corn might have a historical yield record of 4 metric tons (mt) per acre. At 75 percent coverage, the insurance company would indemnify any actual yield of less than 3 mt per acre.

coverage, farmers can also purchase revenue-based insurance, which covers several dimensions of price variability.

The professional literature assessing the innovation and conservation impacts of these insurance programs is still in its infancy. It is likely that new forms of crop insurance will be increasingly important and will likely have powerful incentive effects for farmers, especially within the context of climate variation and change (a topic discussed more thoroughly by Lobell, Naylor, and Field in Chapter 9). Between 2005 and 2012, government budget costs for public-private crop insurance rose sharply, and in 2012, government outlays reached a hefty $14.1 billion (Shields 2012).[8]

Nutritional Assistance

Of all the titles in the Farm Bill, the Supplemental Nutritional Assistance Program (SNAP, also known more generically as "food stamps") is among the most controversial. The wide scope of nutritional assistance within "Farm" Bills often prompts two questions: How did consumer programs become so large? And, why is this safety-net consumer program located within the USDA's budget rather than within the Department of Health and Human Services, the home for most other welfare programs? The answers to both questions are complicated.

The origins of the food stamp program trace to the 1930s and the Great Depression. Under the leadership of Agriculture Secretary Henry A. Wallace (1933–40), the program was originally designed to distribute surplus agricultural commodities and simultaneously to assist poor and needy families.[9] Food distributions were discontinued during the middle of World War II, and it was not until the early 1960s that they resumed. During the 1960 presidential election campaign, West Virginia became a pivotal state for candidate John F. Kennedy's success. The poverty he encountered there during extensive campaigning left a deep impression on him, and served as one of the driving forces for his new "War on Poverty." Food stamps became a part of that war, and Kennedy's first executive order in 1961 focused on them. Later, and as a part of Kennedy's legacy following his 1963 assassination, Congress passed the Food Stamp Act of 1964. The SNAP program, like the 1930s food stamp program, ended up within the USDA, even though its purpose was no longer distribution of farm surpluses. The central role of George McGovern and Herbert Humphrey, two prominent midwestern senators with close ties to agriculture and the Senate Agriculture Committee, also contributed to USDA's control of SNAP. The desirability and political difficulties of changing this arrangement, and other political processes related to food and agriculture, are discussed in a subsequent section of this chapter.

---

[8] In addition to the crop insurance program, the federal disaster relief program provides ad hoc assistance in the wake of weather disasters like droughts and floods. From 1995 to 2012, the federal disaster relief program alone provided more than $22 billion to farmers and ranchers.

[9] See USDA, "A Short History of SNAP," http://www.fns.usda.gov/snap/rules/Legislation/about.htm.

The rapid growth of SNAP payments over the past decade has been a surprise to many. We believe that the expansion is related primarily to four factors. First, while the United States is among the world's richest countries, its personal income distribution is quite unequal, and is becoming more so. For example, in 2011, 17.3 percent of the U.S. population had incomes below 50 percent of the U.S. median income. This measure compares with Norway, Germany, the United Kingdom, and Japan, which have corresponding rates of 7.8 percent, 8.9 percent, 11.3 percent, and 15.7 percent, respectively.[10] In short, there are many poor people in the United States who need assistance in fulfilling their basic nutritional needs (Tiehen et al. 2012).

Second, the SNAP program was designed to be counter-cyclical with respect to growth in the U.S. economy. The expectation was always that SNAP costs would go up when unemployment was high and the country was in recession. Between 2007 and 2010, annual U.S. unemployment rates rose from 4.6 percent to 9.6 percent, and remained at 8.1 percent in 2012.[11] This rise in unemployment was one reason why SNAP expenditures rose by 77 percent during those five years.

Third, food prices have been rising as a consequence of supply shocks, coupled with growing demand as incomes rise rapidly in middle-income countries and from the greatly expanded use of corn for ethanol production. For the period 2005–2012, the all-food component of the U.S. consumer price index rose by 23 percent.[12] Those price increases were then largely translated into additional costs for the SNAP program, which uses retail prices in its benefit calculations.

A fourth explanatory factor traces to the method by which the United States determines its domestic "poverty line," since that line fundamentally determines the number of SNAP-eligible participants. The USDA plays a dominant role in the line-setting process, since it is charged with calculating "thrifty food budgets." These budgets are modest, but they cover basic nutritional requirements and provide for moderate diet diversity as well. The cost of these thrifty budgets is one of just two numbers used in determining the U.S. poverty line. The U.S. Census Bureau calculates the latter simply by multiplying the cost of the thrifty food budget by a multiplier of three.[13] If food costs were lower, or if the multiplier were (say) two rather than three, fewer people would be defined as being in poverty and thus eligible for food stamps.

---

[10] See OECD, "Equity Indicators" http://www.oecd.org/berlin/47570121.pdf.

[11] See Bureau of Labor Statistics, "Annual Unemployment Rates," http://data.bls.gov/timeseries/LNU04000000?years_option=all_years&periods_option=specific_periods&periods=Annual+Data).

[12] See USDA ERS, "Food Price Outlook," http://www.ers.usda.gov/data-products/food-price-outlook.aspx#.Uc23j1PlX34.

[13] See University of Wisconsin, Institute for Research on Poverty, "How is Poverty Measured in the U.S.," http://www.irp.wisc.edu/faqs/faq2.htm. The congressional choice of the multiple "three" indicates that Congress believes that Americans, even poor Americans, should spend no more than 33 percent of their household income on food. If Congress felt that 50 percent were an appropriate amount, the multiple would be two, and many fewer people would be below the poverty line. Interestingly, in most poor countries the poverty-line calculation assumes that 80 percent would be spent on food, making the multiple equal to 1.25.

In 2012, the U.S. thrifty food budget resulted in a poverty line of about $24,000 for a family of four. That year, some 47 million Americans—about 15 percent of the entire nation—received SNAP payments. Support was restricted to those families with net incomes less than the poverty line, with the actual amount of support increasing the further below that line income levels fell.[14] If a hypothetical family of four had zero net income, they would have received $8,000 in SNAP payments for the year; if their net income was $12,000, they would have received payments of $4,000. Overall, payments averaged approximately $1,600 per person for the year.[15]

As a consequence mainly of these four factors, the SNAP portion of the Farm Bill budget totaled $73 billion in 2012. In addition, the budget provided for $19 billion for school lunches; $8 billion in assistance for nursing and pregnant women, infants, and children; and $12 billion in supplemental assistance for new nutrition initiatives.[16] The $112 billion consumer package now constitutes the core of USDA's budget, and not surprisingly, its size and prominence are making the traditional farm constituencies extremely nervous. As discussed subsequently, nutritional costs now seriously threaten the rural-urban political coalition that has been necessary to pass recent Farm Bills.

ADDING UP THE BILL

Some pundits refer to farm legislation as a feed trough. Our preferred metaphor is a cake, which in 2012 represented a multi-layered pastry valued at a whopping $145 billion. Figure 4.2 shows an extremely large portion consisting of nutrition and consumer services. Farm payments—direct, deficiency, and counter-cyclical—form a much smaller second layer. Conservation activities provide an even smaller third ring. And it is important not to forget the icing. Tucked into the language of the thousand-plus pages of the Farm Bill are significant benefits for numerous smaller groups and special interests.

U.S. BIOFUELS POLICY

Carrying the dessert analogy one step further, there is ice cream on the "farm plate" as well. Government support for crop-based biofuels lies largely in the

---

[14] Income eligibility is also defined at 130 percent of the poverty line in terms of *gross* incomes of families. A series of allowable deductions result in the *net*-income definition of 100 percent of the poverty line previously used in the text. See USDA, "Supplemental Nutritional Assistance Program Eligibility," http://www.fns.usda.gov/snap/applicant_recipients/eligibility.htm.

[15] See USDA, "SNAP: Frequently Asked Questions," http://www.snaptohealth.org/snap/snap-frequently-asked-questions/.

[16] See USDA, "Budget Summary and Annual Performance Plan, FY2012," http://www.obpa.usda.gov/budsum/FY12budsum.pdf.

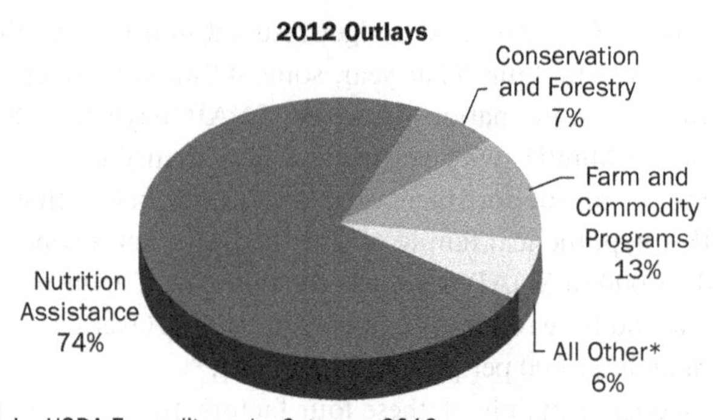

FIGURE 4.2 Major USDA Expenditures by Category, 2012
Source: USDA (2013), http://www.obpa.usda.gov/budsum/FY12budsum.pdf

domain of energy legislation but it also falls within the Farm Bill.[17] It has been a transformative policy in terms of raising crop prices and land values, altering trade, strengthening rural economies, and rebalancing several important programs within food and agriculture legislation.

The partnership that formed in the early 2000s between the agricultural and energy agencies on biofuels policy was consistent with several developing conditions at the time (Naylor 2014). First, with large crop surpluses and low agricultural prices at the turn of the century, policymakers in both the executive and legislative branches were looking for pathways to support rural communities without the burden of huge crop subsidies. Using more corn to fuel cars in a rapidly growing economy was one method of removing surpluses and boosting prices. At the same time, the Second Gulf War served as a stark reminder that the United States needed to wean itself from foreign crude oil sources, especially those associated with rogue governments. Increased recognition of climate change stemming from greenhouse gas emissions presented yet another rationale for moving away from fossil fuels.

Recent biofuels legislation had been built on a history of tax exemptions for production and tariffs (taxes on imports). But the fundamental policy shift was the introduction of mandates for the use of biofuels in transportation—the Renewable Fuels Standard (RFS)—via the Energy Policy Act of 2005 and the Energy Independence and Security Act of 2007 (as discussed also by Lobell and colleagues in Chapter 9). The impacts of biofuels legislation on crop prices, trade, and cropped area were impressive (Beckman et al., 2013). From 2004 to 2012, corn acreage increased 28 percent, mainly by the replacement of other

[17] The Farm Bill component of U.S. biofuels policy (Title IX) covers a variety of initiatives that include, for example, biorefinery assistance, support for advanced biofuels, rural energy programs, infrastructure development, and biomass-based biofuel R&D. The budget expenditures ($320 million) are small relative to the other tiers of the Farm Bill shown in Figure 4.2 and fall into the "other" category. For a cogent summary of the Energy title of the Farm Bill, see:
http://future.aae.wisc.edu/publications/farm_bill/9_Energy_Final.pdf. Accessed July 3, 2013.

crops, such as soybeans, barley, oats, and sorghum, and by taking land out of conservation reserves. Corn prices rose from $2.42 per bushel to almost $7 per bushel during this period. At the same time, the amount of corn used for ethanol increased from about 1.2 billion bushels to about 5 billion bushels. Over 40 percent of annual U.S. corn use went to ethanol in 2012–2013, causing the U.S. share of global corn exports to decline significantly.

Biofuels policy interacts in interesting ways with farm subsidies, the crop insurance system, and nutrition assistance programs. Because counter-cyclical and marketing loan programs are linked to market prices, farm program payments decline when market prices increase. At the same time, the cost of administering the crop insurance program rises as the value of the insured crop increases, because insurance product premiums (subsidized heavily by the federal government) are directly connected to crop values. This dynamic helps to explain why the counter-cyclical and marketing loan programs have provided relatively little support to farmers since the establishment of the RFS, while the cost of the crop insurance program to taxpayers has escalated from $2 billion in 2000 to a record $14 bilion in 2012 (Shields 2012).

The combination of crop insurance and biofuels policy operates de facto as a new safety net for farmers. This producer safety net is matched with a hefty consumer safety net, as SNAP and other nutrition assistance programs are also valued according to current food prices. The decline in direct farm subsidies does not offset the rising cost of the RFS and SNAP programs in budgetary terms. Pitting farmers against SNAP recipients (largely an urban constituency) in a period of fiscal tightening left discussions of the 2013 Farm Bill in turmoil.

U.S. Farm Policy as Viewed Internationally

The structure of the U.S. economy has changed dramatically over the past 100 years, with the agricultural sector declining rapidly in relative importance. This same phenomenon has occurred in other developed countries, including those in the EU, as Johan Swinnen describes in Chapter 5. It is thus instructive to see how varying nations have dealt with their relatively diminished agricultural sectors in terms of policy, and what the spillovers from these policies have been for global food security.[18] Three questions seem crucial: How much support do farmers receive via domestically oriented policies? How much protection does each country accord its rural sector through international trade policy? And, how is international food aid used as a part of both domestic and international food and farm policy?

---

[18] See also C. P. Timmer (2009), "A World Without Agriculture," http://www.aei.org/article/foreign-and-defense-policy/international-organizations/economic-development/a-world-without-agriculture-outlook. Accessed July 10, 2013.

COMPARISONS OF DOMESTIC PRODUCER SUPPORT

When viewed separately, it is easy to conclude that U.S. governmental efforts to assist agriculture have been massive. But viewed in comparative perspective, the support to U.S. farmers appears more modest.

Each year the Organization for Economic Cooperation and Development (OECD)—a grouping of 34 developed countries—calculates various measures of agricultural support for all of its member nations. Using a methodology that is consistent across countries and through time, the OECD provides the most reliable agricultural policy data that are available. The support estimates are broken down by commodity and by various support categories that include product subsidies, input subsidies, and direct payments. These components are summed to provide estimates of total support and the percentage of total farm income that is due to transfers to producers from consumers and the government (OECD 2012). The latter are referred to as producer subsidy equivalents or PSEs.

Box 4.3 summarizes recent comparative data. Interestingly, both the EU and Japan exceeded the total levels of U.S. support to farmers in 2010. In terms of percentage support, only Australia and New Zealand were lower. National differences in support measures notwithstanding, Box 4.3 underscores the widespread global problem of what happens to and for agriculture at upper ends of the development spectrum. The scale of annual global support to agriculture in developed countries reached a gigantic $241 billion in 2010.

Two additional sets of comments are relevant. In terms of the commodity composition of support, the United States is quite diversified. Relative to Japan and Korea, where much of the support goes to rice farmers, U.S. subsidies are more evenly spread across wheat, corn, and milk producers. Second, the estimates shown in Box 4.3 for 2010 are significantly different from those for a decade earlier. Support estimates are made relative to world prices. Since world prices for key commodities have risen, especially since 2005, markets have helped to relieve the cost of the policy "burden" in the United States and other countries. For OECD countries as a whole, the PSE dropped from about 35 percent to 20 percent between 2000 and 2010 (OECD 2012). Even with the lower PSEs, however, it is easy to understand why countries with low levels of support, such as Brazil and Australia, complain bitterly about not having a level playing field for their farmers.

THE MAGNITUDE OF TRADE BARRIERS

One axiom of food policy analysis is that domestic and trade policies are closely related. If a country attempts to keep the domestic price of a commodity significantly above the world price it will no doubt attract imports of that commodity unless some form of tariff (a tax on imports) or other trade restriction is put

BOX 4.3

Government Levels of Agricultural Support in Developed Countries, 2010

Each year the Organization for Economic Development and Cooperation (OECD) produces a comparative analysis of governmental support to the agricultural sectors of developed countries. This box shows the OECD support estimates for 2010 valued in US$, as well as government support as a percentage of total farm income (i.e., producer subsidy equivalents; [PSEs]). Both the total scale of support ($241 billion) and its percentage range (from 1 percent in New Zealand to 61 percent in Norway) are quite remarkable.

Country	Billion US$	Percentage (PSE)
Australia	1.2	3
Canada	7.2	17
Chile	0.3	3
EU	102.4	20
Iceland	0.1	44
Israel	0.9	14
Japan	55.2	52
Korea	17.1	45
Mexico	6.0	12
New Zealand	0.1	1
Norway	3.7	61
Switzerland	5.2	54
Turkey	20.7	26
United States	27.6	8

Source: OECD, *Agricultural Policy Monitoring and Evaluation 2012, OECD Countries*, Paris, 2012.

into place. Those restrictions generally reduce imports and hurt the economies of countries that would otherwise have exported the product. Trade and trade policy thus become integral parts of food and agricultural policy, and are major factors in how food policy activities in one country influence food security in others.

The United States is the world's largest exporter of agricultural products. In 2012 American agricultural exports exceeded $135 billion.[19] Maize, wheat, soybeans, and cotton topped the export listing. That year, agricultural exports comprised almost one-third of total farm earnings. The United States is also a very large importer, with agricultural imports in 2012 totaling slightly more

[19] See USDA "Agricultural Trade," http://www.ers.usda.gov/topics/international-markets-trade/us-agricultural-trade.aspx#.UdBdBFPlX34.

than $100 billion. In value terms, tropical fruits, beer, sugar, and beverage crops such as coffee and tea were among the most important imports.

From the perspective of trading partners, the tax on agricultural imports is a key policy variable. International efforts to reduce those tax and other barriers on all goods and services have long been the focus of the World Trade Organization (WTO). These negotiations take place in various multi-year sessions called "rounds." Because agriculture has been such a politically sensitive sector, it was largely excluded from trade negotiations until the 1994 Uruguay Round and the more recent Doha Round. Global progress has been slow, though there have been some gains, especially in removing import bans and other types of non-tariff barriers. Overall, however, the reduction in trade barriers for agriculture has lagged those in manufacturing. For example, in 2011, the United States imposed average tariffs of 3.5 percent for all goods, but had average tariffs of 4.9 percent for agricultural products.[20]

Comparatively speaking, the U.S. agricultural tariff of 4.9 percent is quite low. The EU maintains an average 13.9 percent tariff on agricultural imports, and Canada, Japan, Switzerland, and Korea impose average agricultural tariffs of 18 percent, 23.3 percent, 43.5 percent, and 49 percent, respectively. Of the major advanced industrialized countries, the U.S. average rate is higher than only New Zealand and Australia, both of which apply an average tariff of 1.4 percent on agricultural products.

The pattern of American agricultural trade policy during the last few decades is one of broad—but not entirely consistent—commitment to reducing trade restrictions. Though recent American agriculture trade policies have mostly been aligned with the goal of advancing the free movement of agricultural goods into the United States, this record is punctuated by several protective actions such as the sugar and cotton programs.[21] Because of these aberrations, the United States is frequently criticized for the negative effects of its domestic farm subsidies on freer agricultural trade.

---

[20] The averages reported here are the "simple average" summary rates published by WTO (WTO, "World Tariff Profiles," http://www.wto.org/english/res_e/booksp_e/tariff_profiles12_e.pdf). Readers should be aware that there is a vast literature on how best to measure "average" protection. While it is easy enough to describe precisely the tariff for a single commodity for a given country, countrywide averages depend on weights given different commodities. For example, protection measures depend in part on how specific "commodity bans" on imports are included in the calculations. Similarly, tariff preferences that apply to some, but not all, trading partners complicate computations, as do "tariff-rate" quotas, whereby a certain quantity of a product enters at a low tariff rate, but quantities about the quota face much higher tariffs. Differing assumptions about how best to calculate averages lead to different countrywide measures of tariff protection. In general, however, the various methods lead to relative rankings among countries that are quite similar.

[21] Brazil, for example, brought a challenge through the WTO dispute settlement mechanism against several elements of the U.S. cotton program. Brazil argued that U.S. cotton payments under the Farm Bill seriously distorted international cotton trade. (See Andersen and Taylor 2009.) After many years of appeals and negotiations, the challenge led to a settlement in which the United States provides $147.3 million per year to a newly created fund called the Brazilian Cotton Institute.

THE ROLE OF FOOD AID

Food aid is a third policy area where comparative perspectives are helpful. Since the mid-1950s, the United States has been the leading supplier of international food aid, almost always supplying more than half of the global total. During 1965, near the height of U.S. government-held surpluses, global shipments of food aid totaled about 20 mmt, with the United States supplying more than 18 mmt (Wallerstein 1980). By 1988, shipments had dropped to about 14 mmt with the United States supplying 8 mmt.[22] And by 2011, the global total had fallen to less than 4 mmt, of which the United States shipped just 2 mmt. Food aid has seemingly gone out of fashion—perhaps in no small part because of U.S. procedures and regulations. The U.S. approach has differed substantially from that of other nations, and seems increasingly at odds with recommendations from the larger development community.

U.S. food aid has been criticized as inefficient, oriented toward domestic political interests over humanitarian needs, and incapable of contributing to sustainable improvements in recipient-country food security. Part of this critique stems from the fact that the U.S. international assistance program was designed originally to provide support to geopolitical allies, develop future markets for domestic agricultural producers, and serve as an outlet for surplus U.S. commodities (Barrett 2005). When President Dwight Eisenhower signed the program into law in 1954, he said the purpose of the legislation was to "lay the basis for a permanent expansion of our exports of agricultural products with lasting benefits to ourselves and peoples of other lands."[23]

Part of the decline in U.S. supplies arose from sourcing difficulties. By law, the vast majority of food aid funding must be used to purchase U.S.-grown commodities; moreover, 50 percent of the volume of commodities financed through the program must be shipped on U.S.-flagged vessels, and 25 percent of any "bagged" food aid must be handled by Great Lakes ports (Bageant et al. 2010; Hanrahan 2014). As a result of having to ship bulky commodities from the United States, very significant amounts of funding are used to transport food across the world (Bageant et al. 2010). Most other countries, by contrast, do not require the purchase of domestic commodities shipped on domestically flagged carriers. Instead, they often deliver aid in the form of cash and purchase food from within the destination region if at all possible.

Rising agricultural prices have also taken their toll on food aid shipments. Since the U.S. government no longer has physical stocks of grain, it must go to the market to purchase the cereals and other commodities to be shipped as food aid. The congressional authorization, however, is for a given dollar amount,

---

[22] See World Food Program, "Food Aid, Quantity Reporting," http://www.wfp.org/fais/reports/quantities-delivered-two-dimensional-report/chart/year.

[23] See "Presidency Project," http://www.presidency.ucsb.edu/ws/?pid=24605.

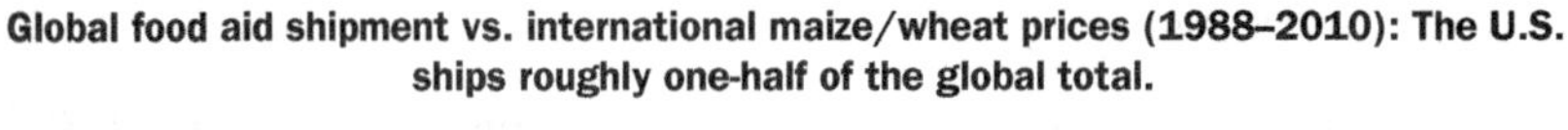

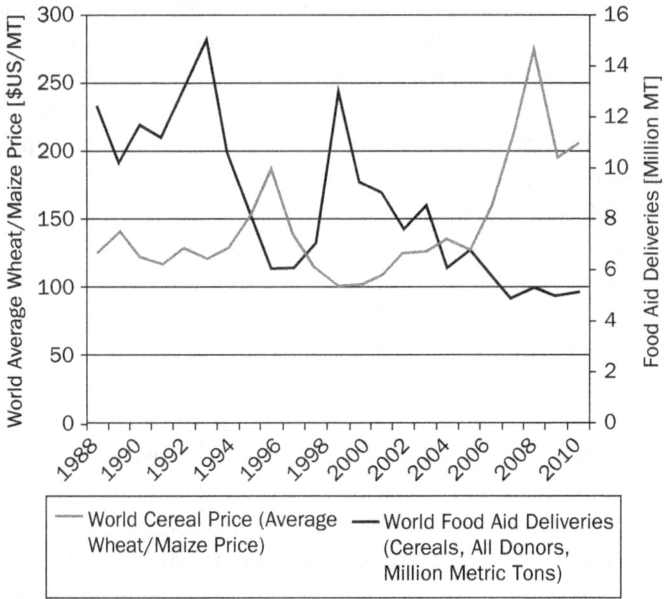

FIGURE 4.3 The Counter-Cyclical Nature of Food Aid

Source: www.imfifs.org, average wheat and maize price; http://www.wfp.org/fais/quantity-reporting, food aid deliveries.

not for a specified tonnage. This arrangement has had a rather cruel effect: the higher the prices for grain—and therefore likely the time of greatest need for food aid—the lower the quantities of food aid that can be provided (Figure 4.3). Because the tonnages have become so low, about 80 percent of food aid funding over the past decade has been provided for emergency relief, even though the legislation indicates that three-fourths of it should be used for longer-run development activities. Finally, a number of humanitarian organizations have strongly criticized the monetization of food aid for project purposes.[24] This has caused one major relief agency, CARE, to withdraw from using U.S. food aid. CARE argues that monetization is inefficient, causes commercial displacement that harms traders and local markets, and undermines the development of local markets.[25]

---

[24] If permitted by the agreement with the recipient country, donated food-aid commodities may be sold into local or regional markets and the proceeds used to finance local development projects. This process is called "monetization."

[25] See Government Accountability Office, "Local and Regional Procurement Can Enhance the Efficiency of U.S. Food Aid, But Many Challenges May Constrain Its Implementation," http://www.gao.gov/new.items/d09570.pdf.

Donor country attitudes toward food aid and to agricultural assistance seem recently to have changed. The food riots that emerged in many countries throughout the world in 2007 and 2008 brought new awareness of the geopolitical risks of food insecurity and resulted in high-level vocal support for agriculture development assistance. At the June 2009 G8 Summit in L'Aquila, Italy, the United States pledged $3.5 billion over three years to a global hunger and food security initiative to address hunger and poverty worldwide. The United States launched an initiative called "Feed the Future" intended to focus the attention of the U.S. government on longer-term investments in international food security. The U.S. initiative is part of a global pledge by the G20 countries and others of more than $22 billion. It remains to be seen whether Congress will provide the significant new food-related resources requested for this initiative, and whether the Feed the Future program can serve as a corrective for the inattention and lack of coordination of foreign agricultural assistance of the past twenty years. We remain skeptical.

Nonetheless, a new bipartisan bill, the Royce-Bass Food Aid Reform Act (H.R. 1947) was presented on Capitol Hill in June 2013, with the intention of improving the efficiency of food aid shipments from the United States and substantially reducing costs. Specifically, the bill sought to eliminate the U.S. procurement and cargo preference requirements and the practice of monetization. It also sought to align non-emergency food aid with the Foreign Assistance Act of 1961. Although it represented a major reconfiguration of historical food aid legislation—from an emphasis on the shipment of U.S. farm commodities to the transfer of cash—it only narrowly failed to pass in the House. In any event, fiscal conservatism seems to forestall any real attention toward food security as a national security concern (a topic discussed by Stephen Stedman in Chapter 13). Whether significant revision of food aid legislation can succeed on Capitol Hill in the near future remains to be seen.

Institutional Dynamics Affecting U.S. Farm Policy

What factors have led the United States to prioritize policies supporting domestic farm incomes and ensuring broad access to nutrition nationally, as opposed to internationally? The United States has far-reaching interests in every region of the world, and food security—or its absence—can affect those interests by altering prospects for development, political security, and political relationships. Why, then, has the United States failed to establish an effective foreign policy agenda or lead an international effort in support of global food security? Even at the national scale, American farm policy is rife with inefficiencies and apparent contradictions in terms of domestic food and nutrition security. What are the political processes at work within the legislative and executive branches of government that perpetuate the influence of special interests and make farm policy so difficult to change?

THE POWER STRUCTURE

Food and agriculture policy in the United States is shaped primarily at the federal level. Congress dominates the policy process—it reauthorizes the Farm Bill approximately every five years, it creates national energy policy, and it provides appropriations for farm, food, and international assistance programs. Legislative power is concentrated in the hands of regionally skewed authorizing committees in the House and Senate, whose members represent domestic producer, consumer, industry, and trade interests. Closely connected, the U.S. Department of Agriculture (USDA) is given broad authority by Congress to implement the Farm Bill. As a result, the USDA oversees a wide range of activities including commodity program supports, crop insurance and disaster relief programs, agricultural research and training, nutritional programs, and many other aspects of farm legislation.

The other components of the executive branch that interact with food and agricultural policy, by contrast, are only loosely coordinated, and policymaking responsibility for food security is diffuse. National and strategic issues with foreign policy imperatives are typically coordinated by the White House, in particular the National Security Council and National Economic Council. For most administrations, agriculture has not been treated as a high enough priority to warrant presidential attention given the many other significant demands of the office. Moreover, there is no single high-profile official within the executive branch with responsibility for overseeing food security policy from a national security perspective. The limited attention that policymakers do devote to food and agriculture tends to be focused on domestic concerns, and particularly on regions where agriculture plays a prominent role. As discussed earlier, the Secretary of Agriculture (a political appointee), in coordination with the White House, often plays an important role in motivating agricultural legislation and implementing policies established by the bills passed by Congress (Merrill and Hickman 2001; Winders 2009).

The Legislative Branch

The Congress works predominantly via a well-defined set of committees and subcommittees. These committees help shield the full legislative chambers from having to learn all the intricacies of particular policies. Additionally, they permit senators and representatives who care about policies for a specific sector, such as agriculture, to focus their energies on particular bills (Weingast and Marshall 1988; Krehbeil 1991). For better or worse, this specialization also permits these groups to exercise outsized policy influence unless offsetting pressures develop; for example, when there is widespread public opposition or an issue-oriented political movement.

The House and Senate agricultural committees thus hold the dominant position in U.S. farm policy. Members of these committees draft the periodic reauthorization of the Farm Bill—a process that requires many months of public hearings around the country, followed by lengthy amendment processes and deliberations in closed committee settings and in open debates on the floors of the Senate and House. Congress typically delegates considerable power to legislative committees if their jurisdiction (in this case, agriculture and food) elicits particularly intense concern from certain regions of the country (Weingast and Marshall 1988). Hence the House and Senate agricultural committees are largely composed of representatives from regions where program commodities are grown—and where Farm Bill payments are concentrated—including the Corn Belt, Southeastern Coastal Plain, Lower Mississippi, and California. These representatives work in a system responsive to concentrated interests, ranging from the American Farm Bureau Federation to commodity associations to large agribusiness. Because of their focus on farm incomes, they often demonstrate dissimilar political views relative to lawmakers from other parts of the country (Winders 2009). Congressional representatives from large urban areas and domestic nutrition interests are also on the committees, mainly to protect low-income consumer constituents.

Though enormously powerful in matters of food and agriculture, the House and Senate agriculture committees are not the only players on Capitol Hill in this space. The agriculture appropriations subcommittees in the House and Senate allocate discretionary funding among food, agriculture, rural development, and conservation priorities. Environment and energy committees (e.g., the Environment and Public Works Committee and the Natural Resources Committee within the Senate, and the Energy and Commerce Committee within the House) help craft energy legislation, including the Renewable Fuels Standard. Other key congressional committees cover the domains of health (including food safety), public finance, infrastructure, trade, food aid, and other forms of foreign assistance. Primary control of the Farm Bill, however, still rests with the agriculture committees in the House and Senate—a structure that reinforces a tight cycle of policy outcomes begetting political representation begetting entrenched policy outcomes.

A classic illustration of this entrenchment occurred in January 2013. At that time a new Farm Bill was being contemplated, in which the program of direct payments to farmers was presumed likely to come under attack from senators wishing to cut budgets. This prospect led to a mini-revolution within the group of agriculturally oriented senators. Senator Thad Cochran from Mississippi, first elected to the Senate in 1978, invoked his "right" of seniority in the Senate to displace Senator Pat Roberts of Kansas as the ranking

Republican member of the Senate Committee on Agriculture.[26] Despite much flowery rhetoric, the timing and purpose of the displacement seemed obvious. As the ranking member, Cochran would have power to help guide the agricultural legislation, thereby protecting the rural interests of his region. (Cochrane's power to influence agricultural legislation was enhanced further by his membership on two other powerful senate committees—Appropriations and Rules.) More specifically, he hoped to retain direct payments in the Farm Bill because of their relative importance to southern farmers. Direct payments per acre for the "northern" crops were quite small, and of less concern to Roberts, whereas for the "southern" crops of peanuts and rice, farmers received payments of $46 and $96 per acre (Ifft et al. 2012). Cochran's new Senate role could thus be seen as a straightforward move designed to help protect Mississippi's interests.

Lobbyists also play important roles in the legislative process. During the debate on crop insurance, for example, the insurance industry lobbied senators aggressively and outmaneuvered the advocates for reform. One senator ruefully approached a group of legislative staffers and was heard to say, "You had better watch out for angry crop insurance agents when you return home to visit your state!"

In addition to the committee structure, there are procedural elements of the legislative process that make significant reform difficult to achieve. The enactment of legislation under the Constitution requires passage by both chambers of Congress in identical form and the signature of the president. The Senate has complicated this process by developing a norm of operating behavior that requires super-majority approval; for most pieces of legislation to pass, at least 60 (of the 100 total) senators must support the bill in order to prevent the minority from blocking legislation using a technique known as the filibuster (the extension of debate through long-winded monologues that can delay or entirely prevent a vote on an issue on the Senate floor).[27] This procedural mechanism is often defended on the rationale that it helps to promote compromise within the Senate, but it also slows the pace of legislation and enables legislative minorities to obstruct the policy process. Moreover, with two senators elected to every state (as opposed to representation in the House, which is proportional to each state's population), power in the Senate is biased in favor of states with

[26] Roll Call, "Cochran Brings Southern Perspective to Senate Agricultural Committee," http://www.rollcall.com/news/cochran_brings_southern_perspective_to_senate_agriculture_committee-224199-1.html.

[27] The super-majority requirement in Senate has developed despite the absence of a filibuster requirement in the U.S. Constitution (Carpenter 2010). The practice traces back at least to the classical Roman Senate, where Senator Cato the Younger was a master of the tactic. Because the Roman Senate required that all debates and votes end by dusk, Senator Cato's long-winded arguments would often prevent a vote from moving forward. The filibuster tactic has increasingly become a tool for political obstructionism within the U.S. Senate too. It has been used more frequently during the administration of Barack Obama than during any other time in U.S. history; for example, from 1917 to 1970, the Senate majority sought to close a vote 58 times, and since the beginning of Obama's first term through 2012, it sought this "cloture" more than 250 times (Klein 2013, p. 24ff).

relatively smaller populations—states that often reflect greater reliance on agriculture as a total portion of state-level GDP.

These various procedural elements favor rural interests and make it difficult for the Congress to change laws or introduce controversial policies relative to those interests. Even the regulatory process makes it hard to change how existing legislation is implemented. Complicated and far-reaching rules—typical of much agricultural legislation—require review and approval by the Executive Office of the President. Once rules are approved, they are subject to an elaborate process of judicial review that can result in the invalidation of many regulatory changes. As a result, it is often difficult for an agency to execute policy changes involving agricultural supports or energy-related agriculture programs even if there is a change in how the majority of representatives views its policy priorities.

The Executive Branch

The presidential veto assures an executive role in the bargaining process over food and agriculture policy, though not necessarily one prominent enough to create fundamental changes in the regionally oriented nature of legislative action on agricultural issues. The President has the largest bully pulpit in terms of setting the legislative agenda, particularly through the yearly State of the Union address and the annual submission of the Presidential Budget. Often the White House also drafts and submits legislation or legislative ideas for Congress to consider. Legislative drafting relies on technical assistance from agencies such as USDA, which gives the executive branch bureaucracy a further opportunity to shape agricultural policy. For these reasons, executive branch priorities can help determine whether domestic food and agriculture priorities are tempered by international concerns. For example, under the administration of President Barack Obama, officials within the State Department, the U.S. Agency for International Development (USAID), USDA, and other agencies have sought to craft a more nuanced position on global food insecurity.[28] Although such interagency coordination is possible and reflects a genuine policy concern over global hunger, most activities related to food and agriculture within the executive branch have historically tended to support the more familiar domestic issues of farm incomes, rural development, and U.S. nutrition assistance.

---

[28] The Obama administration introduced its Global Food Security Initiative in September 2009 in response to rising food prices in international markets, food riots in various countries, and persistent food insecurity (see http://www.state.gov/s/globalfoodsecurity/, accessed June 10, 2013). In announcing this initiative, Secretary of State Hillary Clinton observed: "Food security represents the convergence of several issues: droughts and floods caused by climate change, swings in the global economy that affect food prices, and spikes in the price of oil that increase transportation costs. So food security is not only about food, but it is all about security. Chronic hunger threatens individuals, governments, societies, and borders."

The U.S. Agriculture Secretary is the preeminent cabinet-level decision maker on food and agriculture, and as discussed earlier, has historically set the stage for domestic policy design and implementation within each administration. The Foreign Agricultural Service attempts also to gather data and influence some aspects of international agricultural policy. Nevertheless, the Agriculture Secretary does not participate fully as a so-called "principal" in the National Security Council. As a result, the decisions of the Agriculture Secretary are less subject to pressures and inter-agency scrutiny than are other agencies with shared domestic and international roles, such as Justice, Treasury, and Homeland Security—all members of the National Security Council with oversight from Congressional committees that have strong international interests.[29] By contrast, the USDA is not charged with explicit statutory responsibility to manage international food security concerns; the absence of international focus is how Congress designed the agency, and essentially how the executive branch has left it. A similar situation exists with the Department of Energy (DOE); neither DOE's mandate nor its structure favor concern about the international food security implications of federal efforts to encourage the development of biofuels. By the same token, agencies such as State, Defense, and USAID have reasons to care about international food insecurity—reflected in their institutional structure and their congressional overseers—but have no mechanisms through which to influence core features of domestic agriculture or energy policy that are likely to affect global food supplies and prices.

COLLECTIVE ACTION AND COALITION POLITICS

A constellation of non-profit organizations, trade associations, commodity organizations, think tanks, consultants, diplomats, and other lobbyists seek to influence the various elements of the policymaking process through financial and technical support and political persuasion. They represent a broad range of special interests that includes traditional commodity producers, food manufacturers, anti-hunger advocates, environmental protection groups, and energy production companies. In 2011–12 leading up to the latest presidential election, the agribusiness sector contributed more than $90 million to election campaigns.[30]

---

[29] While these agencies are far from immune to domestic political pressures, they are embedded within an organizational context more favorable to foreign policy and national security concerns. That context makes it enormously difficult for Treasury, for instance, to ignore the implications of financial regulatory decisions to foreign policy or security goals involving economic sanctions or the disruption of terrorist finance (see Cuéllar 2003).

[30] The agribusiness sector includes crop, livestock, and meat producers; poultry and egg companies; dairy farmers; timber producers; tobacco companies; and food manufacturers and stores. Two-thirds of the contributions went to the Republican Party and one-third to the Democratic Party. For more details on political contributions in this and other sectors, see the Center for Responsive Politics, http://www.opensecrets.org/industries/. Accessed June 28, 2013.

The establishment of special interests in American agricultural and food policy reflects what Mancur Olson identified as a "collective action" problem: the less a particular policy mobilizes the mass public across different regions of the country, the more it is driven by organized interests competing over policy outcomes within institutions (Olson 1971). The collective action problem is diminished when majorities emerge in support of policy changes that would eliminate or reduce benefits enjoyed by narrow and concentrated interests. Often, majority-supported positions fail to win in the legislative process because the costs of organizing a dispersed majority are very high, and the prospective gain that any individual within the majority can achieve is very low.

Entrenched support for crop insurance and farm subsidy programs—two major components of successive Farm Bills—illustrates this dynamic. While a majority of knowledgeable policymakers may support reforming these programs in order to reduce budget costs or improve income distribution, the gain to any individual reform-oriented lawmaker (or for that matter, any single taxpayer) is likely to be relatively low. On the other hand, the lost benefit to farmers and crop insurance agents resulting from such policy reform is likely to be very high, resulting in vast disparities in efforts to influence the public debate and legislative process.

The force needed to counter-balance concentrated interests—broad public salience capable of mobilizing voters across regions—has existed in the past, but currently there is much less. During the New Deal period, for example, congressional representatives showed enormous interest in using agriculture policy to address the rural population's exposure to economic risks; the larger public, meanwhile, viewed agriculture policy as a matter of intense national concern. During that era, agriculture was responsible for a much larger share of employment and economic activity. Today, however, with few Americans employed directly in agriculture and less than 2 percent of GDP coming from agriculture, the continuation of New Deal-type farm subsidies resembles a classic collective action problem where the greatest pressures shaping the debate come from organized, concentrated, and entrenched economic actors.

There is another important reason why reform-minded lawmakers may steer clear of challenging entrenched agricultural policies— "logrolling," or the exchange of political favors.[31] Members of Congress often trade votes in order to achieve reciprocal advantage; this practice persists with respect to both the content of particular bills and to the organization of Congress itself. Logrolling especially helps senior members of Congress[32] who can leverage their seniority positions by voting in favor of each other's issues.

---

[31] Logrolling is defined as the informal practice of exchanging favors in politics by reciprocal voting for each other's proposed legislation and originates from the phrase "you roll my log, and I'll roll yours."

[32] Logrolling is most obvious in the Senate where 6-year terms allow legislators to serve for long periods of time and thus develop long-term political relationships.

Arguably, the Farm Bill represents the most prominent example of logrolling and coalition politics in American food and agriculture policy. It exists in its present form largely as a result of the long-standing cooperative arrangement between supporters of domestic nutrition assistance and supporters of domestic farm subsidies in the context of the passage of the Farm Bill. This dynamic has generally entailed an informal understanding whereby members of Congress who support domestic nutrition assistance either vote in favor of, or remain silent on, proposals to subsidize farmers, as long as domestic nutrition assistance programs are also funded adequately. In order to ensure the passage of a bill on either farm subsidies or nutritional assistance, both must be included in the same legislation. The Farm Bill is thus an omnibus piece of legislation requiring the support of both constituencies. What makes this piece of legislation particularly interesting is that the nutritional assistance component provides a safety net for low-income consumers, particularly in times of high or volatile food prices caused in part by agricultural policies like the corn-ethanol program. The convergence of special interests around the omnibus Farm Bill creates a peculiar equilibrium in U.S. food and agricultural policy that is extremely difficult to disrupt.

The coalition between these two very disparate communities around the Farm Bill encourages mutually supportive policy decision making. The process actually works reasonably well under favorable economic conditions, but less well when the growing federal debt and deficit reduction are on the minds of lawmakers, or when political polarization becomes sufficiently widespread to complicate bargains across regions or political parties. The imperative to reduce fiscal deficits—a point of great debate under the Obama administration—limits the amount of money that can be spent through the Farm Bill, placing stress on its "grand coalition." The tightening of available funds encourages each side to seek the best deal they can get individually, regardless of the outcome for the other. The recent weakening of coalition politics and heightened partisanship pertaining to domestic agriculture and nutritional support was almost certainly the cause of the House of Representatives (failed) proposal in 2013 of a Farm Bill without nutritional assistance.

There are two other features at work that diminish the importance of the farm support and food-stamp coalition. The first of these has to do with the economic demography of the country. The agricultural work force has become much smaller than during the first half of the twentieth century; moreover, a sizable portion of that workforce is composed of recent minority immigrants whose politics and interests are different. In addition, new and different critical issues are entering the debate. Genetically modified organisms (GMOs), climate change, and obesity are all examples of hot-button issues that tend to generate political support along non-traditional party and regional lines.[33] Perhaps a combination of these forces will be able to alter the status quo on food-related political coalitions, although the outcome from any such shift is far from certain.

Conclusion: Food Security Implications of U.S. Policy

With its vast resources and long history of agricultural investments, the United States plays a dominant role in international markets, and its farm policies directly influence global food supplies and prices. For decades following World War II, when direct and indirect farm supports fueled domestic crop surpluses, the United States contributed to rising food supplies in international markets at relatively low prices for importing nations. Real (inflation-adjusted) prices for corn, wheat, and soy in global markets declined on trend throughout most of the second half of the twentieth century as shown in Figure 4.4. (An exception occurred in the early 1970s when the combination of high energy prices, agricultural supply disruptions, and shifting policies in the Soviet Union caused a major food price spike.) As a result, U.S. farm policies enhanced global food availability and access, and hence food security as defined by short-run affordability. Over the longer-term, however, surplus production in

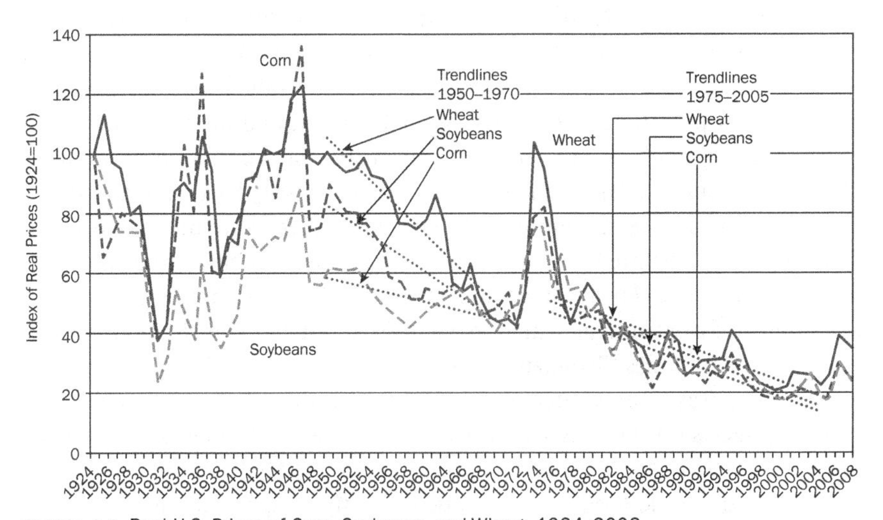

FIGURE 4.4 Real U.S. Prices of Corn, Soybeans, and Wheat, 1924–2008

Note: Trend lines are ordinary least squares fit for respective periods.

Source: J. M. Alston, J. M. Beddow, and P. G. Pardey. eds. 2010. Global patterns of crop yields and other partial productivity measures and prices. Ames, IA: Iowa State University Center for Agricultural Research and Development.

---

[33] See Politics Daily, *Tom Vilsack's Farm Country Tour: A Different Kind of Town Hall* (August 20, 2009) (describing the purpose of the Secretary's "rural listening tour" as being "to hear about the problems facing farmers").

the United States, EU, and other industrialized countries created disincentives for agricultural investments in many developing countries. With the vast majority of the world's poor population concentrated in rural areas of the developing world, lagging agricultural investments translated into low farm incomes and persistent food insecurity in many impoverished regions.

American food aid policy was used in the post-War period to address chronic food shortges in several countries—and more cynically, to increase demand for U.S. farm commodities (Barrett 2005; Riley 2015). However, it also led to a dependence on imports from the United States at below market value, and thus did not solve the longer-run problems of rural poverty and food insecurity. Since the end of the Cold War, U.S. food aid has been aimed increasingly at disaster relief, and only in recent years have political discussions focused on the provision of cash instead of U.S. crop surpluses for food aid. Nonetheless, total food aid shipments remain quite small relative to the chronic shortage of calories and nutrients experienced by over 800 million poor people globally. In essence, food aid addresses short-run famines, not chronic hunger or economic development over the long run.

The direct and indirect effects of U.S. policy on global food prices and food security generally have also taken a significant turn since the establishment of the Renewable Fuel Standard and the escalation of corn-based ethanol production in the mid-2000s (Naylor 2014). Between 2011–2013, over 40 percent of U.S. corn use was devoted to ethanol, more than the share that has gone to livestock feeds. U.S. corn exports fell as a result, particularly relative to other corn producers in the international market, and international prices have remained robust (albeit variable given weather shocks and fluctuations in global feed demand). The impacts on global food security have been mixed: net producers (farmers selling more to the market than they are buying for home consumption) benefitted, while net consumers have suffered from high and unstable prices (Naylor and Falcon 2010). Over the longer term, a high price environment favors agricultural investments that potentially help food security in agrarian economies (Swinnen 2012).

In some ways, the United States has been insulated from the effects of its own policies on food security because of its vast consumer safety net. In other ways—particularly via budget expenditures on SNAP and other nutrition programs—domestic food security is tied directly to U.S. agricultural policy. Despite large expenditures on these programs, one in six American households were classified as having "very low food security" in 2011, and food insecurity increased in most states over the previous decade (Coleman-Jensen 2012, 2013). With the prolonged recession and increasing economic inequality, SNAP payments often fall short of feeding low-income families adequately. And despite major budgetary costs, child nutrition programs fail to cover many children in need, especially outside of school hours. The nutritional quality of these programs is also often inadequate (Pringle 2013).

Domestic nutrition problems related to U.S. farm policy extend far beyond school lunch programs. The focus of agricultural policy on "program commodities" (especially corn, soybean, wheat, rice, peanuts) as opposed to vegetables and fruits has limited crop diversity (in terms of area) and has contributed to greater amounts of starch, meat, and high-fructose corn syrup in American diets. The public health impacts of food policy have become increasingly important with more than one-third of the U.S. population classified as obese, and with medical costs associated with obesity nearing $150 billion in 2008 (Centers for Disease Control and Prevention 2008). Moreover, processed foods are consumed worldwide, and American-style diets have become common in a wide range of countries. As noted in Chapter 1, many developing countries now experience more obesity than hunger.

Although many features of American agriculture policy are deeply entrenched, this domain is far from static. Changing demographics will almost certainly affect the priorities of political leaders in the decades to come. The number of commercial farmers in the United States continues to decrease and is much smaller than it was 50 or even 30 years ago. Demographic shifts may also affect the priorities of those who sit on the powerful House and Senate agriculture committees. These changes could potentially reduce the power of the committees relative to the general chambers or congressional leadership, especially when it comes time to consider policies affecting food prices, environmental benefits, and anti-poverty assistance. The likelihood, however, is that these changes will be both slow and marginal.

The institutions that have long held American food and agriculture policy in place are also beginning to respond to developments in still other domains. Fiscal constraints are already beginning to reshape long-standing agricultural subsidies, and budget pressures are increasing the power of those legislative committees that govern budgets and appropriations relative to those with substantive jurisdiction over food and agriculture. Rising interest in confronting the challenge of climate change (as discussed further by Lobell and colleagues in Chapter 9) and reducing the negative health impacts and medical costs of childhood obesity may create new political pressures and constituencies that impact the future direction of food policy. While global hunger remains primarily a humanitarian issue, a changing foreign policy context may raise the profile of broader security challenges posed by countries with greater vulnerability to food price fluctuations within the security establishment of the United States and other developed nations.[34] (However, the full inclusion of food security into national security is still unlikely, as discussed by Stedman in Chapter 13.) No doubt institutional

---

[34] The $3.5 billion pledge by the United States. toward "Feed the Future" is one such positive example. The problem, however, is whether the pledge will be appropriated. And even if spent, the amount is tiny relative to international needs and to the total size of USAID and USDA budgets. (See U.S. Agency for International Development, Feed the Future: Progress Scorecard; June 2013.)

factors such as the apportionment of the Senate—two senators per state no matter its population—will slow the impact of these changes, and contribute to the continuing entrenchment of some long-standing policies. Still, the structural changes now underway may eventually lead lawmakers and executive branch officials to contemplate modest reforms in executive branch organization, including a revamped National Security Council incorporating USDA, to give food and agriculture their rightful place in discussions of foreign policy issues.

The question for the United States and other advanced industrialized countries is whether any such development will allow the closer alignment of domestic and international concerns that will be needed to improve the prospects for global food security. Most critics of U.S. farm policy focus on specific policy measures that need adjustment. Perhaps they should focus instead on changing the political process that underpins farm policy, which itself has deep roots in U.S. agricultural history.

Coda. On February 7, 2014, President Barack Obama signed the Agricultural Act of 2014. The new legislation made a number of changes, but it was mainly a continuation of earlier approaches described here. In the end, a bipartisan coalition between farm and food stamp supporters continued to hold; 80 percent of projected future expenditures were directed towards nutritional programs; conservation activities were more or less maintained; direct and counter-cyclical payments were abolished, with their risk-reducing objectives covered mainly by expanded use of subsidized crop insurance; and dairy farmers were given a new package of support that protects the margin between the sales price of milk and the cost of feed. Total outlays for the years 2014–18 are projected at $489 billion. For more detail on the 2014 Farm Bill, see http://www.ers.usda.gov/farm-bill-resources.aspx.

References

Anderson, S., and M. Taylor. 2009. "Brazil's WTO Challenge to U.S. Cotton Subsidies: The Road to Effective Disciplines of Agricultural Subsidies." *Business Law Briefs*, *6*(1): 2–10.

Bageant, E., C. Barrett, and E. Lentz. 2010. Food Aid and Agricultural Cargo Preference. *Applied Economic Perspectives and Policy* 32(4): 624-641.

Barrett, C., and D. Maxwell. 2005. *Food aid after fifty years: Recasting its role*. London: Rutledge.

Beckman, J., T. Hertel, F. Taheripor and W. Tyner 2012. "Structural Change in a Biofuels Era," *European Review of Agricultural Economics* 1(39): 137–156.

Benedict, M. R., and O. C. Stine. 1956. *The agricultural commodity programs: Two decades of experience*. New York: The Twentieth Century Fund.

Benson, E. T. 1960. *Freedom to farm*. New York: Doubleday & Company, Inc.

Bureau of Labor Statistics. "Annual Unemployment Rates." Accessed July 9, 2013, http://data.bls.gov/timeseries/LNU04000000?years_option=all_years&periods_option=specific_periods&periods=Annual+Data.

Carpenter, D. 2010. Institutional strangulation: Bureaucratic politics and financial reform in the Obama administration. *Perspective on Politics 8*(3): 825–846.

Center for Responsive Politics. 2010. "Agribusiness Background." Accessed July 9, 2013, http://www.opensecrets.org/lobby/background.php?id=A&year=2012.

Centers for Disease Control and Prevention. Last updated 2012. "Overweight and Obesity." Accessed July 9, 2012, http://www.cdc.gov/obesity/data/adult.html.

Cochrane, W. 1993. *The development of American agriculture: A historical analysis*. Minneapolis: University of Minnesota Press.

Cochrane, W. W., and M. E. Ryan. 1976. *American farm policy: 1948–1973*. Minneapolis: University of Minnesota Press.

Coleman-Jensen, A. 2013. USDA ERS, "Food Insecurity Increased in Most States From 2001 to 2011." Accessed July 12, 2013, http://www.ers.usda.gov/amber-waves/2013-july/food-insecurity-increased-in-most-states-from-2001-to-2011.aspx#.UeRKkD7XjR1.

Coleman-Jensen, A., M. Nord, M. Andrews, and S. Carlson. 2012. Household Food Security in the United States in 2011. Economic Research Report No. (ERR-141) (September).

Cuéllar, M.-F. 2003. The tenuous relationship between the fight against money laundering and the disruption of criminal finance. *Journal of Criminal Law and Criminology 93*: 311–466.

Environmental Working Group. 2013. "2013 Farm Subsidy Database." Accessed July 9, 2013, http://farm.ewg.org/.

Ferguson, E. 2013. "Cochran Brings Southern Perspective to Senate Agricultural Committee," Roll Call, April 30 (1–2). Accessed July, 2013, http://www.rollcall.com/news/cochran_brings_southern_perspective_to_senate_agriculture_committee-224199-1.html.

Gardner, B. 2002. *American agriculture in the twentieth century: How it flourished and what it cost*. Cambridge, MA: Harvard University Press.

Government Accountability Office. 2009. "Local and Regional Procurement Can Enhance the Efficiency of U.S. Food Aid, But Many Challenges May Constrain Its Implementation." Accessed July 9, 2013, http://www.gao.gov/new.items/d09570.pdf.

Hanrahan, C. 2014. International food aid programs: Background and issues. Congressional Research Service.

Ifft, J., Nickerson, C., Kuethe, T., and C. You, 2012. Potential farm-level effects of eliminating direct payments. USDA, ERS, November.

Imhoff, D. 2011. *Food Fight: A Citizen's Guide to the Farm Bill*. Healdsberg, CA: Watershed Media.

Klein, E. 2013. Let's talk: the move to reform the filibuster. *New Yorker*, January 28.

Merrill, T., and K. Hickman. 2001. Chevron's domain. *Georgetown Law Journal 89*: 833.

Naylor, R. L., and W. P. Falcon. 2010. Food security in an era of economic volatility. *Population and Development Review 36*(4): 693–723.

Naylor, R. L. 2014. Biofuels, rural development, and the changing structure of agricultural demand. In Falcon, W. and R.L. Naylor (eds.) *Frontiers in Food Policy: Perspectives on sub-Saharan Africa. Stanford*: Printed by CreateSpace.

OECD. 2011, Equity Indicators. In *Society at a Glance 2011—OECD Social Indicators*, http://www.oecd.org/berlin/47570121.pdf.

OECD. 2012. *Agricultuiral Policies in OECD Countries*. Paris: OECD.

OECD. 2014. "Aid (ODA) by sector and donor [DACS]", *StatExtracts*. http://stats.oecd.org/Index.aspx?datasetcode=TABLE5.

Olson, Mancur Jr. 1971. *The Logic of Collective Action: Public Goods and the Theory of Groups* (Revised edition ed. 1965, 1971). Cambridge: Harvard University Press.

Paarlber, D. 1964. *American farm policy: A case study of centralized decision making*. New York: John Wiley and Sons,.

Pierson, P. 1993. "When Effect Becomes Cost: Policy Feedback and Political Change." Presidency Project. http://www.presidency.ucsb.edu/ws/?pid=24605.

Pringle, P., ed. 2013. *A place at the table: The crisis of 49 million hungry Americans and how to solve it.* USA: Participant Media.

Riley, B. 2015. *The Story of American Food Aid: Past, Present and Uncertain Future* (Forthcoming from Oxford University Press)

"Risk Management Policies," USDA. Accessed July 10, 2013, http://wwris.rma.usda.gov/policies/.

Shannon, F. (1945) 1977. *The farmer's last frontier agriculture, 1860–1897*. New York: M. E. Sharpe.

Shields, D. 2012. Federal Crop Insurance: Background. Washington DC: Congressional Research Service. Accessed July 2013, http://www.nationalaglawcenter.org/assets/crs/R40532.pdf.

Siracusa, J. 2004. *The Kennedy years*. United States: Infobase Publishing.

Swinnen, J., and P. Squicciarini. 2012. Global food supply: Mixed messages on prices and food security. *Science 335*(6067): 405–406.

The Center for Responsive Politics, "Interest Groups." Accessed June 28, 2013, http://www.opensecrets.org/industries/.

Tiehen, L., D. Jolliffe, and C. Gundersen. 2012. Alleviating poverty in the United States: The critical role of SNAP benefits. Economic Research Report No. (ERR-132), April.

University of Wisconsin, Institute for Research on Poverty. "How is Poverty Measured in the U.S."Accessed July 11, 2013, http://www.irp.wisc.edu/faqs/faq2.htm.

USDA, "Agricultural Trade." Accessed July 9, 2013, http://www.ers.usda.gov/topics/international-markets-trade/us-agricultural-trade.aspx#.UdBdBFPlX34.

USDA. 2013. "A Short History of SNAP." Accessed July 10, 2013, http://www.fns.usda.gov/snap/rules/Legislation/about.htm.

USDA. 2013. "Budget Summary and Annual Performance Plan, FY2012." Accessed July 11, 2013, http://www.obpa.usda.gov/budsum/FY12budsum.pdf.

USDA. "SNAP: Frequently Asked Questions." Accessed July 11, 2013, http://www.snaptohealth.org/snap/snap-frequently-asked-questions/

USDA. "Supplemental Nutrition Assistance Program Eligibility." Accessed July 11, 2013, http://www.fns.usda.gov/snap/applicant_recipients/eligibility.html.

United States Department of Agriculture (USDA) Economic Research Service (ERS). "Chart Gallery." Accessed July 5, 2013, http://www.ers.usda.gov/data-products/chart-gallery.aspx#UdNtqFPlX35.

USDA ERS. "Government Payments and the Farm Sector." Accessed July 9, 2013, http://www.ers.usda.gov/topics/farm-economy/farm-commodity-policy/government-payments-the-farm-sector.aspx.

USDA ERS. "Food Price Outlook." Accessed July 9, 2013, http://www.ers.usda.gov/data-products/food-price-outlook.aspx#.Uc23j1PlX34.

U.S. Department of State, Office of Global Food Security. "Women and Agriculture: A Film about Empowering Women to Feed the Future." Accessed June 10, 2013, http://www.state.gov/s/globalfoodsecurity/.

Wallerstein, M. B. 1980. *Food for war–food for peace: United States food aid in a global context.* Cambridge: MIT Press.

Weingast, B., and W. Marshall. 1988. The industrial organization of Congress, or why legislatures, like firms, are not organized as markets. *Journal of Political Economy, 96*(1): 32–163.

Wilde, P. 2013. *Food policy in the United States: An introduction.* New York: Routledge.

Winders, Bill. 2009. *The politics of food supply.* New Haven, CT: Yale.

World Bank. "Data." Accessed July 5, 2013 http://data.worldbank.org.

World Food Programme. "Food Aid, Quantity Reporting." Accessed July 9, 2013, http://www.wfp.org/fais/reports/quantities-delivered-two-dimensional-report/chart/year.

World Trade Organization. Tariff Profiles 2012. Accessed July 10, 2013, http://www.wto.org/english/res_e/booksp_e/tariff_profiles12_e.pdf.

Worster, D. (1979) 2004. *Dust Bowl: The Southern Plains in the 1930s.* New York: Oxford University Press.

WTO. "World Tariff Profiles 2012." Accessed July 11, 2013, http://www.wto.org/english/res_e/booksp_e/tariff_profiles12_e.pdf.

Legislation

The Food, Conservation, and Energy Act of 2008 (Pub.L. 110–234).

Energy Policy Act of 2005 (Pub.L. 109-58).

Energy Independence and Security Act of 2007 (Pub.L. 110–140).

5

Political Economy of EU Agricultural and Food Policies and Its Role in Global Food Security

Johan Swinnen

Since its early days in the post-World War II era, what is today the European Union (EU)[1] has grown from a loosely connected bloc of 6 cooperating countries into a diverse and widely integrated union of 28 member states, representing more than 500 million citizens. Policies designed to protect and subsidize agriculture have been a key part of the EU throughout its history, primarily through its Common Agricultural Policy (CAP). And for much of the past 50 years, critics have argued that the CAP contributes to global poverty and food insecurity by exerting downward pressure on agricultural commodity prices in world markets. As the EU subsidizes its own production and exports of agricultural commodities, it causes a decline in world prices.

Pressure to reform EU agriculture policies increased as the EU grew in size. From the initial six countries that designed the original CAP in the late 1950s, it grew into an economic, regulatory and financial union representing a fifth of the global economy by value. With high subsidies for agricultural production and taxes on imports, imports declined and the EU turned into a major exporter of agricultural products. This reversal also caused budget concerns within the EU with declining revenue from import taxes and growing expenditures on subsidies, and the CAP taking up a major share of the EU budget.

The EU did not only grow in size but also institutionally, and European society became more concerned about the safety and quality of its food, and about

[1] What is now the European Union has gone through many institutional changes. It started as the "European Economic Community" (EEC) of six countries in 1957, building on earlier cross-boundary collaborations as the "European Community of Coal and Steel." Throughout this chapter I mostly abstract from the composition and other institutional changes and refer to the "EU" across the various stages of its institutional development; only at places where it matters do we explicitly refer to the particular institutional design and composition at a given time.

environmental issues broadly. The original European Economic Community was mostly based on economic integration, with the political objective of reducing the chance of another European war by strengthening economic integration and collaboration. In the early years of the CAP, ministers of agriculture met once a year to fix the EU agricultural prices and to decide on the level of the subsidies. All decisions were made unanimously—a difficult task with countries having quite diverse agriculture and food interests. However, since then much has changed. Economic integration has grown to include a common currency used by many member states, and the free flow of workers between countries throughout much of the EU. The EU has also embarked on a process of political integration, including an elected EU Parliament with real decision-making involvement. For the CAP, this means that decisions are no longer made by unanimity but by a voting majority, while major policy issues are set through multi-year policy agreements rather than at annual talks. In addition, a series of food safety scares in the 1990s and growing concern about the environmental impacts of agriculture changed the demands that EU consumers and society imposed on agriculture. All these changes have affected EU agriculture and food policies.

Increasing pressures for policy change from inside and outside the EU led to a series of reforms over the past three decades. These reforms caused a shift in the nature of the subsidies, from taxing imports and subsidizing production and exports to subsidizing first land and animals, and later to subsidizing farms directly. These latter subsidies have a much smaller impact on production and trade, while maintaining support for farmers. At the same time, more stringent food safety and quality standards have been introduced and farm subsidies have been increasingly linked to achieving environmental objectives.

After these reforms of the CAP, the EU now faces a somewhat paradoxical situation: while its agricultural policies no longer push global agricultural and food prices down, global food security problems are now attributed to high food prices (Swinnen and Squicciarini 2012). Surpluses have declined to the extent that there is hardly any food aid being given by the EU to poor countries during the global "food crisis."

The most recent reform was decided in 2013, setting the CAP framework until 2020. There was pressure to re-introduce more market regulation, in response to recent price volatility in global markets. But the 2013 agreement kept the EU largely on its long-term trajectory towards market-based agriculture, while retaining large subsidies to enhance farm incomes directly and to incentivize environmental objectives.

In this chapter I first review the history of the EU's agricultural policy starting with the creation of the CAP and its initial effects,[2] the reforms over the 1980–2010 period and their effects, and the future of CAP. In the final

---

[2] There are some excellent books on the EU's agricultural and food policy, including Grant (1997), Oskam et al. (2011), Ritson and Harvey (1997), and on the earlier periods (Tracy 1989).

section, I discuss other EU policies that also have important implications for agricultural markets and global food security.

The Creation of the Common Agricultural Policy and the Growth of Agricultural Protection in Europe

In 1957 the Treaty of Rome was signed by the six founding EU member states: Belgium, France, Germany, Italy, Luxembourg, and The Netherlands. This treaty provided the foundations of the "European Economic Community." Agriculture was an important sector and element in the initial discussions and was the first sector to have a true "common policy." However, the attempt to work out a common framework to integrate the agricultural policies of the individual countries quickly became too technical and detailed to be negotiated at the level of the heads of state. In the words of Joseph Luns,[3] the Dutch Minister of Foreign Affairs at the time, "We asked the ministers of agriculture to work out some technical issues in a separate working group. . . and we never saw them again."

Those ministers and their staffs remained busy, however, and in 1958, at a conference in Stresa, Italy, they agreed on a set of policies that would be implemented after a transition period of ten years as a "Common Agricultural Policy" (CAP) in all the countries. The policy decisions taken clearly reflected the decision-makers' experience of living with food shortages and upheaval during the then still-recent war years. They also reflected concern about the fast-growing rural-urban income disparities of the 1950s, when economic growth took off and relative incomes in agriculture fell rapidly (Tracy 1989; Swinnen 2009).

The official objectives as stated in Article 33 (39) of the Rome Treaty were to: (1) increase agricultural productivity by promoting technical progress and ensuring the optimum use of the factors of production, in particular labor; (2) ensure a fair standard of living for farmers; (3) stabilize markets; (4) assure the availability of food supplies; (5) and ensure reasonable prices for consumers.

Objective 5 reflects the wartime experience of high food prices and the fact that despite robust growth in the European countries in the 1950s, there remained many poor people for whom food was a dominant budget item. However, this objective arguably soon became the least important in policy-makers' minds. Only the brief upheaval in global food markets in the early 1970s, when prices spiked following the first oil crisis, brought it back to the forefront.

---

[3] Joseph Luns later became NATO Secretary General.

For most of the next 50 years, the European agricultural discussion focused on the protection of farmers and its implications for global markets. The CAP resulted from the integration of various pre-EU member state policies that had been introduced to protect EU farmers' income from foreign competition and market forces. One of the main debates was between the two countries that were the driving forces behind the European integration, France and (then) West Germany. France was a major agricultural producer and exporter while West Germany had no comparative advantage in agriculture since after the war, what had been Germany's most important farmland was now in East Germany and Poland. West Germany's economic growth came from its industry, and it was protecting its smaller farms from international competition by setting high prices and import taxes. The resulting compromise was one where the agricultural prices throughout the CAP were raised close to the German prices. Germany, as the main contributor to the EU budget, paid much of the costs, while France, as the main agricultural producer and exporter, benefitted most from the policy. These differences in costs and benefits associated with common policies have remained important in the political discussions on the CAP to this day.[4]

The mechanism of government support for agriculture was through export subsidies, high import tariffs,[5] and guaranteed prices for products, which were well above world market prices. This system was particularly important for key commodities such as cereals, oilseeds, beef, sugar, and dairy products. While this largely achieved the third CAP objective of a stable market within the EU market, it also created much instability on world markets.[6] The high import tariffs and growing surplus stocks, which were exported with subsidies, caused global agricultural prices to decline. In addition the budgetary cost of the CAP grew as imports declined (bringing in less tax revenue) and surpluses and exports increased (requiring more subsidies). In those years, the EU was described as the land of butter and sugar mountains, wine lakes, and so on, referring to the large and growing surpluses generated by subsidies.

The increase in agricultural protection in Europe in the post–World War II decades is clearly illustrated in Figure 5.1. The nominal rate of assistance (NRA), an indicator that measures how large government support is compared to market revenues for farmers, increased from close to zero in the 1940s to around 80 percent in the mid 1960s when the CAP was implemented. This

---

[4] The "net contribution" status of a country is a highly politically sensitive issue with some countries benefiting more from the CAP support as large agricultural producers and others contributing more to the budget. The issue became especially sensitive after Margret Thatcher insisted, "I want my money back." This resulted in a reduction of UK budgetary contributions to the EU from 1984 onwards—the so-called "UK rebate."

[5] Taxes on imported products are usually referred to as "tariffs" by international economists.

[6] EU import tariffs and export subsidies varied to capture the difference between (fixed) domestic prices and (fluctuating) world market prices. This system of variable tariffs and subsidies ensured stable prices inside the EU, but intensified fluctuations outside the EU since export subsidies would be even higher when world market prices were lower.

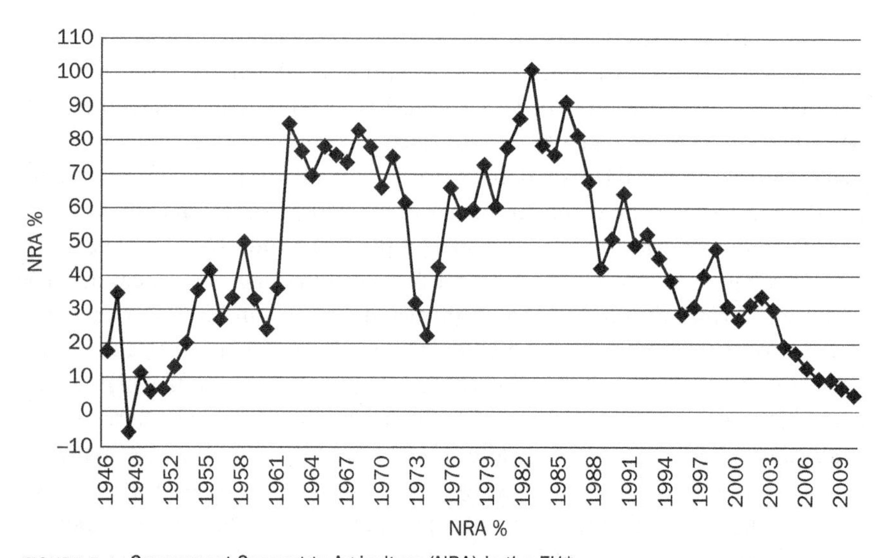

FIGURE 5.1 Government Support to Agriculture (NRA) in the EU*
*NRA = Nominal Rate of Assistance, which is an indicator of government suport to agriculure and is measured as the price of a product in the domestic market (plus any subsidy) less its price at the border, expressed as a percentage of the border price (adjusting for transport costs, quality differences, etc.).
Source: Anderson and Nelgen (2013) and Swinnen (2009)

meant that at the time, farms were receiving almost as much gross revenue through government support as from the market.

I have previously explained this growth in protection by a combination of several factors (Swinnen 2009). First, while farm incomes grew after the Second World War, incomes in the rest of society grew much faster with the rapid economic growth of the 1950s and 1960s. This created a growing urban-rural income gap. Political economists have shown that interest groups turn to governments to assist them when market conditions turn against them and that there are political incentives for decision makers to introduce or adjust policies to assist these groups (for example, Swinnen and de Gorter 1993).[7] Hence, European farmers increasingly pressed their governments to introduce support policies to reduce the urban-rural income gap.

Second, the most important opposition to governments raising prices for farmers came from consumers (workers) and from industry (which was concerned about the inflationary impact of food costs on wages). With economic growth, the share of food in consumer budgets became less important, and thereby the inflationary effect of food on wages was also diminished. These factors reduced the opposition of workers and industry (Swinnen 1994; Swinnen

[7] This so-called "anticyclical policy pattern" is well documented in agri-food markets, both in Europe (e.g., Olper 1998; Swinnen et al. 2001; Swinnen 2009) and elsewhere (Anderson and Hayami 1986; Gardner 1987; Anderson et al. 2013).

et al. 2001). A third factor was the enhanced political organization of farms and agribusiness interests. One important element was the growth of farmer cooperatives as major forces in rural credit, input purchasing, processing and marketing. These cooperatives not only played an important role in improving the economic situation of farmers but also in their political influence. At the same time improvements in rural infrastructure, including in communication, allowed farmers to better organize. Together, these factors led to a strong increase in government support to agriculture, as government tried to protect employment and incomes in a sector in (relative) economic decline from market forces.

Three Decades of CAP Reforms

The EU's impact on the world market increased as it expanded, and as subsidies and tariffs turned the region into a net exporter of food. The EU had previously been a major net importer of agricultural and food products, but the CAP caused a strong reduction in net imports as exports of the agricultural products receiving government support increased. As Figure 5.2 illustrates, the net trade (ratio of exports over imports) increased from less than 40 percent in the early 1960s to more than 80 percent by the end of the 1970s (panel b). While the EU is still the largest agricultural importer in the world, its increasing net trade has been due to a combination of increased exports and falling imports (panel a).

In the 1970s and 1980s, pressure increased on EU policymakers to reduce the CAP distortions. The pressure came both from inside the EU, primarily from ministers of finance concerned about the cost of subsidies, and from outside actors concerned about depressed global prices. The most important outside pressure came from exporting nations such as the United States and Australia, and from developing countries and international organizations that accused the EU of causing poverty and hunger in poor rural households.[8] In response to these internal and external pressures, the EU introduced a series of reforms, spanning three decades, to reduce the impact of its CAP on international markets (Josling 2008; Moehler 2008).

[8] For example, organizations such as the OECD and the World Bank emphasized how the EU (and other countries. including the United States) were hurting the poor by contributing to low agricultural and food prices through their agricultural subsidies: "Many [developed countries]. . . use various forms of export subsidies that drive down world prices and take markets away from farmers in poorer countries. . . . Much of this support depresses rural incomes in developing countries while benefiting primarily the wealthiest farmers in rich countries." (OECD 2003). Non-governmental organizations (NGOs) took the same position. For example, Oxfam International (2005) argued: "US and Europe's surplus production is sold on world markets at artificially low prices, making it impossible for farmers in developing countries to compete. As a consequence. . . farmers are losing their livelihoods." See Swinnen (2011) for more details.

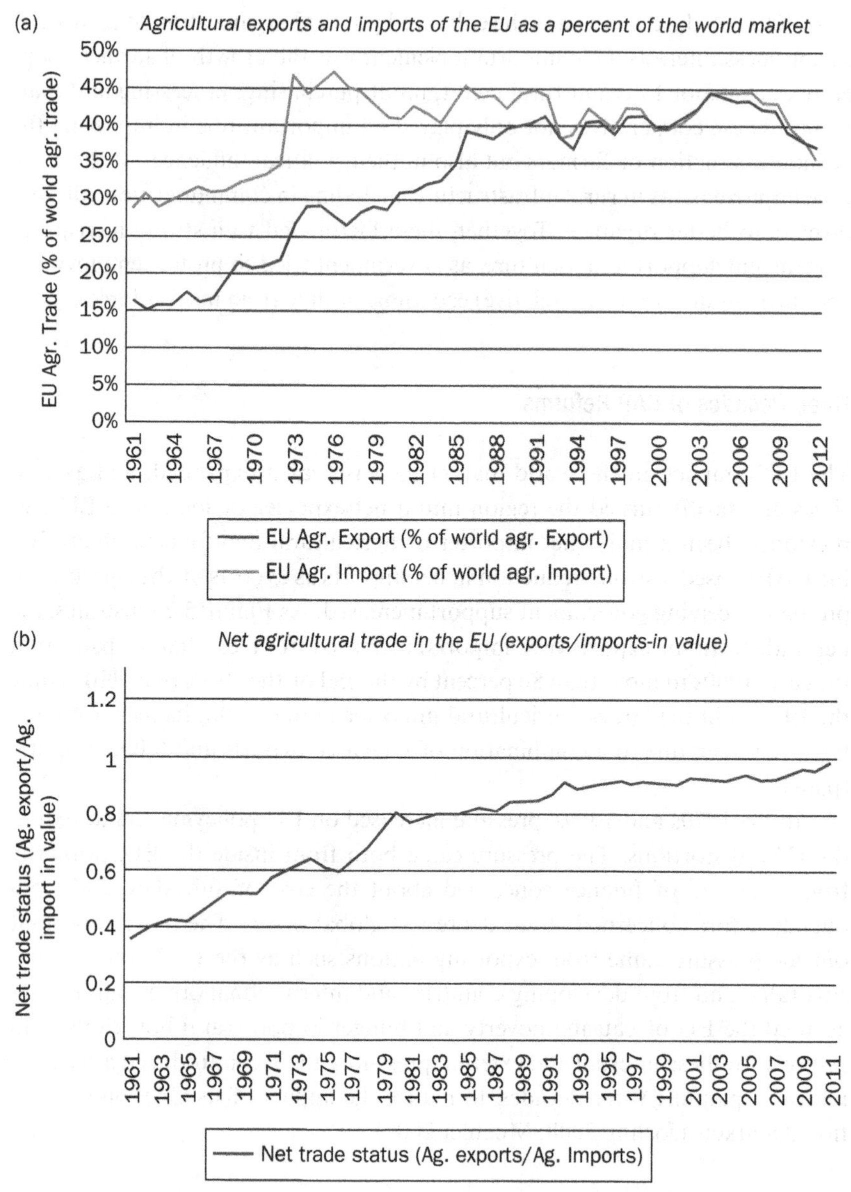

FIGURE 5.2 Agricultural Exports and Imports of the EU as Percent of the World Market
Source: United Nations Food and Agriculture Organization (1960-1990) and World Trade Organization (1991-2012)

The first reforms came in the 1980s with the introduction of production quotas (i.e., maximum quantities that could be produced with subsidies) in the sugar and dairy sectors. These quotas limited the size of subsidies and surpluses, but did not eliminate them. Most production was still subsidized and a substantial amount was still exported with subsidies.

More fundamental problems remained in sectors such as oilseeds and grains where it was more difficult to implement quotas because these products were easier to store and were used in animal feed, complicating the enforcement of supply controls. It took until the early 1990s before policy changes were introduced here: it was decided to replace price support and export subsidies with payments based on the area of land under cultivation, called "direct payments." These reforms were strongly influenced by the GATT[9] negotiations at the time. The GATT negotiations round had been launched in Uruguay several years before and, for the first time, agriculture was included in the discussions—largely because of the impact of the CAP on world markets. The negotiations were long and difficult and ultimately lead to the so-called "Uruguay Round Agreement on Agriculture" (URAA). This agreement imposed limits, among others, on the value of agricultural subsidies that were "trade distorting" (such as import tariffs and production subsidies) and on the volume of subsidized exports. In order to reach an agreement, the EU had to change the nature of the subsidies. Subsidies linked to land (and animals) still stimulated production and exports but less so than subsidies directly linked to production, such as price supports. The compromise in the URAA allowed the EU (and the United States) to continue large subsidy payments linked to land used by farmers, while freeing global markets at least partially from the downward price pressure.

Figure 5.3 clearly illustrates the resulting changes in EU budget expenditures on the CAP: the value of subsidies going to exports and other market support declined significantly in the early 1990s, while the share of "coupled direct payments," the subsidies coupled to land use for specific agricultural production, greatly increased. In Figure 5.3, these are defined as "coupled direct payments." Later, these coupled subsidies would be largely replaced by "decoupled" direct payments, as described later.

EASTERN ENLARGEMENT AND THE CAP

In 1998, I received a job offer from the European Commission, the administration in charge of preparing policies for the EU. They needed advice and understanding about what was going on in Eastern European agriculture.[10] The EU was facing the most challenging extension in its history. After the fall of the Berlin Wall in 1989, Eastern Europe had been able to free itself from the communist system and from Soviet domination. Many of these countries had moved to a democratic political system and a market economy by the 1990s and were keen to join their western neighbors in the EU. The EU had started

[9] General Agreement on Tariffs and Trade (now the World Trade Organization).

[10] After finishing my PhD at Cornell University in 1992 I joined the Leuven Institute for Central and Eastern European Studies to analyze the agricultural transition in the former Soviet bloc. This was an important part of my research during the 1990s, resulting in my book (with Scott Rozelle of Stanford University) "From Marx and Mao to the Market," published by Oxford University Press.

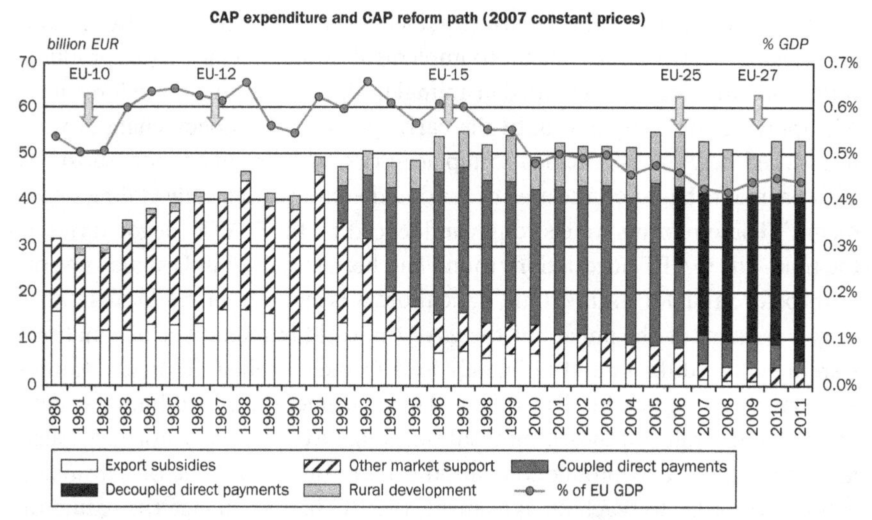

FIGURE 5.3 CAP Budget Expenditures (1980–2012)

Source: European Commission

negotiations with 10 Eastern European countries to become members of the EU (they ultimately joined in two stages, in 2004 and 2007).[11]

Agriculture was a big concern. This "Eastern enlargement" added around 50 percent to the EU's farmland and more than doubled the number of farmers. This created all kinds of challenges for the EU. One challenge was finding money for all the farmers who would become eligible for EU subsidies (Ackrill 2003). Another challenge was that the increase in the total value of subsidies given to the enlarged agricultural sector would put the EU into conflict with the constraints imposed by the URAA. Internally, the old (Western) farms feared that the new (Eastern) farms would flood the newly integrated market with cheap products, while Eastern farmers feared that rich Western farmers and investors would buy up all their land. I discussed the key challenges of Eastern enlargement in my 2001 contribution to the inaugural issue of *EuroChoices* under the title "Will Enlargement Cause a Flood of Eastern European Food Imports, Bankrupt the EU Budget and Create WTO Conflicts?". In this article I argued that there were serious challenges but that many of the fears were unfounded and that some of the most fundamental problems were ignored, such as the structural problems in rural labor markets.

Negotiations on the conditions of accession for the Eastern countries and on CAP reforms were conducted simultaneously. The outcomes included: (1) further reductions in price support and export subsidies for all EU farmers (to satisfy WTO conditions) under the so-called Agenda 2000

[11] The Czech Republic, Slovakia, Poland, Hungary, Slovenia, Estonia, Latvia, and Lithuania joined in 2004, and Bulgaria and Romania joined in 2007.

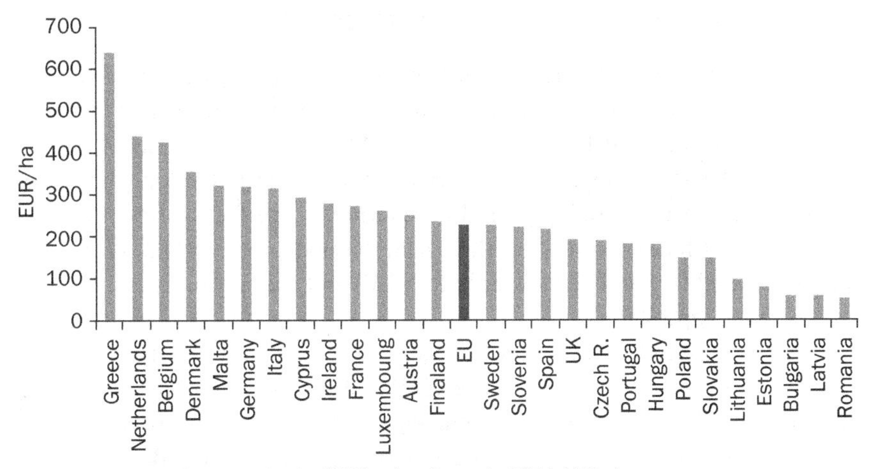

FIGURE 5.4 Direct Payments in the EU Member States in 2011 (€/ha)

Source: Eurostat and European Commission.

reforms; (2) an extension of the direct payments to Eastern farmers but at considerably lower levels than Western farmers received; and (3) a temporary prohibition for Western farmers and investors to buy agricultural land in Eastern countries.[12]

Figure 5.4 clearly shows that even though payments in 2011 have significantly increased compared to the levels when the Eastern countries joined in 2004, Eastern payment levels are still considerably below those in the Western member states. Not surprisingly, one of the demands of the Eastern countries for changes during the recent CAP reform discussions was to make these payments more equitable.

Eastern enlargement also significantly increased the heterogeneity of the EU's agriculture. Despite a "common" agricultural policy, there were always substantial differences in EU agriculture. This applies to everything from the commodities produced and consumed, to farm structures, land use, the importance of agriculture in the regional/national economy, income levels, and more. The "European farm model" concept, of small-scale local family farmers needing government protection from (presumably) "large-scale capitalist farms" was increasingly difficult to maintain. While large-scale farms in France and the UK had always been happy to hide behind this political construct, the integration of the large-scale corporate farms that dominate many regions in the Czech Republic, Slovakia, and the former East Germany made doing so less feasible. The EU's agriculture became much more diverse—and much more publicly so.

[12] The transition period was initially for 7 years, with an exception of 13 years for Poland (Swinnen and Vranken 2009).

THE 2003 "FISCHLER REFORMS": A PERFECT STORM

After three years, I left the European Commission, returning to academic life. However, I was a close observer of what many experts identified as the most radical reform in CAP history: the 2003 CAP reform,[13] presided over (and guided) by then-Commissioner for Agriculture and Rural Development Franz Fischler,[14] This was caused by what I have termed a "perfect storm" (Swinnen 2008). Several factors came together in the period around 2002, creating a strong demand for radical changes in the CAP and causing sufficient pressure to overcome opposition.

Commissioner Fischler in private (and afterwards in public) argued that the CAP was in danger because "the CAP had lost its legitimacy among the EU public." The CAP was seen as hurting EU trade interests, as it had been a major stumbling block in trade negotiations. The EU expected that further cuts in agricultural subsidies would be needed to reach a new trade agreement in the next GATT (what was now WTO) negotiation round—typically referred to as the Doha Round. In addition, there were increasing concerns about the negative effects of agriculture on the environment. When several food safety and animal welfare crises hit the EU in the late 1990s (see further), the CAP not only appeared to be ineffective in addressing the problems and the food safety concerns of EU consumers, but instead to worsen the situation. Hence, when ministers of finance and other members of the European Commission were searching for budget cuts, the CAP immediately came into focus as it was seen as using a disproportionate amount of the EU funds given the problems with the policy. In response, Commissioner Fischler and his team designed a strategy to maintain support for European agriculture by creating a new legitimacy for the CAP by addressing trade, environmental, and food-safety concerns.

The main element of the reform was a shift towards a policy system that continued to support farmers but which created even fewer (or no) distortions in international markets—thereby addressing the trade concerns. The proposal was to "decouple" farm payments fully through a "single farm payments" (SFP) system. Subsidies would no longer be related ("coupled") to what was produced or the land that was used, but were given as a fixed payment to the farm. To address environmental concerns, these payments would be conditional

---

[13] The 2003 reform was initially referred to as the "Mid-Term Review" because it had the original objective to monitor, at the halfway period, the effectiveness of the Agenda 2000 reforms, which had been insufficient in addressing the problems facing the CAP (e.g. Buckwell and Tangermann 1999; Burrell 2000; Núnez Ferrer and Emerson 2000).

[14] In 1995 Austria joined the European Union and Franz Fischler, a then largely unknown Austrian politician, became EU Commissioner in charge of the Common Agricultural Policy (CAP). There was surprise that a new member state had been given the powerful Agricultural Commission chair, but no major expectations surrounded his arrival in Brussels. However, a decade and two tenures later, Fischler left behind a CAP that was dramatically changed from the one he inherited, and he is recognized by friend and foe alike as the architect of the most radical reforms of the CAP.

on farmers addressing certain environmental concerns, such as preventing soil erosion, managing water and taking measures to avoid the deterioration of habitats—the so-called "cross-compliance" requirements. Food safety concerns were addressed in a separate policy (see further).

A key factor in the success of the 2003 CAP reform was that the governing political process had been fundamentally transformed by the 2001 Treaty of Nice, which introduced (qualified) majority voting in the decision-making on the CAP.[15] Previously, decisions had been made by unanimous agreement, effectively giving any opposing member state veto power. The agents involved in agenda-setting and decision-making for CAP had also changed. In 1995, Sweden, Finland, and Austria joined the EU. This reduced the share of the votes of the established players in the EU and brought Sweden, which became a strong voice in favor of more market-based policies, into the decision-making process.[16]

The 2003 negotiations also transformed the politics-as-usual of the CAP. Traditionally, the main pressure group had been the farm unions. Now consumer groups and environmental groups played a more prominent role in the CAP reform debate than previously, and Fischler and his team purposely tried to include them to enhance the broader legitimacy of the future CAP among the increasingly urban EU society. In fact, it appears that farm unions were taken by surprise by this new development and had considerably less influence than they expected in the negotiations. In a related development, the EU's executive body, the European Commission, had also undergone significant changes. Many of the old-style agricultural administrators, whose careers had developed in the early years of the CAP, had retired and were replaced by a younger generation. Thinking within the Commission, which is responsible for administering the EU and preparing policy proposals, was much more open to environmental and economic arguments.

Ultimately though, the combination of Fischler's experience (he was in his second term), his strategic vision, his political tactics, and the Commission officials' effort and preparation led to the successful 2003 Mid-Rerm Review. The reforms were prepared in relative secrecy by a small inner circle of officials, while experts within the Commission administration were calculating the potential effects of the reforms. In-house analyses were prepared by Commission officials to counter critiques with careful arguments. External communications essentially shut down during the spring 2002 French election campaign, since French President Jacques Chirac was a major opponent of the reforms.

---

[15] Following the 2007 enlargement of the EU, the "qualified majority" is 255 votes out of a total of 345 (member states have a number of votes which differs depending on their population, with the winning votes representing at least a bare majority of the member states.

[16] Sweden had gone through a major liberalization of its agricultural policies just a few years before its entry into the EU and was now forced to re-introduce major government interventions under the CAP (Rabinowicz 2003).

The proposals initially faced a strong anti-reform group, led by France, Spain, and Germany, which together controlled a blocking minority of votes. Unexpectedly, the Iraq War played an important role in facilitating the CAP reforms. The March 2003 invasion of Iraq, led by the United States, had split the EU's political unity. Germany and France opposed the invasion, while the UK and Spain supported it. It initially made allies out of Chirac and his German counterpart, Gerhard Schröder, in opposition to the CAP reforms—despite Germany's earlier demand for reforms. But Fischler managed to use the Iraq alliance between the UK and Spain to maneuver Spain out of the anti-reform group, since Tony Blair, the UK prime minister, was strongly in favor of the reform.

Finally, it is important to emphasize that Fischler and his team saw their reforms not as an instrument to reduce the importance of the CAP, but as a way of saving it, and its role in supporting sustainable rural development. They sought bold reforms to reduce its negative effects on the environment, on market distortions and on the WTO negotiations in order to create new support for CAP and reduce the pressure for large budget cuts. Major budget cuts for the next financial period were in fact avoided, something Fischler saw as a major achievement of the reforms. From this perspective, the Fischler reforms contributed to the survival of the CAP, rather than to its demise.

IMPACT OF THE CAP REFORMS

The combined reforms resulted in an overall decline in agricultural support in the EU, and in particular in a strong decline in the use of subsidies affecting production and trade. Figure 5.1 illustrates how the nominal rates of assistance to agriculture (NRA, previously explained) fell strongly, from an average of more than 50 percent in 1991–95 to just 11 percent in 2005–10. The reduction in coupled farm support was especially large. The same conclusion comes from OECD (2012) estimates on agricultural support in the EU. Their measure, the producer support estimate (PSE), declined from an average of 36 percent in the period 1991–93 to 20 percent in 2009–11; the PSE for agricultural support that affects production and trade (the so-called "coupled support") fell below 10 percent.

The dramatic change in the nature of the agricultural subsidies can also be seen from Figure 5.3 which illustrates the effects of changing subsidies change in the nature of the subsidies on the EU budget. From the mid 2000s onwards the vast majority of EU farm support (35 billion euros out of a total of slightly more than 50 billion euros per year) occurs as decoupled direct payments.

After the reforms, prices in the EU have been close to those on world markets and the impact of the current CAP on global prices has been much smaller than in the past. Several studies show the large impact of EU policies on global food markets during the 1980s (e.g. Van Meijl and van Tongeren 2002). Recent studies show that EU policies have no longer had a significant impact on the price volatility of major food commodities (Anderson and Nelgen 2012; Anderson et al. 2012).[17] Unlike countries such as Russia and China, the EU has not introduced export constraints for food during the recent price spikes.

Hence, somewhat paradoxically, after the EU had gone through decades of reforms to reduce the negative impacts of the CAP pushing global food prices downward, the world became concerned with food prices heading in the other direction. After the price spikes of 2007–08, international organizations, NGOs, and many experts pointed at the hunger and poverty effects of high food prices.[18]

The CAP of the 1970s and 1980s would have had a much stronger effect in countering high food prices than the current CAP. The former surplus production and the large food stocks in the EU could have been used to export food, including cereals, and thus to reduce prices when they were rising, both as commercial exports and as food aid. The policy reforms over the past two decades, which have reduced the distortionary effects of EU policies on world food markets, have also reduced the EU's capacity to quickly increase food exports during price spikes. In fact, EU food aid to developing countries was at its lowest in recent years, when food prices were high. See a similar discussion for the U.S. in Chapter 4. Despite high food prices, the EU has not provided more (in-kind) food aid to poor countries. As Figure 5.5 illustrates, food aid declined from a peak of more than 3.5 million tons per year in the early 1990s to close to zero tons in recent years.[19] EU food aid was especially prominent when public stocks were high; with the CAP reforms, agricultural surpluses and food stocks have largely disappeared.

[17] Interestingly, these studies find some impact in the maize market. This is because global maize price increases triggered some policy adjustments, including a reduction in EU import constraints, which contributed to higher world market prices. Hence, even here it is the reduction in import constraints which contributed to higher prices—which in the pre-2008 world would have been considered a positive development.

[18] See Swinnen, Squicciarini, and Vandemoortele (2011) for a political economy explanation.

[19] Instead, in 2008 the EU adopted a regulation establishing a €1 billion "Food Facility" which included measures to improve access to agricultural inputs and services, safety net measures to maintain or increase agricultural production capacity and help meet the basic food needs of the most vulnerable populations, and other small-scale production-boosting measures based on countries' individual needs (microcredit, investment, equipment, infrastructure and storage, vocational training, and support for agricultural professionals). While the EC presented the Food Facility as its highest-profile instrument in development aid, a number of observers, including the European Parliament, questioned the automatic extension of this instrument in times of crisis, as its ability to tackle the structural roots of food insecurity had been rather difficult to assess (European Parliament 2011).

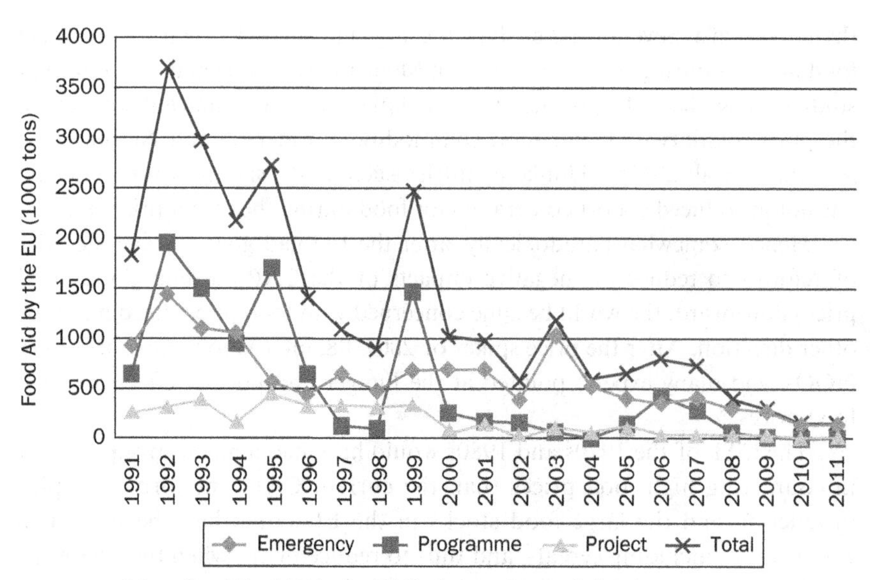

FIGURE 5.5 International Food Aid by the EU (in thousand tons of wheat)
Source: World Food Programme FAIS database.

The CAP of the Future

The food price spikes of the late 2000s coincided with important discussions in the EU on the future of the CAP. The first global price increase coincided with the conclusion of the so-called "Health Check" reforms in 2008. The Health Check reform of the CAP further decoupled support and reduced government intervention in several commodity markets. This included the important decision to continue the gradual abolition of dairy and sugar quotas, and to keep moving toward a more market-based CAP. The aim, according to the European Commission, was to "modernize, simplify and streamline the CAP and remove restrictions on farmers, thus helping them to respond better to signals from the market and to face new challenges" (European Commission 2008).[20]

NEGOTIATIONS ON THE FUTURE CAP: BUDGET PRESSURES AND FINANCIAL CRISIS, HIGH GLOBAL PRICES, AND CHANGED INSTITUTIONS

The CAP's continuing large share of total EU expenditures was one of the main issues in negotiations over the EU budget for the 2013–2020 period. In

[20] Not all farmers benefit from high commodity prices. In particular, EU dairy farmers saw their income falling as feed costs increased. In response, a "milk package" was introduced in 2010. Despite considerable pressure from dairy organizations, the milk package does not include measures that directly intervene in the markets. Instead the policy seeks to improve the functioning of the dairy supply chain, in particular by improving contracts between farmers and dairies and strengthening farmers' collective bargaining power.

2011, the CAP budget amounted to €58.7 billion euros out of a total of €141.9 billion, or around 40 percent.[21] The budgetary discussions were given extra impetus by the global financial and economic crisis, felt keenly in many EU member states. Throughout Europe the economy declined and governments were forced to cut budgets.

As explained earlier, the 2003 CAP decision-making process was fundamentally transformed by the 2001 Treaty of Nice, which introduced qualified majority voting in CAP decision-making. Since then, additional important new institutional changes have taken place. The 2007 Lisbon Treaty gave the EU Parliament "co-decision powers" on CAP and budgetary issues. These co-decision powers were expected to transfer power from the European Commission (which proposes policy reforms to be voted on by the Council of Ministers) to the EU Parliament (which to that point had only a consulting role, with very little effective influence on final decisions) (Crombez et al. 2012).

There was considerable uncertainty about the way the European Parliament would handle this first co-decision experience, and how this would influence the content of the new CAP. Some, such as Roederer-Rynning (2010), argued that the European Parliament would push for a more interventionist policy. This was because interest groups that favor more regulation, such as some farmers associations, were influential in the European Parliament's Agricultural Committee.

The European Commission published its proposal on the CAP for the 2014–2020 period in October 2011. The Commission essentially proposed to maintain the key elements of the CAP as they existed at the time, with some changes in the nature, structure, and distribution of the payments but without a return to market interventions. The changes in the payments can be summarized in three key words: *convergence, greening*, and *capping*. Support is to be more equally distributed (convergence), better linked to environmental objectives (greening), and with a maximum ceiling to any individual producer (capping). The proposals use price volatility as a justification to maintain the CAP direct payments (as a "safety net") to protect farmers against price volatility; they give "basic financial security to farmers, without distorting international markets" (European Commission 2011).[22]

---

[21] European Commission, Financial Programming and Budget, *The 2011 Budget in Figures*, http://ec.europa.eu/budget/figures/2011/2011_en.cfm.

[22] The proposal also included a new "crisis reserve fund" and a "crisis management toolkit." These include funds for crop and weather insurance, and income stabilization to compensate farmers if their income drops by 30 percent or more. The official aim of these instruments is to respond rapidly to an extreme event of price volatility (European Commission 2011). Bureau (2012) concludes that various conditionalities such as maximum quantities on intervention, limits on compensation, and co-financing requirements make these measures consistent with WTO disciplines and limited in practice. In addition, the EC proposes to allocate €4.5 billion for research and innovation on food security, the bio-economy and sustainable agriculture, but the impact is likely modest since a large share of the funding represents a reallocation within the EU Budget (Bureau 2012).

The proposals were amended by the European Parliament and the Council of Ministers (where qualified majority applies). Both had to come to a final agreement on a joint version of the proposals. The amendments did introduce some changes, but overall the reform strategy proposed by the Commission was followed. While the future CAP includes more subsidies that can be linked to production than under the 2003 reforms, the main policy decisions are largely in line with the fundamental long-term strategy towards market-based agriculture. In this latest round of CAP reform, the EU has reaffirmed its move towards an open trade policy, while also underlining the harm done by the restrictive export policies implemented by some countries in response to price volatility. It has also stayed on course with its reform proposals in specific sectors, such as phasing out the quota regime in dairy and sugar, despite a slight change in argumentation (i.e., by linking the motivation to price volatility), as well as to WTO requirements and the other previously stated goals.

POLITICAL ECONOMY CONSIDERATIONS

It is interesting to observe the modest response of EU policies to recent global price volatility in the context of changes in the institutional organization of EU decision-making. As I explained earlier, there are important political incentives for decision makers to adjust policies to changing market conditions. It would therefore be logical to expect much more significant policy adjustments than were observed in response to the dramatic changes in the world food markets. How can the relative lack of action be explained?

First, the lack of major government interventions is consistent with the fact that incomes of EU farmers have increased on average over the 2005–11 period. On average, EU farm incomes in 2011 and 2012 were 25 percent higher than in 2005 and 2006, before the price spikes (Swinnen et al. 2014). Notwithstanding exceptions, including the dairy sector where incomes have fallen with increasing feed costs, most farmers have benefited from higher prices for their products while receiving constant levels of support payments from the EU.

Second, most policy discussions on the CAP in the past years have focused on how to reform the farm payments, as increased pressure from taxpayers and demands from environmental groups challenge the current payment structures. Farm organizations have concentrated more on lobbying to secure the payments rather than on a major shift towards more regulation. They have been supported in these efforts by landowners, who are benefiting from spillover effects of the land-based payments (Ciaian et al. 2010).

Third, the EC and the European Parliament have taken different positions on the CAP, with the latter taking a more interventionist stance. The differences can be related to (at least) two factors. First, the EC has played a leadership role in steering CAP reforms since the 1990s. As a bureaucracy with the right to table policy proposals, combined with strong leadership and a strong capacity in analysis and policy preparation, the EC has been able to steer the reforms through the political process, carefully arranging a qualified majority of votes within its own decision-making apparatus, while largely ignoring the European Parliament. Second, the EC is more sensitive to the international dimensions of the CAP, in particular the WTO constraints, since they have been intensely involved in the ongoing WTO negotiations. (It appears that the EC wants to stay on course in moving the CAP towards more market orientation, continuing its 20-year strategy and legacy.)

In contrast, the European Parliament does not have such a legacy as it is just now becoming involved in the actual decision-making. Moreover, the Agricultural Committee of the European Parliament, where the key positions are prepared, is filled with members who are linked to traditional agricultural interests. This contrasts with the EC's approach in the past decade to broaden the support base for the CAP by reaching out to environmental groups, consumers, and others. Farm organizations have started to target the Agricultural Committee as their key focus for lobbying activities. The WTO agreements do impose real constraints on policy reactions here as well, even though individual actors may not feel them as keenly.

Finally, the multi-annual agreements behind any policies relying on EU budgetary expenditures must be taken into account. The limited response in important EU policies to price volatility is explained at least in part by the fact that the underlying agreements can only be changed after long negotiations. Hence, policy reactions from 2007 to now have been constrained by the CAP and budget agreements covering the 2007–13 period.

Other Policies Affecting Food Security

While the CAP has attracted the most global interest in EU agriculture policy because of its impact on trade and global markets, other EU policies are also important for farmers and food consumers in the EU and globally. In this section I briefly review a few of these.

SOCIAL PROTECTION

European countries are well known for their elaborate social security systems. Social groups that are particularly vulnerable to food costs, such as the elderly, the unemployed and the poor, can draw on social security resources. Over the past decades, government support for poor consumers in the EU has occurred mostly through social spending, not through food market regulations.

Average consumer prices in the EU increased just slightly over the 2005–12 period, with real food prices only 5 percent higher in 2012 than in 2005. An important reason is that the cost of ingredients is a small share of the price of final food products in the EU. For example, the share of ingredients in the cost of bread is merely 5 percent. On average, it is just 20 percent for meat and livestock products (European Commission 2007).

The impact of food price changes on consumer welfare also depends on how much consumers spend on food. European consumers on average spend 15 percent of their household budget on food. Food price changes therefore had a limited impact on the average EU household's overall budget and welfare. However, there are significant differences between and within member states. Poorer families spend a higher proportion of their budget on food. The share of the household budget spent on food varies from 10 percent in the UK to more than 40 percent in Romania.

Unlike CAP subsidies, social policies, such as unemployment benefits, pensions, and disability payments, are still the responsibility of individual member states. The increase in food prices induced pressure from consumers, in particular the poorest, to increase social spending. Increases in other prices, such as those for energy and transport, reinforced this pressure. Even as the financial and economic crisis constrained governments' budgets, social expenditures in the EU increased by approximately 7 percent between 2005 and 2010 (Figure 5.6). Not surprisingly, there are large disparities among member states, but spending on social security benefits increased in almost all of them.

Since 1987, the EU has had a food aid program for the poor and the needy of Europe. Initially, this scheme consisted of the distribution of stocks of surplus food. Due to reduced surpluses from the CAP reforms of the 1990s and 2000s, the food aid scheme was revised in 2008 to buy products on the open

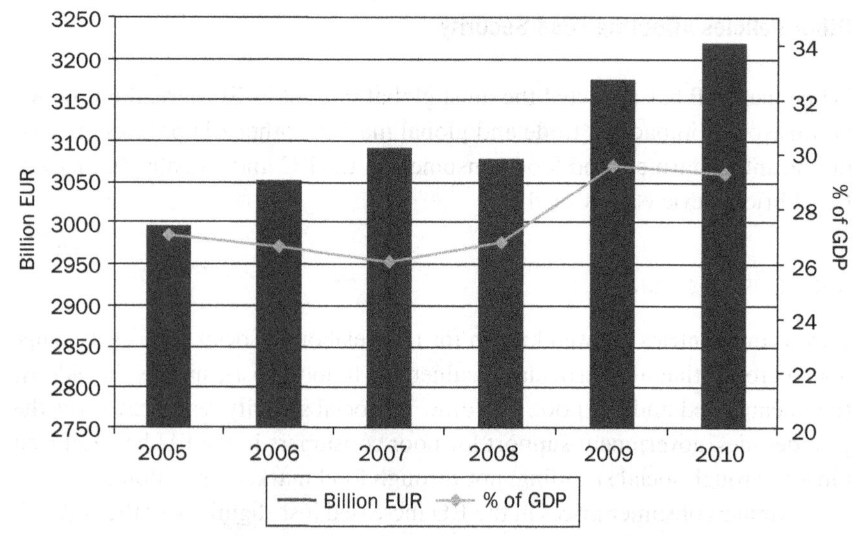

FIGURE 5.6 Evolution of Social Expenditures in the EU27 (in real terms)

Source: Eurostat

market. In 2010, the EC capped food aid spending at €500 million per year. In 2011, under financial pressure, the EC proposed a drastic cut to €112 million. However, aid organizations argued that it was precisely in periods of rising food prices that such programs were most needed. Negotiations took place on the amount of food aid and on (member states') co-financing requirements. Several member states argued that "social policy" is a national competence and should be left to member states. The agreement was reached to maintain the level of funding at €500 million for 2013 only.

FOOD SAFETY POLICIES

EU consumers in the twenty-first century are particularly concerned about the safety and quality of food. These concerns were triggered by the food scares that plagued the EU in the second half of the 1990s, such as bovine spongiform encephalopathy (BSE, or "mad cow disease"), foot and mouth disease (FMD) and episodes of toxic contamination, including the dioxin crises (Bernauer 2003; Scholderer 2005; Graff and Zilberman 2007).

Previously, food safety and quality policies were mainly the responsibility of member states, except for some veterinary directives from the EC. The food safety crises in the 1990s, particularly the emergence of BSE in 1996 and the dioxin contamination crisis in 1998, were crucial in changing this. In 1997, a year after the BSE crisis, the Commission launched a new food safety initiative. This resulted in major legislative initiatives such as the General Food Law Regulation, which was adopted in 2002, and the creation of the European Food Safety Authority (EFSA). The main goal of the new food safety policies is to protect consumer health by introducing a farm-to-fork safety approach, imposing strict traceability requirements throughout EU food chains. At the same time, the policies aim to ensure the smooth operation of the "single market" and to take into account existing or planned international agreements. These include the sanitary and phytosanitary (WTO-SPS) and technical barriers to trade (TBT) agreements under the WTO.

Not only has the public sector responded to the crises, but there has also been a rapid growth in private sector initiatives in the field of food safety and quality standards. These include the GlobalGAP[23] standard which is now used by a large number of the major retailers in the EU (and the world). Furthermore, there is a rise in public-private partnerships in establishing quality assurance schemes.

[23] GlobalGAP is a certification scheme with guidelines for the "Good Agricultural Production" of fresh fruits and vegetables, including food safety requirements and guidelines for reducing environmental impacts and the use of chemical inputs and ensuring a responsible approach to worker health and safety as well as animal welfare.

DECISION-MAKING INSTITUTIONS AND POLICY GRIDLOCK ON GMOS

Heightened food safety concerns have also affected EU regulations on genetically modified (GM) food and feed. The food safety crises of the 1990s contributed to a great wariness about new food technologies, including genetic modification. According to survey data (Eurobarometer 2010), consumers' optimism about biotechnology increased from an all-time low of 24 percent in 1999 to 77 percent in 2010—the highest ever. However, at the same time 58 percent of them still opposed GM food.

Since the end of the 1990s, the EU has followed a precautionary approach in establishing new legislation to regulate GM technology.[24] For some time this led to a de facto EU moratorium on the approval of GM products both for imports and for domestic production. The restrictions on imports have been reduced since 2003, but the staunch opposition of consumers and anti-GM activist groups in combination with the institutional setup of the EU's decision-making procedure on genetically modified organisms (GMOs) have led to something like a regulatory gridlock on GM production in the EU (Swinnen and Vandemoortele 2010).

The current EU authorization process for GMOs consists of several steps, which are explained in Box 5.1. The effect has been that in reality, over the last decade the EU has never been able to make a decision on GMO issues. Votes in the relevant committee (SCoFCAH—see Box 5.1) and in the Council of Ministers typically do not reach a qualified majority on authorizations for GM production despite green lights from the EFSA and the EC. Member state representatives' votes are roughly 50–50 in favor and against. The decision thus has come back to the EC, which according to the EU rules (see Box 5.1) has followed its own proposal and authorized the marketing of some GM crops.[25]

Despite this authorization at the EU level, several member states have banned the cultivation of authorized GM crops. In principle, according to the EU's "free circulation clause," member states may not prohibit, restrict, or impede the placing on the market of authorized GMOs.[26] Following EU legislation, the

---

[24] This is in contrast with the United States, which has chosen to rely on pre-existing laws and agencies, considering GM technology as substantially equivalent to conventional agricultural technology (Sheldon 2002). Interestingly, Vogel (2003) and Cameron (1999) argue that there has been an important shift in regulatory differences between the EU and United States concerning consumer protection policies. From the 1960s through the mid-1980s American regulatory standards tended to be more stringent and comprehensive than in Europe. The United States, more than other countries, used the precautionary principle in domestic law. Vogel (2003) argues that since around 1990 the converse has been true: many important EU consumer and environmental regulations are now more precautionary than their American counterparts, including GM regulation.

[25] An example was the EC's approval of BASF's Amflora potato in 2010.

[26] Member states may install co-existence measures to avoid the unintended presence of GMOs in other products. Additionally, under the EU's safeguard clause, member states may provisionally restrict or prohibit an authorized GMO on grounds of new information that indicates the GMO to be

BOX 5.1

Six Steps for GMO (Dis)approval

1. A company that develops a new GMO must scientifically assess and document the human and animal health effects and environmental safety of the GMO.
2. With these documents, the company has to apply for authorization to a member state, which then passes the documentation on to the European Food Safety Authority (EFSA).
3. The EFSA prepares an opinion, based on a scientific risk assessment conducted by independent experts, and submits this (publicly accessible) report to the European Commission (EC) and the member states.
4. The EC submits a recommendation, either to grant or refuse the authorization, to the Standing Committee on the Food Chain and Animal Health (SCoFCAH). Usually the EC's recommendation follows the opinion of the EFSA.
5. The SCoFCAH—which is composed of member state representatives—may accept or reject the EC's proposal, but only by qualified majority.[27] Three outcomes are possible:

   I. The SCoFCAH accepts the EC's recommendation which then takes effect.
   II. The SCoFCAH rejects the EC's recommendation—the EC then can either rework its proposal and resubmit it to SCoFCAH or submit the initial recommendation to an Appeals Committee (AC)[28]—also made up of member state representatives. The AC decides on the recommendation with qualified majority voting. If the AC fails to reach a qualified majority, either in favor or against the proposal ("no opinion"), the proposal goes back to the EC, which can then adopt its own recommendation.
   III. The SCoFCAH does not reach a qualified majority. In this case, the proposal goes to the AC and follows the AC process as explained in 5.II.

EC then asked for repeal of the member states' bans since the EFSA's opinion was negative. However, in most cases, the Council of Ministers has rejected (with qualified majority) the forced lifting of the provisional safeguard measures. This regulatory gridlock puts the EC in constant violation of both EU legislation and international law (Christiansen and Polak 2009).

To end this regulatory gridlock, the EU is considering several possible institutional reforms, but none of them is obvious. One reform considered is to move from qualified-majority to a simple-majority decision-making in specific cases such as GM regulation. This would allow the committees to reach a

risky. Currently eight member states (Austria, Bulgaria, France, Germany, Greece, Hungary, Italy, and Luxembourg) apply or have applied safeguard measures to GMOs that have been approved at EU level. Member states that invoke the safeguard clause must notify the EC, which then decides on the issue based on the EFSA's opinion. Without exception the EFSA has always opposed the member state's measure because no new information was provided by the member state that challenged the EFSA's prior risk assessment.

[27] According to the definition in the Treaty of Nice, a qualified majority requires the majority of the member states, voting weights (74 percent), and population (62 percent).

[28] Before 2011 the decision moved to the Council of Ministers for Agriculture who had to decide by qualified majority voting; but the process was very similar to the current Appeals Committee.

decision in cases where they previously failed, but whether this reform would lead to more or fewer authorizations of GM varieties remains uncertain.

BIOFUEL POLICY

So far EU biofuel policies have had a limited effect on global food prices, but this could change in the future subject to the outcome of the ongoing policy debate. The debate was triggered by two recent developments. First, while there is disagreement on the size of the impact, biofuels generally have been important drivers of increasing food prices (de Gorter et al. 2013). Second, biofuels were originally thought of as environmentally friendly fuels, due to their decreased carbon impact relative to fossil fuels. However, indirect effects on land use change (e.g., deforestation) may lead to an increase—rather than a decrease—in greenhouse gas emissions (GHGE).

In response to these critiques of biofuels, the EC proposed what the press and the industry have described as a policy U-turn on biofuels. From once strongly encouraging this sector through production targets and blending mandates, the EC is now backtracking on this policy option and seeks to minimize the use of food crop-based biofuels.

The biofuel sustainability requirements of the 2009 Renewable Energy Directive attempted to limit the impact of biofuels on rising food prices (European Commission 2009). The EU biofuel sustainability scheme includes the following criteria: (1) the GHGE saving from the use of biofuels and bioliquids shall be at least 35 percent; (2) biofuels and bioliquids shall not be made from raw material obtained from land with high biodiversity value; (3) biofuels and bioliquids shall not be made from raw material obtained from land with high carbon stock; (4) biofuels and bioliquids shall not be made from raw material obtained from peatland, unless the cultivation and harvesting of that raw material does not involve drainage of previously undrained soil; and (5) agricultural raw materials cultivated in the EU and used for the production of biofuels and bioliquids shall be obtained in accordance with the environmental requirements under the CAP.

In October 2012, the EC published a proposal limiting the use of food crop-based biofuels at 5 percent of consumption of energy for transport in 2020.

Conclusions

The original CAP resulted from the integration of various pre-member state policies, which in turn were introduced to protect farm income from foreign competition and market forces. The mechanism of government support was through high import tariffs, export subsidies, and guaranteed prices for products, which were well above world market prices. While this regime created a stable EU market, it also created much instability on world markets. The high import tariffs and growing surplus stocks, which were exported with subsidies, caused global agricultural prices to decline.

With its expansion, the EU increased its impact on world food markets. In response to internal and external pressures, the EU introduced a series of reforms, spanning three decades, to reduce the impact of its CAP on international markets. The main element of the reform was a shift towards a policy system that continued to support farmers but which created less distortion in international markets. To address environmental concerns, payments were made conditional on farmers addressing certain environmental conditions. Food safety concerns were addressed in a separate policy. Reforms to reduce the CAP's negative effects on the environment, on market distortions, and on the WTO negotiations, and to make it consistent with sustainable rural development, reduced the pressure for large budget cuts and created a new support base for the CAP. From this perspective, the reforms contributed to the survival of the CAP.

The combined reforms resulted in a decline in agricultural support in the EU, and in particular in a strong decline in the use of subsidies that affect production and trade. After the reforms, prices in the EU are close to those on world markets and the impact of the current CAP on global prices is much smaller than in the past. EU policies had no significant impact on the recent price volatility of major food commodities. Hence, somewhat paradoxically, after the EU had gone through decades of reforms to reduce the CAP's (negative) impact on global food prices, the world became concerned with the implications of high food prices. The policy reforms over the past two decades, which have reduced the distortionary effects of EU policies on world food markets, have also reduced its capacity to quickly increase food exports during price spikes.

For the future, the EU has reaffirmed the engagement of the EU towards an open trade policy by underlining the harm done by the restrictive export policies implemented by some countries in response to price volatility; and stayed on course with its reform proposals in specific sectors such as dairy and sugar (phasing out the quota regime), despite a slight change in argumentation (i.e., by also linking the motivation to price volatility).

Other policies are obviously important for farmers and food consumers in the EU and globally. In recent years, government support for consumers in the EU has occurred mostly through social spending, not through food market regulations. European consumers spend on average 15 percent of their household budget on food. However, there are significant differences between and within member states. Unlike CAP subsidies, social policies, such as unemployment benefits, pensions, and disability payments, are still the responsibility of the member states.

The food safety crises in the 1990s were crucial in a new food safety initiative which led to major legislative changes such as the General Food Law

Regulation, including a recast of EU veterinary rules, and the creation of the European Food Safety Authority (EFSA). The new food safety policies introduced a farm-to-fork approach, imposing strict traceability requirements throughout the EU food chains.

The heightened food safety concerns in the EU have also affected EU regulations on genetically modified (GM) food and feed. The EU has followed a precautionary approach in establishing new legislation to regulate GM technology. The staunch opposition of consumers and anti-GM activist groups in combination with the institutional setup of the EU's decision-making procedure on GMOs have led to a "regulatory gridlock" on GM production in the EU.

So far EU biofuel policies have had a limited effect on global food prices but this could change in the future, subject to the outcome of the ongoing policy debate. The EC proposed what the press and the industry have described as a policy U-turn on biofuels. From strongly encouraging this sector through binding targets and blending mandates, the EC is now backtracking on this policy option and seeks to minimize the use of food crop-based biofuels.

References

Ackrill, R. 2003. EU enlargement, the CAP and the cost of direct payments: A note. *Journal of Agricultural Economics*, *54*(1): 73–78.

Anderson, K., and Y. Hayami. 1986. *The political economy of agricultural protection: East Asia in international perspective*. London: Allen and Unwin.

Anderson, K., and S. Nelgen. 2012. Agricultural trade distortions during the global financial crisis. *Oxford Review of Economic Policy*, *28*(1): 235–260.

Anderson, K., M. Ivanic, and W. Martin. 2012. Food price spikes, price insulation and poverty. Paper presented at the NBER Conference on The Economics of Food Price Volatility, August 15–16, 2012, Seattle.

Anderson, K. and S. Nelgen (2013). 'Updated National and Global Estimates of Distortions to Agricultural Incentives, 1955 to 2011'. Washington, DC, June 2013. (Available at www.worldbank.org/agdistortions website)

Anderson, K., G. Rausser, and J. Swinnen. 2013. Political economy of public policies: Insights from distortions to agricultural and food markets. *Journal of Economic Literature*,*51*(2):423–447.

Bernauer, T. 2003. *Genes, trade, and regulation*. Princeton, NJ: Princeton University Press.

Buckwell, A., and S. Tangermann. 1999. The future of direct payments in the context of Eastern enlargement. *MOCT-MOST: Economic Policy in Transition Economies*, *9*(3): 229–252.

Bureau, J. C. 2012. Where is the common agricultural policy heading? *Intereconomics* *47*(6): 316–321.

Burrell, A. 2000. The World Trade Organization and EU agricultural policy. In *Agricultural Policy and Enlargement of the European Union*, ed. A. Burrell and A. Oskam, 91–110. Wageningen: Wageningen University Press.

Cameron, J. 1999. The precautionary principle. In *Trade, environment and the millennium*, ed. G. Sampson and W. B. Chambers, 239–269. New York: United Nations University Press.

Christiansen, T., and J. Polak. 2009. Comitology between political decision-making and technocratic governance: Regulating GMOs in the European Union. *Eipascope 1*:5–11.

Ciaian, P., D. Kancs, and J. Swinnen. 2010. *EU land markets and the common agricultural policy*. Brussels: Centre for European Policy Studies.

Crombez, C., L. Knops, and J. Swinnen. 2012. Reform of the Common Agricultural Policy under the co-decision procedure. *Intereconomics 47*(6): 336–342.

de Gorter, H., D. Drabik, and D. R. Just. 2013. Biofuel policies and food grain commodity prices 2006–2012: All boom and no bust? *Agbioforum 16*(1): 1–13.

Eurobarometer. 2010. Europeans, agriculture and the Common Agricultural Policy. (Special Eurobarometer 336, Wave EB72.5). Brussels: European Commission, Directorate General for Agriculture and Rural Development.

European Commission. 2007. The impact of the developments in agricultural producer prices on consumers. CM/WM/PB 34703, Brussels.

———. 2008. The Health Check of the Common Agricultural Policy. Proposal for a council regulation establishing common rules for direct support schemes for farmers under the common agricultural policy and establishing certain support schemes for farmers, on modifications to the common agricultural policy by amending Regulations (EC) No 320/2006, (EC) No 1234/2007, (EC) No 3/2008 and (EC) No [. . .]/2008, amending Regulation (EC) No 1698/2005 on support for rural development by the European Agricultural Fund for Rural Development (EAFRD), amending Decision 2006/144/EC on the Community strategic guidelines for rural development (programming period 2007 to 2013), Brussels.

———. 2009. Directive 2009/28/EC of the European Parliament and the Council of 23 April 2009 on the promotion of the use of energy from renewable sources and amending and subsequently repealing Directives 2001/77/EC and 2003/30/EC, Brussels.

———. 2011. Press Release, Commissioner Dacian Ciolos, Member of the European Commission Responsible for Agriculture and Rural Development, Delivering sustainability and resource efficiency in Europe's farms, fields and forests. Koli Forum Joensuu, 15/09/2011, Brussels.

European Parliament. 2011. Resolution on an EU policy framework to assist developing countries in addressing food security challenges. (2010/2100(INI)), Brussels.

Gardner, B. L. 1987. *The economics of agricultural policies.* New York: Macmillan.

Graff, G. D., and D. Zilberman. 2007. The political economy of intellectual property: Reexamining European policy on plant biotechnology. In *Agricultural biotechnology and intellectual property: Seeds of change*, ed. J. P. Kesan, 244–267. Cambridge, MA: CAB International.

Grant, W. 1997. *The common agricultural policy*. New York: MacMillan Press.

Josling, T. 2008. External influences on CAP reforms: An historical perspective. In *The perfect storm: The political economy of the Fischler reforms of the Common Agricultural Policy*, ed. J. Swinnen, 57–75. Brussels: Centre for European Policy Studies.

Moehler, R. 2008. The internal and external forces driving CAP reforms. In *The perfect storm: The political economy of the Fischler reforms of the Common Agricultural Policy*, ed. J. Swinnen, 76–82. Brussels: Centre for European Policy Studies.

Núnez Ferrer, J., and M. Emerson. 2000 Goodbye agenda 2000, hello agenda 2003. CEPS Working Document No. 140, CEPS, Brussels.

OECD. 2003. Cancún and the Doha agenda: The key challenges. September 10–14, 2003 [Also repeated in the Declaration by the Heads of the IMF, OECD and World Bank,

September 4, 2003] http://www.bfsbbahamas.com/photos/old_images/Declaration.pdf.

———. 2012. Producer and consumer support estimates database. Paris: Organisation for Economic Co-operation and Development.

Olper, A. 1998. Political economy determinants of agricultural protection levels in the EU member states: An empirical investigation. *European Review of Agricultural Economics 25*(4): 463–487.

Oskam, A., G. Meester, and H. Silvis. 2011. *EU policy for agriculture, food and rural areas*. Wageningen: Wageninge Academic Publishers.

OXFAM International. 2005. International celebrities get dumped on at the WSF. November 1, 2005. http://www.oxfam.org/en/node/283.

Rabinowicz, E. 2003. Swedish agricultural policy reforms. Paper presented at the Workshop on Agricultural Policy Reform and Adjustment, Imperial College, Wye, October 23–25, 2003.

Ritson, C., and D. R. Harvey. 1997. *The Common Agricultural Policy*. Wallington: CAB International.

Roederer-Rynning, C. 2010. The Common Agricultural Policy: The fortress challenged. In *Policy-making in the European Union*, 6th ed., ed. H. Wallace, M. A. Pollack, and A. R. Young, 181-205. Oxford: Oxford University Press.

Scholderer, J. 2005. The GM foods debate in Europe: History, regulatory solutions, and consumer response research. *Journal of Public Affairs 5*(3): 263–274.

Sheldon, I. 2002. Regulation of biotechnology: Will we ever "freely" trade GMOs? *European Review of Agricultural Economics 29*(1): 155–176.

Swinnen, J. 1994. A positive theory of agricultural protection. *American Journal of Agricultural Economics 76*(1): 1–14.

———. 2008. "The perfect storm". Brussels: Centre for European Policy Studies.

———. 2009. The growth of agricultural protection in Europe in the 19th and 20th centuries. *The World Economy 32*(11): 1499–537.

———. 2011. The right price of food. *Development Policy Review* 29(6): 667–688.

Swinnen, J., and H. De Gorter. 1993. Why small groups and low income sectors obtain subsidies: The "altruistic" side of a "self-interested" government. *Economics and Politics 5*(3): 285–296.

Swinnen, J., A. Banerjee, and H. de Gorter. 2001. Economic development, institutional change and the political economy of agricultural protection: An empirical study of Belgium since the 19th century. *Agricultural Economics 26*(1): 25–43.

Swinnen, J., and L. Vranken. 2009. *Land & EU accession: Review of the transitional restrictions by new member states on the acquisition of agricultural real estate*. Brussels: Centre for European Policy Studies.

Swinnen, J., and T. Vandemoortele. 2010. Policy gridlock or future change? The political economy dynamics of EU biotechnology regulation. *AgBioForum 13*(4): 291–296.

Swinnen, J., P. Squicciarini, and T. Vandemoortele. 2011. The food crisis, mass media and the political economy of policy analysis and communication. *European Review of Agricultural Economics 38*(3): 409–426.

Swinnen, J. and P. Squicciarini. 2012. Mixed messages on prices and food security. *Science 335*(6067): 405–406.

Swinnen, J., Knops L., and K. Van Herck. 2014. Food price volatility and EU policies. In P. Pinstrup-Anderson. *Political Economy of Food Price Policy*, forthcoming.

Tracy, M. 1989. *Government and agriculture in Western Europe 1880–1988*. New York: Harvester Wheatsheaf.

Van Meijl, H., and F. van Tongeren. 2002. The Agenda 2000 CAP Reform, world prices and GATT–WTO export constraints. *European Review of Agricultural Economics 29*(4): 445–470.

Vogel, D. 2003. The hare and the tortoise revisited: The new politics of consumer and environmental regulation in Europe. *British Journal of Political Science 33*(4): 557–580.

PART THREE

Challenges for the Poorest Billion

6

Creating Synergies between Water, Energy, and Food Security for Smallholders

Jennifer A. Burney

The Connections between Poverty and Food Security

When considering the world's poorest populations, *food insecurity* and *poverty* are often used interchangeably. This is understandable given that economic access is a key component to overall food security. The overlap between poor households and food-insecure households is nearly complete. The poor (living on less than $2.50 per person per day) and extremely poor (< $1.25 per person per day) for the most part live in rural areas, earn most of their income from agriculture, depend heavily on food they have produced themselves, and still spend most of their money on food. These households are also more likely than wealthier ones to be food insecure in several ways: they are typically net calorie deficient, have lower overall diet diversity, and suffer from "hidden hunger"—an insufficient supply of micronutrients (Wang 2009; FAO 2012). At larger spatial scales, the world's poorest households are found in low income food deficit countries (LIFDCs), nations in which agriculture is a greater percentage of GDP (Figure 6.1a), the food supply is more likely to depend on food imports and food aid (Figure 6.1b), and in which poor governance hinders development.

These gross patterns nevertheless mask a tremendous heterogeneity among smallholders. Poverty is often defined by one number—daily income—or perhaps a few, if assets and housing material are included. Food security—as discussed earlier in this volume—comes in many forms and is defined by many measures. Household surveys reveal that, even among extremely poor rural households, some are net consumers and some are net producers of food. Further divisions exist. Among the net consumers, some households do not sell any of their products, while others do. Among net producers, some grow one main cash crop but then purchase all of their food while others are marginal net producers across an array of food crops, some of which they eat and some of which they sell. Moreover, household behavior varies from year to year based on factors like weather and amount of money available for investment in the current crop (Jayne et al. 2003; Jayne et al. 2010; FAO 2012). Because they are poor, these households make money and labor allocation decisions under severe resource and energy constraints, and must often choose between different objectives, such as caloric sufficiency, diet diversity, and other non-food expenditures. For this reason, net food-producing households may still suffer from caloric, protein, or micronutrient insufficiencies: they may decide to purchase higher-cost calories/protein/micronutrients instead of lower-cost starchy staples, or they may put their limited resources towards non-food expenditures or investments (Banerjee and Duflo 2007; Naylor and Falcon 2010).

Poverty and food security for smallholder farmers also interact over time, complicating measurement of these problems (Carter and Barrett 2006). Measuring income (or consumption expenditure) at one point in time may be misleading, because some households—particularly agricultural households—are frequently subject to bad years and good years that make them look more or less poor than they actually are. Measuring household assets adds a dimension to the poverty picture: how much of a savings buffer do households have?

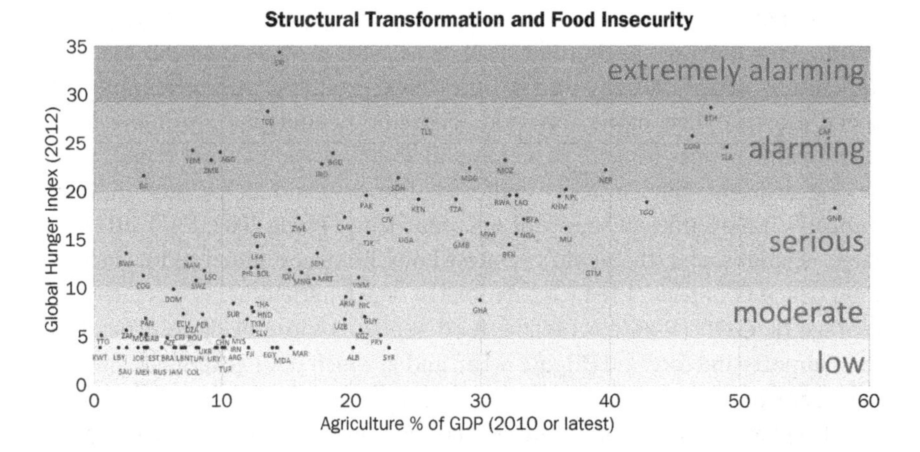

FIGURE 6.1 Hunger, Agriculture as a Percentage of GDP, and Reliance on Food Aid
Figure 6.1a shows the relationship between the 2012 Global Hunger Index and Agriculture (Value Added) as a share of GDP for most countries in the world. At the micro level, most of the world's poor households are agriculture-dependent yet food-insecure; this is reflected at the macro level as well. Note that several important food-insecure countries are not represented in this figure as they do not have a 2012 GHI due to lack of data: These include (but are not limited to) Somalia, Myanmar, South Sudan, DRC, Papua New Guinea, and the Occupied Palestinian Territories. Additionally, several countries (most notably Haiti) lack reliable data on agriculture as a percentage of GDP.

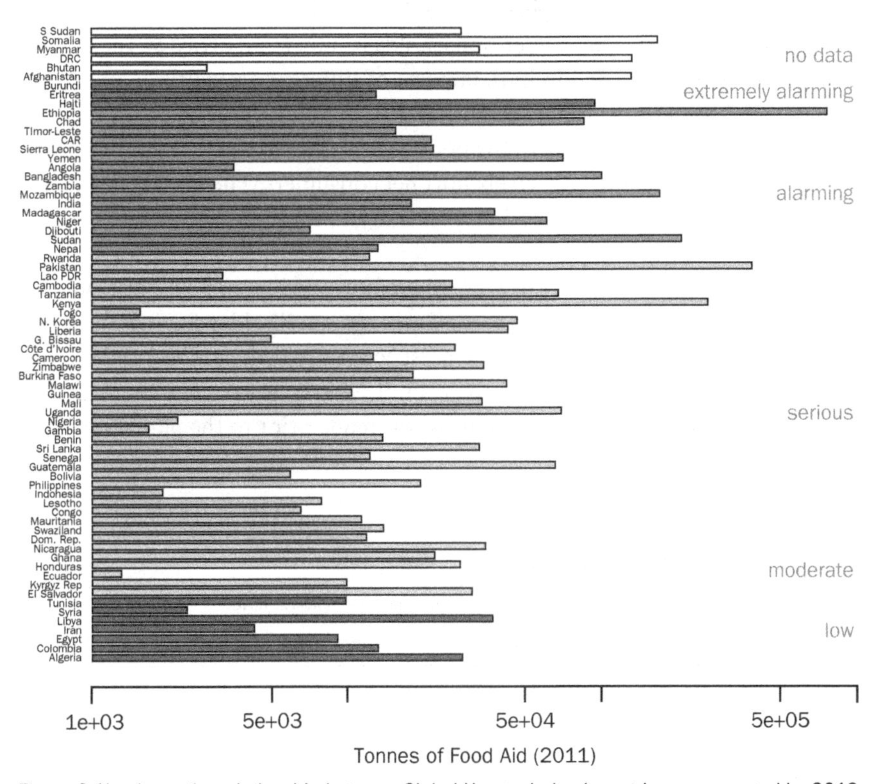

Figure 6.1b shows the relationship between Global Hunger Index (countries are arranged by 2012 GHI on the vertical axis) and Food Aid Shipments in 2012 (note the logarithmic horizontal axis). Countries without a 2012 GHI are shown in white at the top of the chart. Colors/categorizations for Figures 6.1a and 6.1b follow those defined by the GHI.

Source: Data are from the International Food Policy Research Institute's Global Hunger Index website (International Food Policy Research Institute 2012) and the World Bank Indicators data portal (World Bank 2012).

However, looking at assets at one point in time can be similarly misleading, as it gives no indication of whether assets are stable, accumulating, or being depleted. Ideally, both consumption and assets should be measured over time to provide a more complete picture of how the poorest households are actually faring.

When thinking about smallholder farmers through this lens of dynamic poverty analysis, an even more complex web of connections between poverty and food security emerges. For most smallholders, agricultural production is the main source of food and income. But the connection is deeper: the second source of income for these households is usually agricultural labor on other farms. These families most often do not have access to traditional bank accounts; instead, they accumulate assets in the form of livestock (the rural savings account). Their well-being is thus even more deeply tied to the fate of the local agricultural sector in a given year and over time. Moreover the mix of production, sales, labor, and assets varies substantially across households within a given village, and especially across countries and years.

This diversity in smallholder households results in an array of very different risk profiles and varied welfare impacts of different agricultural and development policies. For example, studies have shown that government-sponsored cereal price supports might help the poorest households who are net producers of those crops, but could conversely hurt net consumers. The impacts of such a change on caloric sufficiency and diet diversity, among other indicators, would likely be mixed (Hertel et al. 2010).

Poverty and food insecurity are therefore overlapping but not entirely equal concepts, and the notion of the "average smallholder farmer" is somewhat limited, from a macroeconomic perspective. Indeed, national-level trends indicate a long-term reduction in the overall poverty headcount over the past 15 years, although most of that progress was made prior to the global financial

TABLE 6.1

GINI Index for Low-Income Countries

	GINI Index (Most Recent Value)		
Country	1980–1989	1990–1999	2000–2009
Bangladesh	28.5	33.46	33.22
Burkina Faso		46.85	39.79
Burundi		42.39	33.27
Cambodia		38.28	36.03
Central African Republic		61.33	56.3
Ethiopia	32.42	39.96	29.83
Gambia, The		50.23	47.28
Guinea		44.87	39.35
Guinea-Bissau		47.84	35.52
Kenya		42.51	47.68
Kyrgyz Republic	26.01	35.98	36.19
Madagascar	46.85	41.81	47.24
Malawi		50.31	39.02
Mali		50.56	38.99
Mozambique		44.49	45.66
Nepal	30.06	35.23	43.83
Niger		41.53	34.55
Rwanda	28.9		53.09
Tajikistan		29.01	30.83
Tanzania		33.83	37.58
Uganda	44.36	43.07	44.3

National-level income inequality data (GINI Index) for low-income countries. For many developing countries—even those with strong macroeconomic growth—inequality has increased or plateaued over the past several decades.

Source: World Development Indicators. Countries included are those with data for more than one of the time periods.

crisis, (starting in 2007), and levels have held constant since then (FAO 2012). At the same time, macroeconomic data from sub-Saharan Africa and Asia have remained tremendously promising. These numbers hide a painful truth, however: the poorest smallholder farmers across the developing world seem to be perpetually poor and food insecure, and are being left behind in the growth process (Table 6.1). Why and how do so many smallholder farmers remain in poverty, despite an overall context of economic progress?

The Low-Productivity Trap

Put simply, smallholder farmers are very often caught in a type of vicious cycle called a poverty trap. The basic structural feature of a poverty trap is that growth is much more difficult at low levels of development. Below a certain threshold of well-being it is almost impossible to overcome all of the constraints that would facilitate growth, and therefore stagnation, or even decline, occurs. Only beyond that critical threshold level does self-sustaining growth become possible. There are many possible mechanisms for poverty traps, including low investment in human capital, low savings rates, endemic poor health, and more. All of these mechanisms in turn lead to more poverty (Easterly 2009). Most important, this unfortunate feedback can happen across scales, forming vicious cycles from household to local to regional to national levels, that are self-similar and self-reinforcing (Barrett and Swallow 2006). For example, local poverty often leads to poor governance, lack of infrastructure, and environmental degradation at larger spatial scales, which in turn conspire to prevent individuals and households from escaping poverty.

As previously noted, the notion of an "average" smallholder farmer may be misleading given the diversity in agroecological zones, production practices, market interactions, risk profiles, and institutional contexts represented by this group. Nevertheless, some instructive commonalities exist even within this heterogeneous group, and these structural traits help explain why many smallholder farmers and their households remain food insecure in various ways across space and time.

Notably, many of the world's poor smallholder farmers face a particular type of poverty trap that is characterized by low agricultural productivity. What are the structural features of this low-productivity trap? First, most smallholder farmers, by definition, do not have much land, and often do not have clear title rights (as documented by Smith and Naylor in Chapter 8). They typically own or lease plots up to one or two hectares in size—though again, this varies widely. Most often, they use that land to grow starchy staple crops, including cereals like rice, corn, sorghum, millet, and wheat, as well as roots and tubers like cassava, yam, or potato. Typically their only source of water is seasonal rainfall. They may also own some poultry or livestock, which as mentioned often serve as a savings account, as well as a source of protein. These animals also depend on rainfall, directly for water and indirectly for growth of the forage crops.

The rain-dependence of these systems imposes some important constraints on household food security. First, rain-dependent farms in the poorest developing communities are typically low-yielding, and they are also extremely sensitive to weather shocks. This means that farmers may make the decision to grow a lower-yielding, but more drought-resistant crop (e.g., sorghum instead of maize) as a risk-management strategy. They may also avoid experimentation with high-yielding varieties and increased fertilizer use as uncertain rainfall poses an enormous, uncontrollable risk for such investments. In such an environment, a sub-par rain year can very easily lead to overall calorie and income shortages that have economic, health, and human capital repercussions for years into the future. Families may simply not have enough food to eat; as a result, they may not be able to purchase protein or micronutrients, they may deplete assets (such as livestock) to cover their shortages, they may pull children out of school due to inability to pay fees or a need for additional labor income, and they may be unable to make investments in their own farming systems (e.g., purchase inputs) in the following year, potentially creating a vicious cycle.

A second feature of rain-dependence is that production is confined to the rainy season or seasons. In the dry tropics and monsoonal areas with only one main rainy season, households may depend on a single season for their entire annual income. This often results in a "hungry season"—the time of year when households have depleted or almost depleted their stores from the previous harvest, or have spent their incomes and are waiting for the next harvest. Prices for calories, protein, and micronutrients can be several times higher than average during the hungry season. Some households have no choice but to go hungry, and even the slightly more well off face difficult choices between hunger and asset depletion to purchase additional food. Thus, in addition to shortfalls in potential annual income and calories on an annual time scale, many smallholder farmers and their households also face seasonal food security challenges, which can similarly have lasting repercussions.

Perhaps most important, this seasonality in food and income availability can leave farmers eager to sell their crops immediately upon harvest—when local prices are at annual lows. Low institutional capacity across scales deepens the problem, making lasting food security and escape from poverty extremely difficult for the world's poorest smallholder farmers. These farmers often lack the resources (including information and education) to seek better prices at market, for example by storing or transporting crops, leaving them vulnerable to large local price slumps when many farmers in the same region harvest similar crop mixes at the same time. Worse, the price elasticity of demand of starchy staple crops is relatively low, meaning people do not consume much more even as supply increases and prices fall. The poorest smallholders thus end up desperately needing to sell as soon as they harvest, and doing so for the lowest possible price. These low returns leave farmers and their families with less surplus income to purchase other calories, protein, and micronutrients, as well as fewer overall resources to invest in health, education, and income-generating activities. Most perversely, these price slumps ultimately act as a disincentive to adopt productivity-enhancing technologies.

In this way, low-productivity smallholder systems form a particular kind of poverty or food insecurity trap, illustrated in greater detail in Box 6.1. This vicious cycle at the micro level—low productivity leads to poverty and food insecurity, which leads to low productivity—is reinforced by the local and regional institutional contexts surrounding smallholder farmers. For example, the literature on nutrition shows that low levels of educational access (for children and for adults, via extension services) mean that farmers are less likely to maximize yields for their production systems. And given a lack of nutritional education, they are even less likely to have enough diversity in their diets or to recognize the signs of malnutrition (as discussed by Rozelle et al. in Chapter 2). Furthermore, they are less likely to have access to market information, which may be lacking entirely, or inaccessible due to illiteracy, that would allow them to maximize their sales price. Low profitability then leaves these households with fewer resources to invest in their children's education. Similarly, the lack of infrastructure and market connectivity plaguing many poor rural areas makes

BOX 6.1

An Illustration of the Low-Productivity Trap, along with Leverage Points

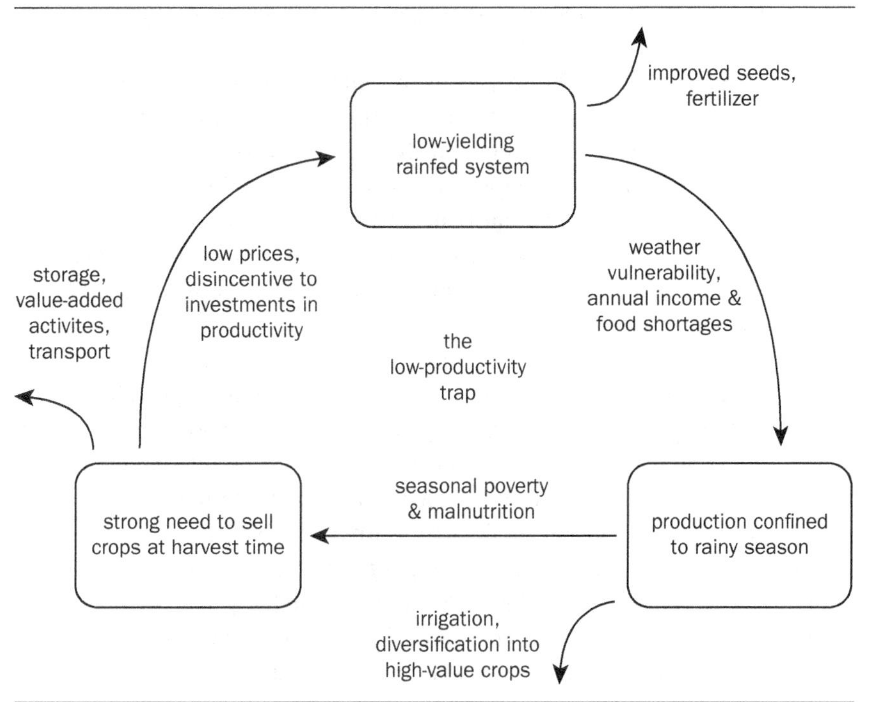

access to inputs expensive if not prohibitive, leads to higher hungry-season prices for calories, protein, and micronutrients, and intensifies the harvest-time price slumps, given the price inelasticity of staples.

The cartoon schematic in Box 6.1 illustrates the low-productivity trap, along with leverage points for breaking the trap. The Sudano-Sahel region of West Africa provides an ideal illustration of these mechanisms. Across the region, most smallholder farmers produce some crop mix of corn, sorghum, millet, and yams, with cassava serving as an important secondary crop. Planting of the cereal crops and cassava is timed with the onset of the rainy season (in May or June), and yams are planted earlier in the year. Harvest of grains is typically in fall (October or November), as the dry season begins again. The cassava and yams may be left in the ground longer and harvested gradually, depending on household food needs, local prices, and weather.

These smallholder production systems are a cornerstone of the persistent poverty found in the rural Sudano-Sahel. Most smallholder staple systems have low yields, and producers are very vulnerable to weather shocks. On an annual basis, this can result in caloric and income deficiencies. In addition, because production is limited to the rainy season, there is a strong seasonal component to nutrition and income, as households must stretch their stores through the beginning of each rainy season to the next harvest. Fruit and vegetable crops are typically only grown in the rainy season and are essentially unavailable in rural areas during the dry season. And while households may purchase some legumes (peanuts and soy) grown locally, as well as milk products produced by local cattle-owners, availability of these products also dwindles during the dry season.

The existence of the "hungry" season results in an often-urgent need for smallholders to sell their crops at harvest time. Since seasonal needs affect farmers across the region simultaneously, local price slumps result, due to both the inelastic nature of staples markets and lack of market connectivity. Prices for staples and micronutrient crops can vary by factors of three or more between seasons. Most perversely, these price slumps can be a disincentive to adoption of productivity-enhancing technologies, thus completing the low-productivity trap.

Smallholder irrigation projects in this region have helped farmers break out of poverty by facilitating year-round production of high-value crops. Farmers using solar-powered drip irrigation systems in northern Benin, for example, have increased household consumption of both staples and nutrients, and have begun putting their new income towards longer-term investments like education and assets (Burney et al. 2010; Burney and Naylor 2012). This type of production is critically dependent on energy and water access, however. Fuel supplies in the region are highly variable, with volatile prices. Groundwater depths and flow rates vary dramatically. Farmers who successfully cultivate fruit and vegetable crops need to be able to access markets and information to fetch the best prices, and doing so typically requires using cell phones. Those who do not have access to such technology typically sell "over the back fence" for lower prices than at regional markets.

The vicious cycle continues to propagate up to larger scales. The persistent low profitability of rural regions leaves governments less likely to invest limited infrastructure funds on extending roads and communication services to these areas. Low-productivity areas tend to lack political clout because of their poverty, and cash-strapped governments may reap higher (and quicker) returns on infrastructure investments in urban manufacturing or wealthier cash crop-producing areas. Low returns to production also create disincentives for private-sector investment in supply chains and financial services. Lack of access to credit, insurance, and savings accounts renders the entire low-productivity cycle more susceptible to biophysical and economic shocks alike. In short, the high vulnerability of poor smallholder farmers across larger regions creates an environment of coordinated risk that leaves governments and financial institutions loath to extend such services to smallholders.

Macro Pressures Exacerbating the Low-Productivity Trap

Two important trends overlay this multi-scale low-productivity trap and further threaten the prospects of even the most successful smallholder farmers. The first is population growth. Global population is expected to peak at 9–10 billion at mid-century, which implies an overall need for more food worldwide. But essentially all the additional population growth by mid-century is expected in the developing world and thus presents an additional constraint to smallholder farmers. Large swaths of the developing world still feature high fertility and mortality rates; they have not yet undergone the demographic transition where rising standards of living lead to a decrease in the mortality rate followed closely by a decrease in the fertility rate. As such, per capita landholdings are declining in many rural agricultural areas, as small farms are subdivided among large families. This is particularly prevalent in sub-Saharan Africa, where per capita landholdings are declining fastest (Figure 6.2). Areas such as eastern Africa that were not considered land-constrained two decades ago now very much are. These trends are most devastating for the poorest households as they often further divide land among the next generation, making lasting food security and escape from poverty ever more difficult from a shrinking land base. And the longer the rural poor remain poor, of course, the longer the demographic transition is delayed and the higher peak population estimates grow. This trend is also exacerbated by so-called "land grabs"—demand for larger-scale plots of land by domestic or foreign investors (see Chapter 8 and Deininger 2011; Hertel 2011).

The second major trend is climate change. Given that rural poor smallholder farmers depend on rain-fed production of staple crops, as well as livestock for

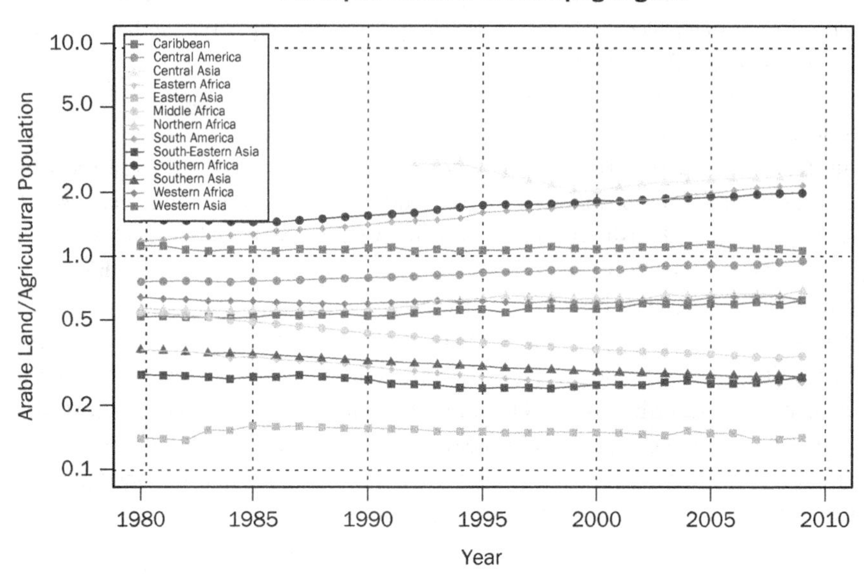

FIGURE 6.2 Per Capita Land Holdings Macro trends in arable land and agricultural population across developing regions of the world. Across much of sub-Saharan Africa and Asia, per capita arable land is holding constant or declining.

Source: Data are from the FAO Statistical Database (Food and Agriculture Organization of the United Nations).

both food and assets, trends in temperature and precipitation have important food security ramifications. Declining yields due to climate change have already been documented (Lobell et al. 2011). Some of the most important impacts have been felt in key food security regions, such as India. Global-scale trends and trends in developed countries are also important, given the relatively small proportion of food that is traded worldwide. At the household level, absent adaptation, yields of many key staple crops are expected to decline further in the next two decades as temperatures rise, with important impacts coming in regions that are already largely food insecure. Furthermore, precipitation patterns in the dry tropics and monsoonal regions have in many cases changed dramatically in recent decades (e.g., northern India, the Sahel, northeastern Brazil). Fairly rapid changes in the onset and length of growing seasons (with rainy seasons occasionally disappearing entirely) pose obvious problems for farmers trying to allocate very limited resources as efficiently as they can. Surveys indicate that smallholder farmers sense these changes to differing degrees but have severely limited capacity for adaptation (Brondizio and Moran, 2008 and Mertz et al. 2009).

Connections to Energy and Water Security

The low-productivity trap helps illustrate the connections between food, energy, and water security across space and time. On the water side, against a backdrop of weak infrastructure and poor market connectivity, the ability of both net consuming and net producing smallholders to access sufficient calories and protein throughout the year is driven to a large extent by rainfall. In addition, rain replenishes seasonal or annual surface sources, fills catchments and reservoirs, and recharges groundwater supplies (Taylor et al. 2012). Access to an ample and clean drinking water supply leads to a healthier population—a population more likely to benefit from available food and have the greater physical capacity to grow food and engage in other income-generating activities (as discussed by Davis et al. in the following chapter).

Smallholder productivity is similarly dependent on energy in numerous ways. Farmers who might want to use technologies such as plows and tractors are limited by manpower, animal power, or money to buy fuel—often at high prices and from unreliable suppliers. Even households with access to supplemental water supplies (e.g., tubewells or seasonal or annual surface sources) may be limited by the energy to lift and move that water. They will either need to pay for pumping energy or use human labor. The power output of a healthy well-fed human is around 80W—enough to power a quality treadle pump to draw water from up to ~10 m deep and irrigate an area of 1–2 hectares (ha). However, human power output is often constrained by hunger, poor health, and infectious disease (as shown in Chapter 7). At larger scales, energy costs drive up input prices (due to transport costs), and contribute to harvest-time price slumps as getting harvest products to larger regional markets is often prohibitively expensive. Finally, regional-scale energy poverty in rural regions means that farmers have less access to communications technologies (although this is rapidly changing due to cellular technology), cold storage for higher-valued products, or processing technologies that might add value to harvested products and/or help prevent post-harvest losses.

At larger scales, it is important to recognize that most food-insecure households are also water- and energy-insecure (Figure 6.3). This results in detrimental feedbacks across scales, as smallholders without sufficient access to these critical resources are forced to engage in behaviors that further threaten food security. For example, water-insecure households have no choice but to allocate much of their labor energy to hauling water, often over a distance of several kilometers. In most areas, this burden is born primarily by women and girls, and the criticality of water often means that girls cannot attend school at all, or are frequently absent. Meanwhile, their labor is also unavailable for agriculture or other economic activities. Even then, the water fetched may not be clean, leading to increased household disease burden, a reduced capacity for labor and income-generating activities, and lower nutritional status (see Chapter 7).

A similar story can be told on the energy side: nearly 3 billion people worldwide depend on unprocessed biomass for their cooking and space heating needs. This group includes most of the rural agricultural poor. Because they need to gather fuelwood, crop residues, and dung, smallholder households

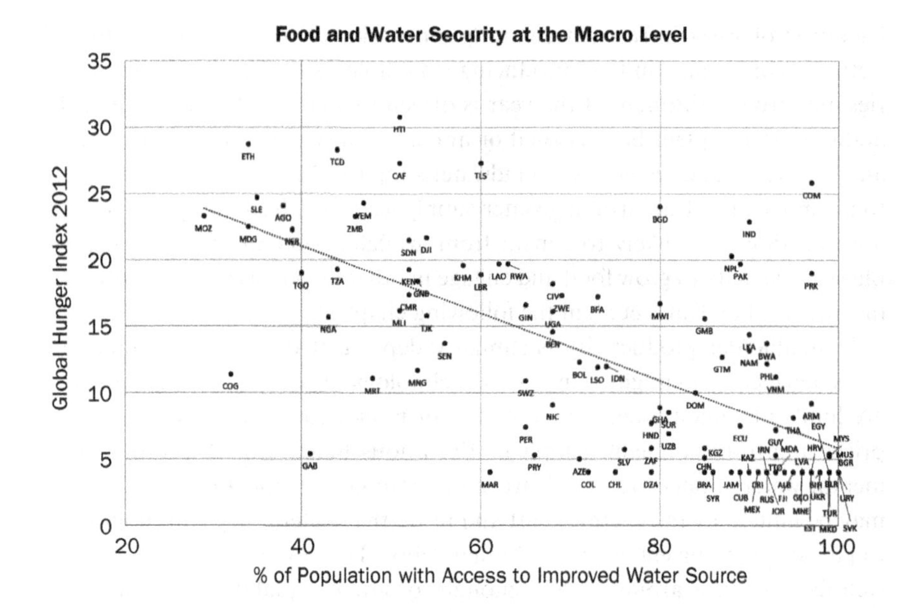

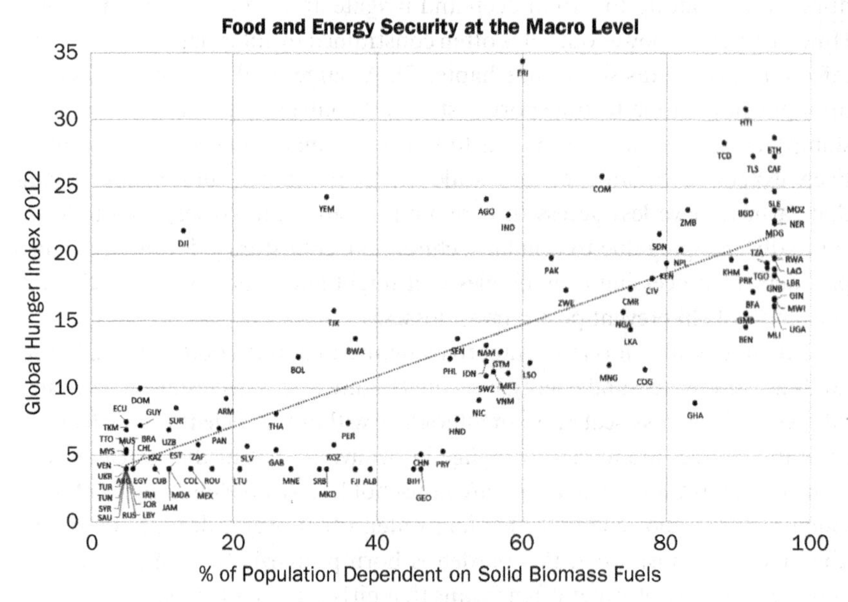

FIGURE 6.3 Water-, Energy-, and Food-Security Linkages at the Macro Level
Figure 6.3a shows the relationship between Global Hunger Index (GHI) and the percentage of a country's population with access to an improved water source, by country.
Figure 6.3b shows the relationship between GHI and the percentage of a country's population dependent on solid biomass fuels for cooking and heating.

Note: The biomass fuels variable does not include significant populations (e.g., rural China) dependent on solid unprocessed coal as a main cooking and heating fuel.

Source: Data are from the Millennium Development Goals website and the World Bank.

divert labor energy away from agricultural production, other income-generating activities, and schooling. More perversely, the extensive use of fuelwood leads to deforestation, which in turn affects water availability, as deforested areas are more susceptible to erosion and are less able to hold soil moisture. Deforestation also affects rainfall patterns at the regional scale by removing a major terrestrial source of water vapor. Once degraded, these deforested lands also become less suitable for agricultural production than they would have been under more careful management of native habitat. In addition, the burning of biomass leads to the emission of pollutants, including aerosols and ozone precursor compounds. Emissions of these pollutants are responsible for over 2 million premature deaths each year (concentrated in the developing world, and among women and children), millions of tons of crop loss, and the weakening of important seasonal precipitation patterns, including Sahelian rainfall, the Indian monsoon, and the North-South China rain pattern (UNEP 2008; UNEP 2011). Moreover, these short-lived climate pollutants (SLCPs), including black carbon, tropospheric ozone, methane, and HFCs, are responsible for roughly half of the global anthropogenic climate warming to date (with greenhouse gases making up the remainder).

Breaking the Low-Productivity Trap

Smallholder farmers and their families are perpetually confronted by difficult trade-offs about how to allocate limited resources—land, labor, and money—to meet their food, water, energy, health, education, and other needs. Compounding matters, optimal allocation of these resources can often be impeded by lack of education and by cultural/gender norms (Udry 1996). So what can be done? Any successful strategy will necessarily improve returns to land, labor, and inputs. As seen in Box 6.1, there are several entry points for breaking the low-productivity trap, which fall into three general categories: improving productivity of starchy staple crops, facilitating diversification into high-value crops and year-round production through irrigation, and improving post-harvest performance and storage. In line with the literature on poverty traps, each of these types of strategies relies on some sort of input to raise productivity to the critical threshold level, at which point self-sustained growth becomes possible.

(1) The first set of strategies involves improving yields of rainfed starchy staple crops and enhancing the ability of farmers growing those crops to handle weather shocks. In places like sub-Saharan Africa, this effort amounts to bringing the Green Revolution to the regions it has not yet reached. This approach includes development and dissemination of access to improved cultivars and drought- and heat-tolerant varietals.

The effort to breed such crops requires coordinated science, germplasm collection and storage, and enhanced extension services, and builds on the basic model spread across Asia in the 1960s and 1970s (as discussed in Chapters 2 and 3). A main example of success in this realm has been the New Rice for Africa (NERICA) project, developed through the Africa Rice Center, one of the research centers of the Consultative Group on International Agricultural Research (CGIAR). The NERICA project is producing higher-yielding, more resistant, higher protein upland rice varietals for countries in sub-Saharan Africa, many of which had previously purchased imported milled Asian rice for consumption. The NERICA project has expanded since its inception in the early 1990s to include lowland rainfed and irrigated rice systems as well (Somado et al. 2008). Between 2008 and 2013, the Alliance for a Green Revolution in Africa (AGRA) engaged smallholders in "farmer participatory variety selection" across 13 countries in sub-Saharan Africa and produced 332 improved varieties of various crops. Over half of these seeds have been commercialized and are being sold (The Alliance for a Green Revolution in Africa 2011). A majority of these have been maize varietals, more than 10 other crops are represented in the mix (reflecting the diversity of African production). Well-established breeding programs in other parts of the world (e.g., Brazil) are focused on reaching smallholder farmers via extension services to ensure their access to improved seeds. In regions where farmers already have access to improved seeds, efforts have turned to boosting resilience via access to other inputs.

(2) The second set of strategies involves diversification into production of high-value crops and breaking the seasonality of production, income, and malnutrition. During the Green Revolution in Asia, governments across East, South, and Southeast Asia worked to install irrigation infrastructure for farmers growing staple grains. Similar transformations have taken place in Central Asia, the Middle East, and North Africa. These systems have helped improve yield growth over the past half-century and have also helped make farmers more resilient to climate shocks. They have also enabled farmers to produce more than one crop per year, breaking the harsh seasonality of income and nutrition of rainfed production. For these reasons, development of irrigation infrastructure remains a goal for countries, particularly in sub-Saharan Africa, where very little cropland is irrigated. Community-scale schemes have had a significant positive impact on smallholder farmers in Mali (Dillon 2008) and Ethiopia (Tesfaye et al. 2008), among others, but the construction of these types of projects lags far behind demand in most countries.

Extensive surveys of smallholder farmers in East and Southern Africa have shown that diversification of crop production is inversely proportional to land holdings (Jayne et al. 2010). As plot sizes shrink, families cannot earn enough from their land to feed and support their families by growing only low-value (and price-inelastic) staple crops. As a result, the most land-constrained farmers by necessity try to diversify their crop production, thus increasing returns to land. If they do not manage to do this, smallholders may be pushed out of agriculture altogether, typically into low-value rural or urban activities (as opposed to being pulled by higher wages or other inducements). This has been seen in India and other parts of South Asia, and across sub-Saharan Africa (Badiane 2014; Binswanger-Mkhize 2014). Diversification into high-value crop production is a survival strategy, but it can also be a positive growth strategy if farmers are able to make the necessary investments.

The end results of diversification are often much better than the alternative push out of agriculture. Net producer smallholders tend to be net producers of many crops (Wang 2009). There are several reasons why this is the case: fruits and vegetables (and even legumes) tend to fetch higher market prices than staple crops, local demand for such products tends to be much more elastic than for staples, and increased production of such crops can help address the nutritional dimension of food security in rural areas (Jayne et al. 2003). Many programs to address iron-deficiency anemia and vitamin A deficiency across the developing world have focused on household gardens as tools for improved nutrition security (Burney and Naylor, 2012). As will be discussed later in the chapter, diversification is closely tied to access to water for irrigation, and access to energy to move that water.

(3) The final set of strategies involves improving post-harvest performance. Crop storage to prevent post-harvest loss (see Chapter 2) is critical in regions where limited infrastructure hampers efforts to get crops to market. Even more effective might be combinations of storage and credit—allowing farmers to borrow against their stored crops at harvest time so they can sell them later for higher prices. Value-added activities including milling, canning, and juicing not only bring more income to farmers, but also are a form of storage, avoiding crop loss. Even access to information about markets via cell phone can help farmers make better decisions about where to sell their products (and, over time, what to grow and when). This has been shown to stabilize prices in grain markets throughout the year (Aker 2010). Cold storage plus transport (a "cold chain") for produce, eggs, meat, and other perishable agricultural products gives farmers both a mechanism and time to move their products to the best markets.

These three groups of strategies are complementary: In the immediate term, solutions such as storage and value-added activities help farmers deal with poor local institutions and infrastructure in relatively low-cost ways. There is a strong role for the private sector in this realm, as large market potential exists for both low-cost post-harvest technologies (e.g., peanut shellers, dehydrators, juicers, storage bags) and introduction of new services (e.g., credit, insurance, storage services, market information, "middlemen" purchasers). It is tempting to emphasize post-harvest activities at the expense of other strategies, but the need for large-scale coordinated efforts to increase yields of staple crops remains urgent in both the short and long term. Starchy staples provide the bulk of calories consumed, and market connectivity is not high enough simply to allow smallholders to sell cash crops and purchase staple calories. Over the long term, smallholders will need access to cultivars that are able to thrive in a world of hotter temperatures and more extreme precipitation and drought events.

Perhaps most important, all three strategies should be pursued simultaneously because access to energy and energy services—directly and indirectly—is a limiting factor for each. Post-harvest activities require local energy services (to run shellers or juicers or to operate a cold chain), as well as fuel for transport. Introduction of new products and services similarly requires sufficient energy and transport infrastructure to keep prices accessible for farmers. Even information-based strategies (like market pricing or SMS-based extension services) require functioning cell phone networks and/or local power sources for phones and computers.

Beyond such direct energy dependencies, indirect linkages to energy also loom large. Research on the Green Revolution in Asia indicates that the successes in yield growth from the 1960s forward were roughly attributable in equal parts to better seeds (breeding), increased use of fertilizers, and irrigation (see Chapters 2 and 3). Indeed, research suggests that many agricultural soils in sub-Saharan Africa are being depleted of nutrients (Sanchez 2002; Sanchez 2002), and only 4 to 5 percent of cultivated lands in sub-Saharan Africa are irrigated (FAOSTAT). International organizations and national governments have also sought to improve fertilizer access and development of water resources for agriculture, along with improved seeds, to boost overall agricultural productivity.[1] Maximizing performance of new high-yielding varietals requires the synergies of seeds, water, and fertilizer. For example, it takes approximately 30,000 Btu of energy to produce a kilogram of nitrogenous fertilizer (kg N). Corn

[1] For example, the government of Malawi began a large fertilizer subsidy program in 2006 (after a food crisis in 2005). In a few short years, Malawi became a net exporter of maize (a similar story exists for Zambia). Nevertheless, such subsidy programs are complicated (see Chapter 2), expensive, and not clearly effective in reducing rural poverty: despite the large-scale successes, household-level impacts of the subsidy program have been mixed (Mason et al. 2011).

grown in the midwestern United States requires on average 100kg N per hectare per year, or an equivalent of ~880 kWh of energy per hectare per year. Energy requirements for pumping of irrigation water vary by depth of wells and type of equipment, but are of a similar order of magnitude on a per-hectare per-year basis. Although the input use in sub-Saharan Africa would be much lower, longer-term strategies for improved fertilizer and water access must include more regional fertilizer production (perhaps using renewable energy sources) and lower transport costs, as well as use of renewable energy sources for pumping water.

Efforts to extend electrical grids and build roads, crop breeding programs, and input subsidies are by definition out of the hands of individual smallholder farmers. They require governments, donor organizations, and private firms to coordinate work across large spatial scales, and the smallholder does not really enter the equation until s/he is opting to purchase or use the finished product. While critical, they harness individual smallholder entrepreneurialism only at the end of a large-scale coordinated effort. As such, complementary strategies and interventions that reach smallholder farmers directly are also needed. Moreover, almost all of the more immediate solutions outlined (irrigation, diversification, market information, storage, value-added activities, improved transport) require stable and secure energy access, whether in the form of an extended grid, access to fuel at reasonable and reasonably stable prices, or distributed energy sources (solar panels, generators, batteries, etc.). Of these technologies, the one that offers perhaps the strongest potential for lifting smallholder farmers out of poverty is irrigation. It also perhaps best illustrates the interdependence of food, water, and energy.

Irrigation as an Illustration of Food-, Water-, and Energy-Security Linkages

Across temporal and spatial scales, irrigation technologies offer several different potential entry points for breaking the low-productivity trap. Irrigation, however, can mean many things, from access to government-run, infrastructure-scale flood irrigation to privately purchased drip or sprinkler irrigation systems run out of private tubewells or surface sources. In staple crop production, access to irrigation water several times during the growing season can provide a buffer against climate shocks and can mitigate some heat stress; it may also facilitate the addition of a second growing season. Access to a regular water supply can facilitate a transition by farmers into production of higher value crops including fruits and vegetables, which require regular water application. This diversification of production, as described earlier, results in greater returns to land (and potentially labor, depending on the type of irrigation). Cultivation of high-value crops also allows for sophisticated crop calendar planning, allowing farmers to produce high-demand products when supplies are lowest in the counter-season.

These benefits have been documented in different regions around the world (Comprehensive Assessment of Water Management in Agriculture 2007). But overall success depends on both technical functionality of the irrigation system and the surrounding institutional context (see Thompson, Chapter 11). Large-scale irrigation infrastructure is costly to build and maintain, and because so many coordinated parties are involved in construction and oversight, they may be more susceptible to corruption in areas with weaker institutions. Moreover, by definition, large-scale centralized systems can only impact smallholder farmers within the irrigation network. Private or smallholder-based systems can be designed to meet individual farmers' needs, and are owned and maintained by the farmer. They therefore obviate the need for oversight agencies like water user associations. Although returns on investment are high for both types of irrigation, smallholder systems yield even greater gains (Jones 1995; Inocencio et al. 2007).

One can think of any smallholder irrigation-based production system as consisting of three main components: a water access technology, a water distribution technology, and a productive application (Burney and Naylor 2012). Water access technologies range from buckets to pumps of all varieties, and facilitate movement of water from source to field. Water distribution technologies range from hands to watering cans to furrows to sprinklers to drip irrigation tubes, and facilitate more efficient distribution of water. Productive applications refer to cropping systems that render a smallholder irrigation system profitable. All three components are complementary, and synergies arise from simultaneous uptake and use. Nevertheless, research has shown that farmers may not elect to adopt complementary technologies, particularly if credit is involved; even if they do adopt all three, the institutional backdrop may still preclude success.

In particular, water access—both from institutional and technological perspectives—is perhaps the most critical element for successful smallholder irrigation. Numerous studies have shown that suboptimal management of water access can undermine the poverty-alleviation effects of irrigation. These potential failure modes make sense intuitively. On the institutional side, even if farmers have access to irrigation technologies (including pumps and drip kits), they may not reap the benefits if they do not have secure access to a water source. On the technological side, farmers with institutional access to water but no improved access technology (i.e., they carry water by hand) may abandon their irrigation systems if they feel the effort doesn't produce enough of an improvement. Finally, within a given irrigation basin where farmers have both technical and institutional access to irrigation water, upstream households may still fare better than downstream households, as they are more likely to get sufficient water when supplies are too limited to benefit all smallholders in the network.

Nevertheless, having physical and economic access to water is a key starting point. So what does water pumping cost, in terms of energy? The main goal of irrigation is to replace soil water lost through evaporation and transpiration from plant leaves. Across the dry tropics, this evapotranspiration can easily exceed 7 mm of water per day during the dry season. A smallholder farming 1 ha of vegetables will require around 70 m^3 of water per day to replace evapotranspiration losses from the soil. Motorized pumps can obviously achieve sufficiently high pumping rates most easily, but they are costly to maintain and operate, especially in remote areas where fuel supplies are unreliable. Smallholder farmers in Niger using gas-powered pumps from shallow (~ a few meters depth) borewells to grow tomatoes and other vegetables reported needing an average of ~0.15 L of fuel per m^3 water pumped. Considering that fuel prices are often ~$1/L, gas- or diesel-powered pumps can cost several dollars per day to operate. At the other end of the fuel-use spectrum, even top-of-the-line human-powered pumps, like the KickStart Super Money Maker Pump (www.kickstart.org), are only capable of pumping perhaps to ~5 m^3 per hour. Fuel costs and human energy output thus quickly limit plot size for smallholder irrigated systems. These limitations increase with the depth of the water table, and treadle pumps face a physical suction limit of ~8 m at sea level.

These fuel and human effort costs of irrigation mean that a premium is placed on efficient water use. In this regard, efficient, effective distribution technologies are also critical for the economic viability of irrigation systems. In particular, drip irrigation helps increase crop yields, promote uniform distribution, facilitate efficient fertilization (by using soluble fertilizer), reduce weeds (by watering only plant locations as opposed to the entire bed), and reduce overall water use. Side-by-side comparisons between drip irrigation and conventional hand-watering have validated these impacts in the Sudano-Sahel (Woltering et al. 2011). Fortunately, a large area is not needed to make a good income when growing crops like vegetables. Vegetable yields vary by type, of course, but are usually in the range of several kg/m^2, and vegetable crops can be grown in multiple seasons in an irrigated system. In this way, irrigation dramatically increases returns to land and water.

Labor must also be considered for both access and distribution. On the access side, while treadle pumps may be an option in some locations, water depth or cultural considerations—for example, women who wear ground-length skirts—may preclude their use in others. In such locations, farmers who wish to irrigate have only two options: haul water by hand or invest in a motorized pump. On the distribution side, drip irrigation has been shown to be labor saving compared to traditional hand-watering. Most of those labor savings come in the form of water hauling, but some are also from reduced time spent weeding. Since drip irrigation delivers water directly to plant roots, neighboring weeds are less likely to be watered and fertilized in the process (Woltering et al.

2011). However, earlier studies have shown that the labor savings of drip irrigation alone may not be enough for farmers to adopt the technology if they still have to haul water by hand (Moyo et al. 2006). Economists historically viewed the rural agricultural developing world as a vast pool of underutilized labor, but there is inevitably an opportunity cost to water hauling or treadle pumping. In a solar-powered drip irrigation project in Benin, West Africa—where farmers were relieved of both the physical burden of water hauling/pumping and the economic burden of volatile fuel supplies—farmers immediately invested the time saved irrigating into other income- and human capital-generating activities (Burney et al. 2010). Beyond addressing their own food security needs with the income generated from irrigated horticulture, farmers in this project opened small businesses, and invested in assets. They also paid previously unaffordable school fees for their children, in many cases sending their daughters to school for the first time.

Photographs in Box 6.2 show before and after images from a community garden using solar-powered water pumps for irrigation of vegetables in Benin. Women's agricultural groups in these villages previously cultivated beds of 10 to 20 square meters and watered them by hand; they persisted with this laborious task in the dry season for both nutritional reasons (local sources of micronutrients can become very sparse) and financial reasons (prices in the off-season for vegetables can be several times higher than in the rainy season). Now, a 2-kw photovoltaic pumping system pumps water to a concrete reservoir, which then

BOX 6.2

Smallholder Irrigation and Food-Energy-Water Linkages

A solar-powered drip irrigation project in Benin.

Source: Center on Food Security and the Environment

gravity-distributes water to a pressure-regulated drip irrigation system watering the 100 to 200 square-meter plots of 30 to 40 women and their families. Access to these systems has brought many of these women and their families above the extreme poverty line.

When farmers are able to invest in the future, the poverty trap can be broken, and positive feedbacks can replace the vicious cycle feedbacks of low productivity. In the same Benin project, schools autonomously developed a curriculum comparing different horticultural production practices. Students did the labor and cost calculations for hand-watered versus drip-irrigated plots and visited the project sites to learn about solar power and irrigation. Farmer groups using the technology registered as NGOs within the country and secured deeds to their land to safeguard their investments. More than one ton of food per week from the gardens flowed into mostly local markets. This increase nowhere near satisfied demand, but it has had a measurable impact on overall consumption (Burney et al. 2010). In this way, irrigation generated positive institutional feedbacks and has helped promote food security across scales.

Vulnerabilities, Past-Present-Future

Taking irrigation as an example, but thinking in general about investments in smallholder agriculture, what would it take to make results like those described above the norm? Above all, farmers need to have access to improved technologies so that they can leverage their own way out of the low-productivity trap. Improved governance, institutions, and infrastructure will all help—and are necessary for long-term, large-scale success—but as noted, growth in the rural agricultural sector itself can help drive those improvements. Nevertheless, there are three main hurdles—economic, institutional, and environmental—standing in the way of widespread success.

(1) Economic. Farmers need access to financial services appropriate for the rural agricultural sector. Microfinance institutions (MFIs) have mainly focused on very small loans for non-agricultural enterprises in urban and peri-urban areas. MFIs are reluctant to extend services to the agricultural sector because loan sizes will need to be larger, and lenders face coordinated risks (e.g., in the event of a drought, all farmers in the drought area would likely do badly). Since irrigation reduces weather vulnerability, investments in irrigation might be more appealing to lenders than, say, loans for fertilizer or improved seeds for rain-fed crops. Furthermore, the collateral for loans (driplines, pumps, solar panels, reservoirs, etc.) are physically present and visible. However, irrigation and other agricultural technologies might also raise smallholder risk profiles in unforeseen ways. For example, higher productivity and

better market connectivity may expose farmers to new unknown markets and different price regimes. This reinforces the importance of sequencing of policies raised in Chapter 2. As former net consumers become net producers, they will need to be able to access markets and plan their cropping calendars, transport, and so on carefully to maximize returns. The importance of sound macro-level economic policy emphasized in Chapter 2 holds true here as well: the private sector needs to be able to extend into rural low-income areas with innovative products and services.

(2) From an institutional perspective, rural farmers need to secure land tenure rights to safeguard investments they make in their own productivity. They also need to secure water rights, which may become more difficult because of climate change and population growth. In regions like sub-Saharan Africa, farmers and pastoralists must coexist peacefully under ever-tighter resource constraints. Ultimately, infrastructure—access to roads, markets, fuel, and the electrical grid—may make or break the rural sector. The deep ties between food, energy, and water security at all scales underscore the need for development of functional institutions to support rural development. The private sector will play a critical role in creation of new energy and water services models for this vast market. Finally, research is needed to better understand the links between the rural poor and urban poor, who are often net food consumers and most vulnerable to high prices. As development of the rural sector continues, food security strategies for the newly urban, not-quite-poor population will be needed to safeguard the vibrancy and growth stemming from urbanization.

(3) Finally, environmental safeguards need to be put in place as the rural sector develops worldwide. Perhaps most important, policies across scales need to promote agricultural productivity growth through intensification, as opposed to land expansion. This is because agricultural land use change is a major source of global greenhouse gas emissions (see Chapters 8 and 9), and native habitats (especially forests) play crucial roles in the hydrological cycle. Education and expansion of local extension services can help educate farmers about optimal fertilizer use and help reduce downstream impacts of overfertilization (see Chapter 10). Integrated pest management can help offset some of the need for inorganic pesticides.

Across scales, water for irrigation must be used efficiently. The evolution of irrigation in more developed countries offers instructive lessons, as discussed further in Chapter 11. Differences in management and pricing of water in these systems, as well as large-scale irrigation schemes all over the world (see Chapter 11), have led to sub-optimal management, salinization, and water-based conflicts (Postel 1999). In some places, such as northern India and the western United States, farmers turned to private tubewells and borewells. While private groundwater extraction is beneficial to the individual farmers, at larger scales this type of use has resulted in unsustainable aquifer drawdown in key food security regions (Tiwari et al. 2009). Looking toward the future, many regions in sub-Saharan Africa, for example, have ample renewable water resources—the problem is not physical access, but economic. Nevertheless, emphasis should be placed on using any groundwater efficiently, which would prioritize drip irrigation over other, more water-intensive forms, and would minimize the risk of soil salinization. Studies have shown that flow rates from borewells in sub-Saharan Africa would be more amenable to smallholder schemes as opposed to community-scale irrigation infrastructure (MacDonald et al. 2012), which supports the idea of smallholder irrigation as a development priority, but points to the clear need for coordinated oversight.

Concluding Thoughts

After looking in detail at the microscale low-productivity trap, it is clear that there are many technologies, strategies, and policies that might help households break out of the vicious cycle of low productivity. In the long run, both macro-scale crop breeding programs and micro-level innovations and strategies will play critical roles. Across scales, energy poverty and food

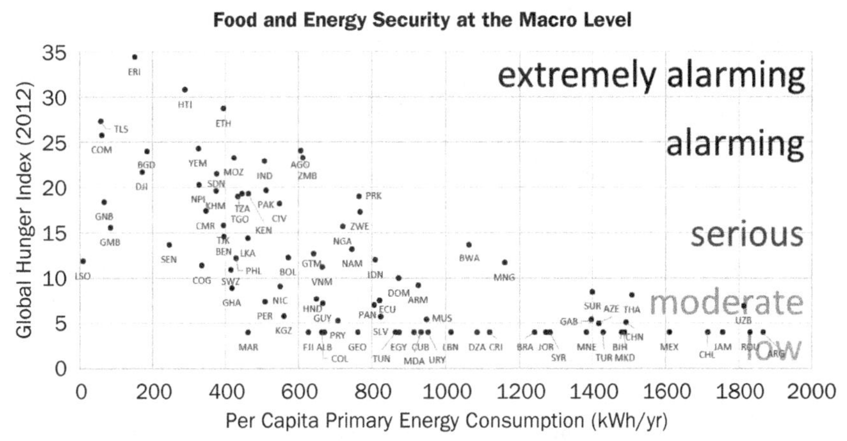

FIGURE 6.4 Energy Consumption and Human Development Global Hunger Index (GHI) and per capita equivalent Energy Consumption.

Source: Data are from the International Food Policy Research Institute's Global Hunger Index website (International Food Policy Research Institute 2012) and the World Bank Indicators data portal (World Bank 2012).

insecurity go hand in hand. Figure 6.4 shows national-level food insecurity (Global Hunger Index) as a function of per capita primary energy consumption. Beyond a level of about 100 kWh (equivalent) per person per year, countries are largely food secure. This cutt-off is the equivalent of about $10 in U.S. electricity; or 11 percent of the average U.S. household's monthly use in 2011. It is also roughly equivalent to the amount of energy provided per person per year by the photovoltaic pumping system pictured in Box 6.2. Although causality obviously runs in both directions in this relationship—energy services facilitate development and more developed countries can afford more energy services—it provides a reminder that promoting efficient energy inputs into the agricultural sector is likely to have huge payoffs for food security and beyond. Investments in diversification, irrigation, breeding, and improved post-harvest performance will be key to transforming rural smallholder farm life from a vicious cycle to a virtuous one.

References

African Development Bank Group. 2012. Benin Country Strategy Paper 2012–2016. http://www.afdb.org/fileadmin/uploads/afdb/Documents/Project-and-Operations/Benin%20-%20Country%20Strategy%20Paper%202012-2016.pdf.

Aker, J. C. 2010. Information from markets near and far: Mobile phones and agricultural markets in Niger. *American Economic Journal: Applied Economics* 2(3) (July): 46–59.

Badiane, Ousmane. 2014. "Agriculture and structural transformation in Africa." In Falcon, W. P. and R. L. Naylor (eds.), *Frontiers in Food Policy: Perspectives on sub-Saharan Africa*. Stanford, CA: Printed by CreateSpace.

Banerjee, A. V., and E. Duflo. 2007. The economic lives of the poor. *The Journal of Economic Perspectives* 21(1): 141.

Barrett, C. B., and B. M. Swallow. 2006. Fractal poverty traps. *World Development* 34(1): 1–15.

Battisti, D., and R. L. Naylor. 2009. Historical warnings of future food insecurity with unprecedented seasonal heat. *Science* 323(5911): 240.

Binswanger-Mkhize, H. P. 2014. "India 1960–2010: Structural change, the rural non-farm sector, and the prospects for agriculture." In Falcon, W.P. and R.L. Naylor (eds.), *Frontiers in Food Policy: Perspectives on sub-Saharan Africa*. Stanford, CA: Printed by CreateSpace.

Brondizio, E. S., and E. F Moran. 2008. Human dimensions of climate change: The vulnerability of small farmers in the Amazon. *Philosophical Transactions of the Royal Society B: Biological Sciences* 363(1498): 1803–1809.

Burney, J. A., and R. L. Naylor. 2012. Smallholder irrigation as a poverty alleviation tool in Sub-Saharan Africa. *World Development* 40(1): 110–123.

Burney, J., L. Woltering, M. Burke, R. L. Naylor, and D. Pasternak. 2010. Solar-powered drip irrigation enhances food security in the Sudano-Sahel. *Proceedings of the National Academy of Sciences* 107(5): 1848–1853.

Carter, M., and C. Barrett. 2006. The economics of poverty traps and persistent poverty: An asset-based approach. *The Journal of Development Studies* 42(2): 178–199.

Comprehensive Assessment of Water Management in Agriculture. 2007. *Water for food, water for life: A comprehensive assessment of water management in agriculture*. London and Colombo: Earthscan and the International Water Management Institute (IWMI).

Deininger, K. 2011. Challenges posed by the new wave of farmland investment. *The Journal of Peasant Studies* 38(2): 217–247.

Dillon, A. 2008. Access to irrigation and the escape from poverty: Evidence from Northern Mali. IFPRI Discussion Paper. International Food Policy Research Institute.

Easterly, W. 2009. Can the West save Africa? *Journal of Economic Literature* 47(2): 373–447.

Food and Agriculture Organization of the United Nations (FAO). 2012. *The state of food insecurity in the world 2012: Economic growth is necessary but not sufficient to accelerate reduction of hunger and malnutrition*. Rome: Food and Agricultural Organization of the United Nations.

Food and Agriculture Organization of the United Nations (FAO). Food and Agriculture Organization of the United Nations Statistical Database (FAOSTAT). http://faostat.fao.org/.

Hertel, T. W. 2011. The global supply and demand for agricultural land in 2050: A perfect storm in the making? *American Journal of Agricultural Economics* 93(2): 259–275.

Hertel, T. W., M. B. Burke, and D. B. Lobell. 2010. The poverty implications of climate-induced crop yield changes by 2030. *20th Anniversary Special Issue* 20(4): 577–585.

Inocencio, A., M. Kikuchi, M. Tonosaki, A. Maruyama, D. Merrey, H. Sally, and I. De Jong. 2007. *Costs and performance of irrigation projects: A comparison of sub-Saharan Africa and other developing regions*. International Water Management Institute.

International Food Policy Research Institute. 2012. "2012 Global Hunger Index." http://www.ifpri.org/ghi/2012.

Jayne, T. S., N. Mason, N. Sitko, D. Mather, N. Lenski, R. Myers, J. Ferris, A. Chapoto, and D. Boughton. 2010. *Patterns and trends in food staples markets in Eastern and Southern Africa: Toward the identification of priority investments and strategies for developing markets and promoting smallholder productivity growth*. Michigan State University International Development Working Paper. East Lansing: Michigan State University Department of Agricultural, Food, and Resource Economics.

Jayne, T. S., T. Yamano, M. T. Weber, D. Tschirley, R. Benfica, A. Chapoto, and B. Zulu. 2003. Smallholder income and land distribution in Africa: Implications for poverty reduction strategies. *Food Policy* 28(3): 253–275.

Jones, W. I. 1995. *The World Bank and irrigation*. Washington, DC: The World Bank.

Lobell, D. B., M. B. Burke, C. Tebaldi, M. D. Mastrandrea, W. P. Falcon, and R. L. Naylor. 2008. Prioritizing climate change adaptation needs for food security in 2030. *Science* 319(5863): 607.

Lobell, D. B., W. Schlenker, and J. Costa-Roberts. 2011. Climate trends and global crop production since 1980. *Science* (May 5).

MacDonald, A. M., H. C. Bonsor, B. É. Ó Dochartaigh, and R. G. Taylor. 2012. Quantitative maps of groundwater resources in Africa. *Environmental Research Letters* 7(2): 024009.

Mertz, O., C. Mbow, A. Reenberg, and A. Diouf. 2009. Farmers' perceptions of climate change and agricultural adaptation strategies in rural Sahel. *Environmental Management* 43(5): 804–816.

Meze-Hausken, E. 2004. Contrasting climate variability and meteorological drought with perceived drought and climate change in Northern Ethiopia. *Climate Research* 27(1): 19–31.

Mitchikpe, C. E., R. A. Dossa, E. A. Ategbo, J. M. Van Raaij, and F. J. Kok. 2009. Seasonal variation in food pattern but not in energy and nutrient intakes of rural Beninese school-aged children. *Public Health Nutrition* 12(3)(March): 414–422.

Moyo, R., D. Love, M. Mul, W. Mupangwa, and S. Twomlow. 2006. Impact and sustainability of low-head drip irrigation kits, in the semi-arid Gwanda and Beitbridge districts, Mzingwane Catchment, Limpopo Basin, Zimbabwe. *Physics and Chemistry of the Earth* 31(15–16): 885–892.

Naylor, R. L., and W. P. Falcon. 2010. Food security in an era of economic volatility. *Population and Development Review* 36(4): 693–723.

Sanchez, P. A. 2002. Soil fertility and hunger in Africa. *Science* 295(5562): 2019–2020.

Somado, E. A., R. G. Guei, and S. O. Keya. 2008. "NERICA: The new rice for Africa: A compendium." Africa Rice Center (WARDA). http://tropicalmedicinecentre.com/downloads/NERICA%20Compendium.pdf.

Talukder, A., L. Kiess, N. Huq, S. de Pee, I. Darnton-Hill, and M. W. Bloem. 2000. Increasing the production and consumption of vitamin A-rich fruits and vegetables: Lessons learned in taking the Bangladesh homestead gardening programme to a national scale. *Food & Nutrition Bulletin* 21(2) (June): 165–172(8).

Taylor, R., M. Todd, L. Kongola, L. Nahozya, H. Sangha, and A. M. MacDonald. 2012. Evidence of the dependence of groundwater resources on extreme rainfall in East Africa. *Nature Climate Change* 3: 374–378.

Tesfaye, A., A. Bogale, R. E. Namara, and Dereje Bacha. 2008. The impact of small-scale irrigation on household food security: The case of Filtino and Godino irrigation schemes in Ethiopia. *Irrigation and Drainage Systems* 22(2): 145–158.

The Alliance for a Green Revolution in Africa. 2011. *AGRA in 2011: Investing in Sustainable Agricultural Growth: A Five-Year Status Report.*

Tiwari, V. M., J. Wahr, and S. Swenson. 2009. Dwindling groundwater resources in Northern India, from satellite gravity observations. *Geophysical Research Letters* 36(18): L18401.

Udry, C. 1996. Gender, agricultural production, and the theory of the household. *The Journal of Political Economy* 104(5): 1010–1046.

UNEP. 2008. "Atmospheric Brown Clouds: Regional Assessment Report with Focus on Asia." United Nations Environment Programme.

———. 2011. *Integrated Assessment of Black Carbon and Tropospheric Ozone: Summary for Decision Makers.* United Nations Environment Programme and World Meteorological Organization.

Vedwan, N., and R. E. Rhoades. 2001. Climate change in the Western Himalayas of India: A study of local perception and response. *Climate Research* 19(2): 109–117.

Vitousek, P. M., R. Naylor, T. Crews, M. B. David, L. E. Drinkwater, E. Holland, P. J. Johnes, et al. 2009. Nutrient imbalances in agricultural development. *Science* 324(5934): 1519.

Woltering, L., A. Ibrahim, D. Pasternak, and J. Ndjeunga. 2011. The economics of low pressure drip irrigation and hand watering for vegetable production in the Sahel. *Agricultural Water Management* 99(1): 67–73.

World Bank. 2012. Agriculture, Value Added (% of GDP). http://data.worldbank.org/indicator/NV.AGR.TOTL.ZS.

www.sima-niger.net. "SIMA NIGER." *Système d'Information Sur Les Marchés Agricoles.* http://www.sima-niger.net/index.php.

7

Health and Development at the Food-Water Nexus

Jennifer Davis, Eran Bendavid, Amy J. Pickering, and Rosamond L. Naylor

Food security in the world's poorest regions is as closely linked to water and infectious disease as it is to agriculture. Without water for irrigation, crop production is confined mainly to the rainy season, which lasts only a few months in many tropical and sub-tropical countries, with timing and duration that vary from year to year. As described by Burney in Chapter 6, farmers without irrigation are less likely to cultivate fruits and vegetables that improve household nutrition and generate income for the family. Many households in low-income countries also have limited access to domestic water supplies, often spending dozens of hours per week hauling small volumes of water from distant sources for their drinking, cooking, and washing needs. Inadequate water supply and sanitation services are key contributors to the spread of diarrheal disease, which in turn limits people's ability to absorb calories, protein, and essential micronutrients. The malnutrition resulting from repeated or prolonged diarrheal illness profoundly compromises the human immune system, and can accelerate the progression of other diseases such as HIV/AIDS. A pernicious cycle arises, in which the physical impacts and medical costs of malnutrition and frequent illness reduce labor productivity and farm output, limit household income, and prevent individuals from making biological use of the food they are able to obtain. Effective interventions to reduce food insecurity thus need to address the intersection of these water, food, and health constraints, as well as the fundamental economic conditions that cause families to live in persistent poverty.[1]

[1] For a discussion of poverty traps and persistent poverty, see Easterly (2009) and Barrett and Carter (2012).

This chapter focuses on the water-food-health nexus as a root of food insecurity among the world's poorest billion. Sub-Saharan Africa (SSA), in particular, is a region plagued by the complex interface of limited food and water availability, along with high rates of preventable infectious disease. Two-thirds of SSA's economically active population earns its income from agriculture. With less than 5 percent of the continent's agricultural land irrigated, the vast majority of households rely on rainfed farming systems. Indeed, despite significant increases in average per capita gross domestic product (GDP) and other human development indicators since the turn of the century, more than a quarter of the African population remains undernourished (UNDP 2012). Rates of chronic malnutrition, as measured by the percentage of children under the age of five who are stunted, barely budged from 43 percent in 1990 to 41 percent in 2010. By comparison, Asia reduced its rate of preschool stunting from 49 percent to 28 percent during the same 20-year period.

Four in ten SSA households also lack access to an improved domestic water source, such as a yard tap or borewell with a handpump. Most survive on fewer than 25 liters of water per person per day for all of their domestic needs, compared to the 50–100 liters recommended by the World Health Organization for meeting basic needs and protecting public health (WHO 2010a). By contrast, most households in Europe use 200–300 liters per person per day, while a typical American uses 300–400 liters.

SSA also has the highest child mortality rate in the world. Chronic malnourishment and limited access to water for both productive and domestic uses increases vulnerability to communicable diseases, including three of the leading causes of morbidity and mortality in SSA: HIV, tuberculosis, and infectious diarrhea. One in eight children dies before reaching the age of five, often from infectious diseases such as diarrhea and pneumonia (UNICEF 2011, Liu et al. 2012). Global mortality due to HIV is concentrated in SSA, severely curtailing average life expectancy (Lopez et al. 2006).

Addressing these interdependent problems, whether at the policy level or within international development agencies, raises unique challenges. At the institutional level, groups responsible for agriculture, health, or water supply programs are typically "siloed" from one another, with little or no overlap in responsibility or coordination of effort. Even if greater coordination were to develop, identifying the most effective entry points for intervention is complicated by the lack of clear and consistent causal relationships within the water-food-health nexus. The nature of interactions among water quantity and quality, food security, and infectious disease is typically shaped by site-specific factors. The time, effort, and money needed for field research, outreach, and the scaling up of locally appropriate interventions can be prohibitive for resource-limited development organizations. Moreover, a coordinated approach to solving water, food, and health problems requires a strong political constituency advocating for such solutions. Such constituencies are a rarity in the world's lowest-income countries.

Detrimental cycles within the water-food-health nexus are not always obvious on the ground, and they tend to masquerade as the status quo in SSA. Individuals may deliberately conceal their HIV status because of social stigma, complicating detection and tracking of disease progression. Malnutrition is most noticeable in its extreme forms—distended bellies of children caused by severe protein deficiencies (*kwashiorkor*) or skeletal forms of starvation. It is not uncommon for parents, especially those who lack education or nutritional training, to overlook signs of malnutrition in their children. Recurrent mild or moderate diarrhea is frequently considered to be a normal rite of passage for children, rather than a concerning symptom. It is also difficult, even for public health professionals, to attribute malnutrition to a specific cause such as drinking contaminated water. Politicians, development specialists, and citizens alike seem to have grown accustomed to the chronic confluence of illness, malnourishment, and poverty that characterizes much of the sub-Saharan African region.

Our objectives in this chapter are to shed light on the web of water-food-health impediments that prevent impoverished communities from achieving food security, and to develop a framework for analyzing and addressing these interconnected constraints. We begin with a trip to western Kenya where (in 2012) we launched a collaborative research project with the Kenya Medical Research Institute and the Centers for Disease Control (KEMRI-CDC).[2] This work provides real-life context, a research framework, and a foundation for the following sections of the chapter that trace the linkages between water access, food production, infectious disease and malnutrition. We conclude with observations on the institutional challenges of solving this complex array of problems, and suggestions for moving forward.

The tension between improving water quality versus water quantity is highlighted in this chapter as a key development issue for low-income countries. In sub-Saharan Africa, where the population is projected to double from 900 million in 2012 to 1.8 billion in less than 30 years (UNDP 2012), the trade-off between water availability and water quality will have increasingly stark consequences. Natural resource scarcity associated with continued population growth underscores the need for family planning services, which play critical roles both in improving maternal and infant health, and in reducing population pressure in the future. Investments in family planning clinics can serve as a valuable complement to agricultural investments, as shown by Falcon

---

[2] This project, "Rural Health and Development at the Food-Water Nexus," is supported by the Environmental Ventures Program of the Woods Institute for the Environment, and by the Global Underdevelopment Action Fund of the Freeman Spogli Institute for International Studies, both at Stanford University. The field research began in 2012 and remains in an early phase at the time of writing. Research articles on our findings will be published in peer-reviewed journals after data collection and analysis are complete. This project follows many years of fieldwork by Davis, Bendavid, and Pickering on water, sanitation, and health issues in several sub-Saharan African countries.

in Chapter 2. The links between natural resources, the environment, and food security are explored in greater depth in subsequent chapters of this volume.

A Trip to Kisumu and Asembo in Western Kenya

Kisumu is a rapidly growing city located on the northeastern shore of Lake Victoria (Figure 7.1). Home to roughly 500,000 residents in 2010, it is the third largest city in Kenya and the capital of Nyanza province. Kisumu is an important trade center for the Lake Victoria region, filled with government offices, banks, universities, hospitals, and research institutions. Walking through Kisumu, one is struck by the uncontrolled and haphazard growth, pot-holed roads, and poor sanitary conditions. The informal economy flourishes: countless stalls are scattered in the city center selling rubber sandals, M-pesa (mobile money) cards, and secondhand clothing. The main roads are lined with men on motorcycles offering to give rides for 20 Kenyan shillings (less than US$0.25). Just beyond the main business district lie dozens of slums, housing 60 percent of the city's residents. The majority of Kisumu's residents live on less than US$2 per day, and roughly half are classified as "food poor" by the municipal council (UN-HABITAT 2005).

Leaving Kisumu by car, the urban landscape quickly gives way to green, gently rolling hills dotted with small villages. Following the all-weather highway westward along the northern shore of Lake Victoria, the village of Asembo lies an hour's drive from the bustling city. Along the village's dirt paths, rows of maize,

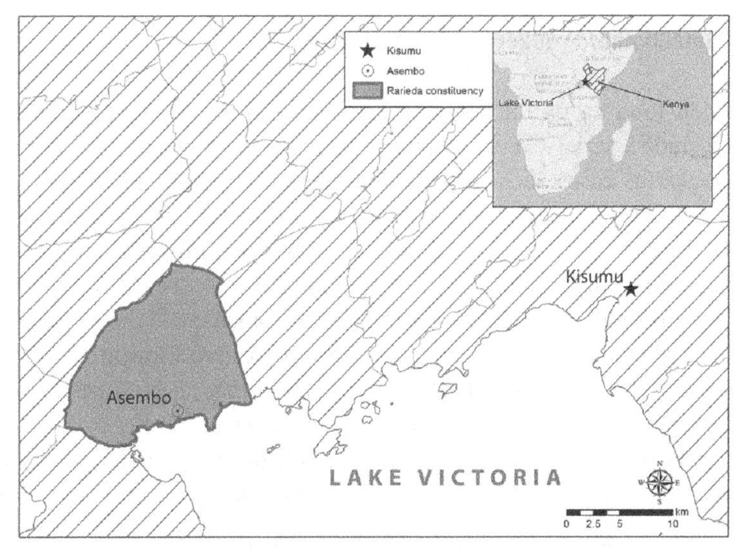

FIGURE 7.1 The Kisumu/Asembo Region, Western Kenya

Source: Jeff Ho.

beans, millet, and sorghum are cultivated and harvested by families. Although this region has reasonably good soil, agricultural productivity is constrained by the absence of irrigation infrastructure, commercial fertilizers, and credit. Households survive mainly on a diet of starchy staples, with limited protein and micronutrient intake. Women and children spend up to two hours per day fetching water from public taps, ponds, and streams, carrying 45-pound containers of water back to their homes. More than 70 percent of the households in Asembo live below the Kenyan poverty line, and one in five children dies before the age of five. The 2008 Demographic and Health Survey (Kenya DHS Final Report 2008) estimated that roughly 27 percent of children in Nyanza province are "stunted," or have height-for-age measurements that are two standard deviations or more below normal. Against this backdrop of privation, St. Elizabeth's Mission Health Centre (a donor-supported clinic in nearby Lwak village) is an oasis of support, overflowing with children and mothers waiting to receive medical treatment.

The people of Asembo are caught in a complex cycle of morbidity and mortality, driven by food insecurity and variable access to water and health care. In an effort to understand and measure the relationships among water access, food production and consumption, and health, we developed a framework for data collection and analysis (Box 7.1). Among the 2,700 households we visited in the first year of our field study, half reported that they had completely run out of food at least once in the preceding month, and one third reported going to bed hungry at least once during the same time period. Our survey thus revealed evidence of considerable food insecurity that often goes unnoticed in this bustling region of western Kenya.

Water, Smallholder Production, and Nutrition

The staple food crops for regional consumption in much of sub-Saharan Africa include a variety of cereals, roots, and tubers—mostly maize (corn), but also indigenous crops such as tef and sorghum. Virtually all of these crops are rainfed, with irrigation covering less than 5 percent of total arable land in SSA (Burney et al. 2013). With added pressures from pests, crop diseases, and low soil fertility, average maize yields in Africa are below 2 tons/hectare (ha)—less than half the mean yields in Asia (4.4 tons/ha) and the world (5 tons/ha), and far below yields in North America and Western Europe (8 to 10 tons/ha) (FAOSTAT 2012).[3] Such low yields reflect, in large part, farmers' limited access to high-yielding varieties, synthetic fertilizers, and credit. Even when these inputs are available, farmers without access to irrigation may be reluctant to adopt new technology because of the risks of crop failure and poor nutrient

[3] These data refer to the most recent update by the Food and Agriculture Organization (FAO) at the time of writing, covering the 2008–2009 period.

BOX 7.1

Studying Water Access, Infectious Disease, Nutrition, and Economic Development in Rural Kenya

The figure below depicts the linkages between water, food, and health that can influence food insecurity, infectious disease burden, and morality rates at the household or community level. Our research team, which includes members from Stanford University and the Kenya Medical Research Institute/Centers for Disease Control visited 2,700 households in Asembo in order to collect information about each construct in the schematic below.

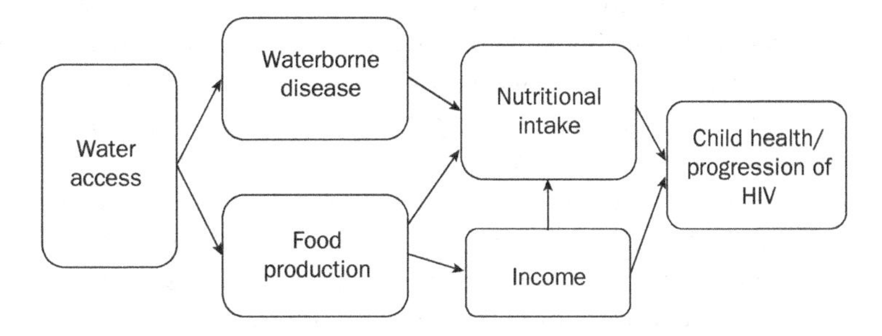

The specific data collected from participating households include:

- *Water access:* Distance to water sources; volume of water used for domestic and productive activities (including irrigation); microbial water quality of source and stored water.
- *Waterborne disease:* Diarrhea prevalence among children under age five and women of childbearing age.
- *Food production*: Type and number of crops grown; cropping intensity; use of irrigation.
- *Nutritional intake*: Dietary diversity of children under age five and women of childbearing age.
- *Income*: Income-generating activities; household wealth index; crop sales.
- *Child health and progression of HIV:* Height, weight, and upper arm circumference of children under age five and women of childbearing age; change in CD4 counts among HIV-positive women.

Household data on water; food, and health collected at a given point in time permit analysis of associations between variables, but causal relationships remain obscure. Without targeted interventions and randomized trials, it is difficult to pinpoint, for example, precisely how water quantity or quality influences a family's nutritional status and vulnerablilty to disease (Banerjee and Duflo 2011). When experimental interventions and randomized trials are not feasible, careful design of the survey sample can help elucidate whether certain relationships exist. For example, data can be compared across four types of households, each with different water access status: 1) households with adequate access to both domestic and productive (irrigation) water: 2) households with access to domestic water supply but not productive water; 3) households with access to productive water but not domestic water supply; and 4) households without access to either domestic or productive water. This scheme allows for an analysis-of-variance (ANOVA) approach to dissecting key relationships in the water-food-health nexus.

uptake during dry periods. Since the turn of the twenty-first century, SSA remains the only major region of the world that has repeatedly experienced famines, the majority of which have been initiated by drought and extreme weather events, and often compounded by misguided government interventions or lack of response (Devereux 2009).

As noted by Burney in Chapter 6, irrigation is critical to diversifying smallholder agricultural production, in particular for enabling the cultivation of market garden crops (fruits and vegetables) and contracted export crops (e.g., green beans, cashews, cotton) that earn higher returns in the market. Irrigation can also help increase cropping intensity by allowing for year-round agricultural production. Public investment in large-scale irrigation infrastructure or centralized small-scale systems has been minimal in most SSA countries, however, and returns on investment for large-scale systems have been relatively low. Families growing garden crops might carry water from the river or a dug well to their plots for hand-watering, but such practices are time intensive and physically demanding. In villages that have access to piped water from a handpump or a tap, smallholders will often use these domestic sources to irrigate small areas of vegetables or fruits, or to provide water for livestock.

The use of domestic water supplies for productive purposes can be controversial. Piped water systems in rural communities are typically designed to deliver a "basic needs" volume of water on the order of 20–50 liters per capita per day. If households regularly withdraw more water in an effort to support "unplanned" productive uses such as irrigation, the result could be a reduction in the lifespan of both the infrastructure and the water source. At the same time, the use of domestic water supplies for productive purposes has been shown to improve nutrition and increase income among smallholder households in many low- and middle-income countries (Moriarty et al. 2004). In Kenya, for example, households using piped water for smallholder agriculture and dairy farming earned an average of US\$16–\$18 per cubic meter (m^3) of water, respectively, for which they paid only US\$1/m^3 (Davis et al. 2011).

Given the dearth of investment in large-scale irrigation infrastructure, along with the growing body of evidence that small-scale irrigation can be an important force for improving nutrition and reducing poverty in rural SSA (Burney and Naylor 2012; Burney et al. 2013), it would seem that an opportunity exists to re-orient the design of community water systems to accommodate both domestic and productive needs. Such efforts would challenge current institutional arrangements in the water sector of many countries, however. Historically, the planning, financing, and operation of domestic water systems have been implemented in isolation from that of productive infrastructure. In addition, a "combined use" village water system—with increased storage capacity and a broader distribution network that delivers water closer to households and agricultural plots—inevitably costs more to construct than a simple borehole with a handpump. The capital cost-sharing requirements that most

SSA governments and international donors place on so-called beneficiary communities could make such systems unaffordable for the rural poor.

Until access to irrigation expands in sub-Saharan Africa, the bulk of smallholder farm production will likely be comprised of staple cereals and tubers, and relatively little additional area will be devoted to the cultivation of more nutritious vegetables and fruits. Even if these low-income households can overcome basic calorie deficits, their diets will still be lacking in protein, fat, and essential micronutrients (UNDP 2012). Dietary diversity, one important metric of malnutrition (Headey and Ecker 2013), varies widely across SSA countries; on average, however, more than a third of rural households in SSA experience low dietary diversity, as compared to 11 percent of urban households (Smith et al. 2006).[4] Malnutrition, in turn, compromises individuals' ability to fend off or slow the progression of infectious disease. Both hunger and infectious disease may reduce one's ability to irrigate crops, especially when physically demanding manual technologies such as treadle pumps are used. Similarly, those affected by malnutrition and illness are less able to travel to water sources, or to carry heavy loads of water for household or agricultural use.

Effects of Water Quality versus Water Quantity on Infectious Disease

Even if adequate quantities of water are available for diversified food production, poor water quality can lead to reduced nutritional uptake. Diarrheal illnesses remain a major cause of morbidity and mortality in developing countries, responsible for 700,000 deaths annually, principally among young children (Walker et al. 2013). Between 25 and 33 percent of growth failure observed in children is thought to be caused by diarrheal disease (Brown 2003), and acute diarrheal infections in early childhood can impair cognitive function and physical fitness much later in life (Guerrant et al. 1999; Niehaus et al. 2002). In sub-Saharan Africa, the health burden of diarrhea is particularly large; each year, half of child deaths and almost 8 percent of all deaths are related to diarrheal illness (Kosek et al. 2003).

Infectious diarrhea—which is characterized by the passing of three or more loose stools within a 24-hour period—can be caused by a variety of pathogens, including bacteria, viruses, and parasites. Transmission of diarrheal illness is typically through a fecal-oral route, in which a susceptible host ingests fecal material containing infectious organisms. Fecal contamination of drinking water is all too common where sanitation is poor and water sources are

[4] The metric of dietary diversity is based on the number of food groups consumed within the household and is derived from household surveys. Typical food groups include: (1) cereals, roots, and tubers; (2) pulses and legumes; (3) dairy products; (4) meat, fish, seafood, and eggs; (5) fats and oils; (6) fruits; and (7) vegetables.

inadequately treated. A single gram of human feces can contain up to 10 million viruses, 1 million bacteria, 1,000 parasite cysts, and 100 parasite eggs.

Even when a child is not suffering from diarrhea, if she lives in an environment with extensive fecal contamination she may develop a condition from this exposure known as environmental enteropathy (EE). Individuals with EE suffer from poor nutrient absorption in the small intestine, caused by changes in the function and physiological structure of the gut (Lunn 2000). It is hypothesized that EE may contribute significantly to child growth faltering, particularly in populations with relatively low rates of diarrhea (Humphrey 2009; Dewey and Mayers 2011).

Individuals can be exposed to fecal pathogens in a variety of ways (Figure 7.2). Pathogens and fecal contamination are common in stored drinking water and on the hands of mothers and children among households lacking in-home water supplies (Pickering et al. 2010). Household soil has also been found to contain high levels of fecal bacteria and diarrheal pathogens, as we have seen in Tanzania (Pickering et al. 2012). Microbes can also be detected on common household surfaces and objects, such as dishes and toys (Pickering et al. 2012). In the public health arena there is little consensus about the relative importance of these different exposure pathways to the global burden of diarrheal illness.

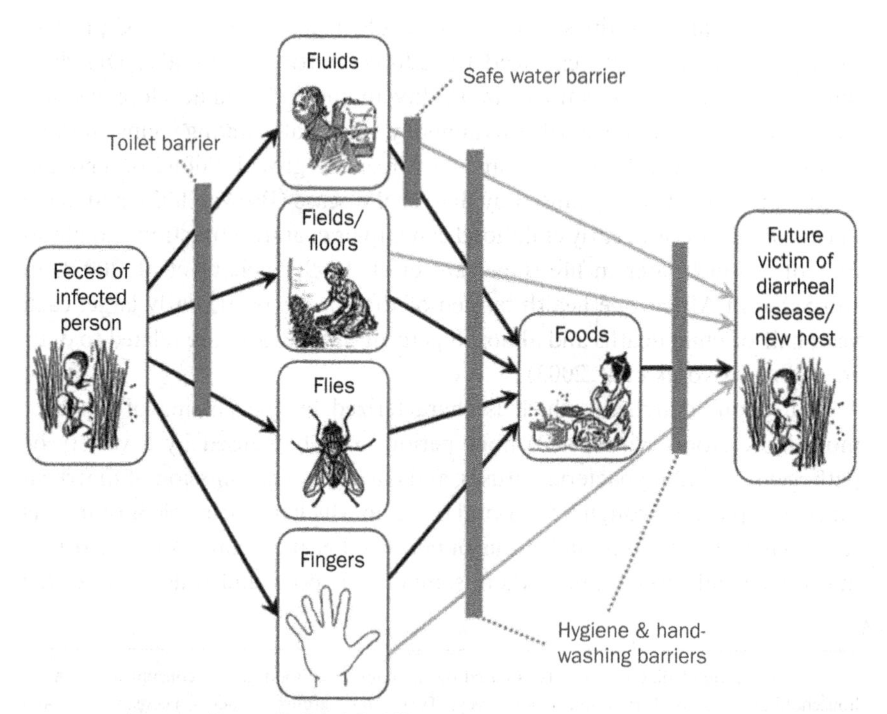

FIGURE 7.2 Exposure Routes for Fecal-Oral Illnesses

Source: Marla Smith-Nilson/Water1st International.

This dearth of knowledge has important implications for the design of interventions in the water sector. For example, it is not uncommon for households that rely on shared point sources (such as borewells with handpumps or public taps) to use only 10–20 liters of water per person per day. Limited water use constrains opportunities for personal hygiene, including hand-washing after defecation and before preparing food. Increasing the volume of water available to a household, for example by installing water points closer to the home, can enable more frequent hand-washing and thus reduce transmission of pathogens *via* hands (Curtis and Cairncross 2003; Waddington and Snilstveit 2009). Hand-washing interventions have been shown to reduce the incidence of diarrhea by 30–40 percent (Ejemot et al. 2008).

Alternatively, supplying households with technologies to improve the quality of water at the point of use can reduce the incidence of illnesses such as typhoid and cholera, which are contracted by ingesting contaminated water. Reductions in diarrheal disease of up to 50 percent have been documented among households that receive interventions such as household water filters and chlorine products (Fewtrell et al. 2005; Arnold and Colford 2007). At the same time, serious challenges have been documented in sustaining use of point-of-use water treatment products beyond the initial project or study period.

Diarrhea can impair growth through multiple forms of malnutrition. Diarrheal illness can decrease appetite, resulting in reduced caloric intake; it can reduce absorption of nutrients in the intestine; and it can increase anabolic requirements and/or catabolic losses (for example, to support the immune system as it fights against infection). Although quantitative assessments are rare, earlier research suggests that children suffering from chronic diarrhea take up roughly 20 percent fewer calories from the food they consume as compared to healthy children (Martorell et al. 1980).

Water access also affects nutritional status through the energy demands of water collection, which can be particularly large for rural households using sources far from their homes. More than 40 years ago, the *Drawers of Water* study found that rural water carriers in East Africa spent an average of 240 calories per day collecting water, which was equivalent to 12 percent of their daytime energy expenditures (White et al. 1972). In arid zones where people walked long distances and/or traversed steep terrain to access water, up to one quarter of their daytime energy was expended on water collection. A subsequent review found that little had changed in SSA; water carriers in sub-Saharan Africa spent an average of 134 minutes per day collecting water, the equivalent of 8 to 10 percent of their daily caloric intake (Rosen and Vincent 1999).

Not surprisingly, water-related energy expenditure falls disproportionately on females. Analysis of representative data from 25 sub-Saharan African countries found that women and girls were the main water haulers in 71 percent of all households lacking in-home water supplies (JMP 2012), and that the mean round-trip fetching time to collect water was 30 minutes. In these 25 countries, it is estimated that women and girls spend a combined total of at least 16 million hours every day collecting water. The water-fetching burden makes it harder for them to eat enough food to maintain good nutritional status. It also reduces the time available to women for childcare, with significant implications for child health. In a study of 26 sub-Saharan African countries, Pickering and Davis (2012) found that just a 15-minute decrease in one-way walk time to a water source is associated with a 41 percent average relative reduction in diarrhea prevalence, and an 11 percent relative reduction in under-age-five child mortality.

In sum, water constraints can lead to serious health and developmental problems not only in the current generation, but for the next generation as well. Myriad connections and feedbacks among water availability, health and nutritional status, and food security have been identified. But much remains to be learned about the full costs and benefits, as well as the sustainability challenges, of improving water quality versus increasing water quantity for low-income households.

The Evolving Story of Food Insecurity and Infectious Diseases

The burden of life-threatening infectious diseases in addition to diarrheal illnesses—principally HIV, tuberculosis, and malaria—is also high in sub-Saharan Africa. Many diseases that have been successfully controlled in Europe and North America through sanitation, vaccination, and elimination of disease vectors (such as malaria-carrying *anopheles* mosquitoes) continue to cause substantial disease burdens in SSA's low- and middle-income countries. When we visited South Africa in 2000, two out of every three patients in Cape Town's Groote Schuur hospital had HIV. In the United States, by comparison, the HIV epidemic has never been a dominant cause of disease burden outside the one epidemiologic group at highest risk—men who have sex with men (MSM) between 25 and 49 years old. The confluence of HIV, tuberculosis, and child infectious diseases such as diarrhea in SSA's low- and middle-income countries raises questions about the causes of such pervasive challenges to development, especially the relationship between disease, nutrition, and access to food and water.

The difference in infectious disease burden between sub-Saharan Africa and wealthier nations reflects SSA's slow progress within the "epidemiologic transition"—a shifting of disease burden associated with the process of economic development (Omran 1971). This progression starts with a proliferation of infectious diseases ("the age of pestilence") at low levels of development, moves through a stage of receding pandemics, and proceeds to the dominance of chronic and non-communicable diseases (e.g., heart disease, cancer) at higher levels of per capita income. North American and European countries had undergone this transition by the first half of the twentieth century, while most East and Southeast Asian countries, with the exception of India, largely transitioned out of "the age of pestilence" by the 1990s.

Why do millions in sub-Saharan Africa die every year from diseases that are entirely preventable and commonly treatable in other parts of the world? The main correlates of sub-Saharan Africa's persistently high infectious disease burden include limited access to clean water, widespread malnutrition, a lack of capacity to control many vaccine-preventable illnesses, a stubbornly active malaria epidemic, and HIV. The HIV epidemic is relatively new, having spread across large portions of the continent during the 1990s. By the first part of the twenty-first century, HIV had reduced life expectancy in the southern cone of Africa to the same level it was in 1960, effectively erasing 50 years of health gains.

Africa's HIV epidemic has many faces and manifestations. In Kenya, only 1 percent of the population in the rural North Eastern Province has HIV, while the prevalence is more than 16 percent in Nyanza Province (Kenya DHS Final Report 2008). Across SSA countries, the variation in HIV incidence is even more striking, ranging from about 1 percent in Senegal to over 40 percent in Swaziland among pregnant women visiting antenatal clinics. Several factors are thought to play an important role in the extremely high HIV rates found in some areas, including low rates of male circumcision, as well as trucking routes and concurrent sexual networks that allow the rapid spread of the virus along a "transmission superhighway." At the same time, food access, nutrition, and water quality and quantity remain critical variables underpinning the HIV epidemic throughout the continent.

The Intersection of Food Insecurity and HIV

The role of food security in the spread of HIV is embedded fundamentally in the socio-behavioral aspects of HIV transmission. In Botswana and other hyperendemic countries, control efforts have identified the role of food security in sexual behavior (Weiser et al. 2006; Gourvenac et al. 2007; Carter et al. 2007a; Agency 2008). Women, in particular, engage with older and wealthier men in sexual networks to gain access to food and other basic amenities (Lagarde et al. 2001; Kalichman et al. 2007; Carter et al. 2007b). These transactional relationships have been observed to increase with rising food prices in several African countries.

The role of food insecurity is also important in the context of HIV care (Weiser et al. 2006, 2009). Treatment for HIV, currently best accomplished with antiretroviral therapy (ART), requires complex partnerships between infected individuals and the health care system. In the early days, ART meant taking upwards of 40 pills per day at four scheduled periods (Mills et al. 2006). This kind of pill burden required a high level of discipline and motivation, especially as symptoms decreased over time. But safe drinking water supplies and nutrition were also significant success predictors, helping to moderate the drugs' gastrointestinal side effects and to enable absorption and metabolism of the drugs (Monforte et al. 2000; Clark 2005). Guidelines for the use of some antiretroviral therapies also indicate that a minimum intake of calories and fat is required in order for the drugs to be metabolized and absorbed properly. Low-fat meals may decrease the absorption of the drugs, leading to increased toxicity and decreased efficacy. Most modern ART regimens are much simpler—one or a few pills daily—but the side effects and interplay with meal calorie and fat composition remain important factors for the health of HIV-infected patients.

Beyond the daily task of taking medicines, care for HIV patients requires frequent visits to a health care facility, regular blood draws, and monitoring (World Health Organization 2010a,b). The transition of HIV from an acute to a chronic disease depends on the substantial and sustained burden of self-care required of affected individuals. In rural African settings, where population is sparse, road conditions are poor, and transportation is rudimentary, the time cost of making regular clinic visits is high. When a large portion of every day is spent caring for crops and livestock, and collecting water and fuel, the time cost of seeking HIV care—irrespective of the monetary cost of receiving medical attention and antiretroviral drugs—may be prohibitive (Rosen et al. 2007; Brinkhof et al. 2009). Among individuals who start antiretroviral therapy in Africa, only 50–80 percent remain in care and return to the clinic within the first year (Fox and Rosen 2010; Rosen and Fox 2011). Women, in particular, identify food insecurity as a major cause of failing to show up in the clinics and adhere to ART. It is not surprising, then, that clinics in Botswana that offer food in addition to HIV care achieve higher rates of retention and adherence than those that do not (Weiser et al. 2006).

Just as food insecurity influences HIV treatment, the impacts of HIV on mortality and morbidity also affect food insecurity in many communities. Projections indicate that the nine SSA countries with the highest HIV burden will lose between 13 and 26 percent of their potential agricultural labor force to HIV-related illness by the year 2020 (Drimie 2002). The demographic strata most highly affected by HIV are women between the ages of 20 and 45, followed by men in the same age group (Larson et al. 2009). This concentration of disease burden on young women and men has led to the creation of a generation of orphans—upwards of 10 million by most estimates—who have lost at least one parent to HIV (Guest 2003). Such mortality compromises the capacity of farm households to produce food in the short run, as well as the transmission of farm knowledge from older to younger generations, potentially lowering productivity in the longer term.

The impact of infectious disease on knowledge transfer and capacity building also extends into agricultural sector institutions. In Kenya's Ministry of Agriculture, for example, it is estimated that 58 percent of staff deaths have been caused by HIV (UNAIDS 2001). Almost one in five staff members of Malawi's Ministry of Agriculture and Irrigation are living with the disease (UNAIDS 2001). Throughout SSA, up to 50 percent of staff time for agricultural extension workers has been lost to HIV-related challenges, such as worker illness, caring for sick family members, and organizing and attending funerals (Qamar 2003).

The relationship between agricultural labor productivity, HIV, and ART is complex. A longitudinal study of Kenyan tea growers demonstrated that farm labor productivity decreased with HIV disease progression; once ART was initiated, labor productivity improved but never returned to baseline levels (Larson et al. 2012). That is, the recovery of immunity and health status did not lead to full recovery of farming labor productivity, partially because of the significant time and resource cost needed to maintain good health while on ART, including regular clinic visits, medication refills, and laboratory tests.

Other infectious diseases have similar effects on food access and labor productivity. Household members suffering from infectious diseases regularly lose time from work, and have reduced capacity for labor when in the fields (Antle and Pingali 1994). A study in Vietnam found that the lost income of those suffering from diarrhea was equivalent to 64 percent of the per capita monthly non-food expenditures for the poorest fifth of households (Luong et al. 2007). In Ethiopia, research has shown that healthier farmers produced more crop per unit of input, earned more income, and supplied more labor than farmers reporting higher rates of illness (Ulimwengu 2009). As crop production declines in a household with ill members, so too does household income, and thus the ability to secure food in sufficient quantity and diversity (Keverenge-Ettyang and Ernst 2010). In short, it has become widely apparent that just as malnutrition is a clear contributor to illness, poor health of agricultural laborers is a major cause of chronic malnourishment in SSA (U.S. Government Accountability Office 2008).

Re-thinking the Structure of Institutions

The depth and complexity of the interconnections between water, food production, and health open up opportunities for synergistic policies and programs across these domains. At the same time, they also present a substantial challenge for organizing the kinds of multi-sectoral, multidisciplinary institutional approaches required for effective interventions. The experience of donor-supported "integrated rural development" programs in sub-Saharan Africa during the 1970s and 1980s suggests that such institutional impediments are considerable. The establishment of new governance structures for such programs (for example, project management units) allowed for smooth administration of funds and avoided traditional bureaucratic roadblocks within responsible ministries. But these governance structures also prevented widespread institutionalization of a multi-sectoral approach to development.

More recently, the Millennium Villages Project aimed to directly address the interconnected problems characterizing poverty in rural sub-Saharan Africa. This project instituted a holistic approach to underdevelopment, concurrently addressing major challenges such as health, nutrition, and access to clean water (Sachs et al. 2004; Friedrich 2007). To date, however, evaluation of these investments shows only a modest decline in child mortality. This experience suggests that even well-funded, multi-sectoral programs come up against barriers to development that are not entirely explained by existing models of self-sustaining poverty (Pronyk 2012; Pronyk et al. 2012).

Many of these same tensions are currently playing out in the water sector across SSA, as decision makers and donor agency staff reconsider the institutional silos that separate planning and policy for domestic water supplies from that of irrigation infrastructure. Community-level experience in many countries suggests that the rural poor would be better served by water systems designed to meet both domestic and productive demands, instead of the "basic needs" infrastructure that is typically provided.[5] Combined systems give rise to some engineering challenges, including potentially large seasonal variation in demand and the related implications for system sizing. These challenges, however, are readily identified and generally manageable. Far more difficult are the institutional complexities associated with assigning rights to use water resources (typically organized by domestic versus agricultural applications); financing, including requirements for capital cost sharing by communities; water pricing schedules, which often differ for domestic and commercial versus agricultural and industrial users; and responsibility for the operation, maintenance, and replacement of infrastructure.

In Kenya, for example, there are at least six different administrative bodies that play some role in the allocation of water use rights, including the Water Services Regulatory Board (WSRB), the Water Resource Management Authority (WRMA), Water Service Boards (WSBs), Water Service Providers (WSPs), Catchment Area Advisory Committees (CAACs), and Water Resource

---

[5] Basic-needs water supply investment strategies have dominated the rural sub-sector for decades, resulting in the vast majority of those households with access to "improved" water supplies being served by shared community taps or borewells. Indeed, the share of households with individual water connections in sub-Saharan Africa has increased by just one percentage point (from 15 percent to 16 percent) between 1990 and 2010 (JMP 2012). In Kenya, the share of rural households with individual water connections also increased by one percentage point (from 11 percent to 12 percent) during this period.

User Associations (WRUAs). Prior work by our team in 50 rural communities across Kenya's Rift Valley, Central, and Eastern provinces found that permits for withdrawing raw water had been granted that exceeded the available flow in nearby rivers by threefold. In addition, the National Water Act (2002) acknowledges the role that water can play in supporting rural livelihoods, but emphasizes the provision of "basic needs" supplies of 20 liters per capita per day as a first priority for sector investments. Ensuring universal access to basic service is an understandable political imperative; however, it arguably results in the installation of water infrastructure that makes a negligible contribution to food security, health, or poverty alleviation.

Conclusion

Reducing the high rates of morbidity and mortality from malnutrition and infectious disease in sub-Saharan Africa requires a multidisciplinary mindset and a suite of compatible interventions that take the interconnected nature of the water-food-health nexus into account. Whereas our experience and the available data suggest that integrated approaches may be more likely to achieve meaningful progress on health and development, there are very few examples of successful programs in the African context from which to draw inspiration.

Instead, major infectious disease initiatives are often launched in regions where food insecurity abounds, and where malnutrition compromises people's ability to engage in or respond to drug treatment. Should investments in ART be targeted toward food-insecure populations in the absence of complementary nutrition or safe water interventions? Is it worth funding a large malaria intervention if most of the intended beneficiaries are likely to die of diseases related to hunger, or if stunting is so severe that population survival is likely to be curtailed by infectious diseases such as tuberculosis? Similarly, in an area with chronic micronutrient and protein deficiencies, is it worth funding a program to install shared community water points that do not help households grow crops in the long dry season, and that confer large time and physical burdens upon women and girls?

Governments, development agencies, or even individual communities cannot tackle all of these challenges at once. As clearly described by Falcon in Chapter 2, solving these problems at the early stages of development requires careful attention to sequencing. History has shown that grand integrated development plans have failed time and time again (Easterly 2009), and it seems clear that at best, two to three major initiatives can be successful at one time. Otherwise, competing agencies end up in power struggles, funds run out, the administrative and intellectual capacity of agencies is too limited to engage numerous problems in tandem, and there is rarely the buy-in that is needed to support these programs over time. Instead, the keys to success seem to be figuring out the right "first steps," and coordinating efforts across ministries or agencies to ensure that progress in one area is not thwarted by inaction or even counterproductive activities in another.

A first-order question remains: Where should this coordination take place, and by whom? Should the coordination be centralized within the office of the president, prime minister, or perhaps a finance minister who holds most of the cards in the policymaking process? Or should the coordination take place at the local level under the leadership of a governor or provincial chief? A long history of development experience suggests that, although establishing semi-autonomous regional bodies for coordination and action can be effective initially, this model is not sustainable once foreign donors go home. Whether the overall approach is centralized or decentralized, the relevant ministries must typically be meaningfully engaged in the process to ensure long-run development success.

So what are the best "first steps" to intervention? The answer will always depend to a large extent on local conditions, yet there appear to be a few good candidates that should be considered in any situation. First, investments in local health and family planning clinics, especially those targeting children two years of age or younger, as well as pregnant and lactating women, are important for reducing rates of stunting and acute malnutrition. For example, Indonesia's major success in improving food security (described in Chapter 2) depended not only on investments in agriculture, but also on the establishment of rural health and family planning clinics throughout the country. Second, bringing water closer to households is critical for women's time management, food production, and health. Even small incremental volumes of water can have a high payoff when used to irrigate vegetables and fruits with high micronutrient content, or to raise livestock with high protein content. Safe water for drinking and washing, and a diet rich in protein and micronutrients, will play a central role in preventing childhood morbidity and mortality from diarrheal illness and respiratory infections.

Irrespective of the starting point, it is clear that food security for the poorest billion requires investments beyond agriculture alone. Food insecurity for many households is not just about food on the table. It is also about the struggle against HIV, the hours spent collecting water that is still not safe to drink, and the missed opportunities for stunted children whose malnourishment originated in the womb.

References

Agency, N. A. C. 2008. Botswana Campaign on MCP. UNAIDS/Soul City Regional Meeting on Multiple Concurrent Partnerships, Johannesburg, South Africa.

Smith, L. C., H. Alderman, and D. Aduayom. 2006. Food Insecurity in Sub-Saharan Africa: New Estimates from Household Expenditure Surveys. Research Report 146. Washington, DC: International Food Policy Research Institute.

Arnold, B. F., and J. M. J. Colford. 2007. Treating water with chlorine at point-of-use to improve water quality and reduce child diarrhea in developing countries: A systematic review and meta-analysis. *The American Journal of Tropical Medicine and Hygiene* 76(2): 354–364.

Bannerjee, A. and E. Duflo. 2011. *Poor Economics: A Radical Rethinking of the Way to Fight Global Poverty*. New York: Public Affairs (Perseus Book Group).

Batliwala, S. 1982. Rural energy scarcity and nutrition: A new perspective. *Economic and Political Weekly* 17(9): 329–333.

Berio, A. J. 1984. The analysis of time allocation and activity patterns in nutrition and rural development planning. *Food and Nutrition Bulletin* 6(1): 53–68.

Black, R. E., et al. 2010. Global, regional, and national causes of child mortality in 2008: A systematic analysis. *Lancet* 375(9730): 1969–1987.

Boschi-Pinto, C., L. Velebit, and K. Shibuya. 2008. Estimating child mortality due to diarrhea in developing countries. *Bulletin of the World Health Organization* 86(9): 710–717.

Brinkhof, M., M. Pujades-Rodriguez, and M. Egger. 2009. Mortality of patients lost to follow-up in antiretroviral treatment programmes in resource-limited settings: Systematic review and meta-analysis. *PLOS One* 4(6): e5790.

Brown, K. H. 2003. Diarrhea and malnutrition. *The Journal of Nutrition* 133(1): 328S–332S.

Burney, J. A., R. L. Naylor, and S. L. Postel. 2013. The case for distributed irrigation as a development priority in sub-Saharan Africa. *Proceedings of the National Academy of Sciences of the United States of America*.

Burney, J. A., and R. L. Naylor. 2012. Smallholder irrigation as a poverty alleviation tool in Sub-Saharan Africa. *World Development* 40(1): 110–123.

Carter, M. W., J. M. Kraft, T. Koppenhaver, C. Galavotti, T. H. Roels, P. H. Kilmarx, and B. Fidzani. 2007a. A bull cannot be contained in a single Kraal: Concurrent sexual partnerships in Botswana. *AIDS and Behavior* 11(6): 822–830.

Clark, R. 2005. Sex differences in antiretroviral therapy-associated intolerance and adverse events. *Drug Safety* 28(12): 1075–1083.

Curtis, V. and S. Cairncross. 2003. Effect of washing hands with soap on diarrhoea risk in the community: A systematic review. *The Lancet Infectious Diseases* 3(5): 275–281.

Davis, J., R. Hope, and S. Marks. 2011. Assessing the link between productive use of domestic water, poverty, and sustainability. Report to the World Bank.

Devereux, S. 2009. Why does famine persist in Africa? *Food Security* 1(1): 25–35.

Dewey, K. G., and D. R. Mayers. 2011. Early child growth: How do nutrition and infection interact? *Maternal & Child Nutrition* 7: 129–142.

Drimie, S. 2002. The impact of HIV/AIDS on rural households and land issues in Southern and Eastern Africa. A Background Paper prepared for the Food and Agricultural Organization, Sub-Regional Office for Southern and Eastern Africa. Pretoria, South Africa: Human Sciences Research Council.

Easterly, W. 2009. Can the West save Africa? *Journal of Economic Literature* 47(2): 373–447.

Ejemot R., J. Ehiri, M. Meremikwu, J. Critchley. 2008. Hand washing for preventing diarrhoea. *Cochrane Database System Reviews*.

Food and Agriculture Organization of the United Nations (FAOSTAT). 2012. "FAO statistical databases." http://faostat.fao.org.

Fewtrell, L., R. B. Kaufmann, D. Kay, W. Enanoria, L. Haller, and J. M. Colford Jr. 2005. Water, sanitation, and hygiene interventions to reduce diarrhoea in less developed countries: A systematic review and meta-analysis. *The Lancet Infectious Diseases* 5(1): 42–52.

Fox, M. P., and S. Rosen. 2010. Patient retention in antiretroviral therapy programs up to three years on treatment in sub-Saharan Africa, 2007–2009: Systematic review. *Tropical Medicine & International Health* 15: 1–15.

Friedrich, M. 2007. Jeffrey Sachs, PhD: Ending extreme poverty, improving the human condition. Interview by MJ Friedrich. *Journal of the American Medical Association* 298(16): 1849–1851.

Gourvenac, D., N. Taruberekera, O. Mochaka, and T. Kasper. 2007. Multiple concurrent partnerships among men and women aged 15–34 in Botswana. Gaborone, Botswana, PSI-Botswana.

Guerrant, D. I., S. R. Moore, A. A. Lima, P. D. Patrick, J. B. Schorling, and R. L. Guerrant. 1999. Association of early childhood diarrhea and cryptosporidiosis with impaired physical fitness and cognitive function four to seven years later in a poor urban community in northeast Brazil. *American Journal of Tropical Medicine and Hygiene* 61(5): 707–713.

Guest, E. 2003. Children of AIDS: Africa's orphan crisis. London: Pluto Press.

Humphrey, J. 2009. Child undernutrition, tropical enteropathy, toilets, and handwashing. *The Lancet* 374: 1032–1035.

Headey, D., and O. Ecker. 2013. Rethinking the measurement of food security: From first principles to best practice. *Food Security* 5(3): 327–343.

Kalichman, S. C., D. Ntseane, K. Nthomang, M. Segwabe, P. Phorano, and L. C. Simbayi. 2007. Recent multiple sexual partners and HIV transmission risks among people living with HIV/AIDS in Botswana. *Sexually Transmitted Infections* 83(5): 371–375.

Kenya DHS Final Report. 2008. Demographic and Health Surveys. ICF International. Accessed December 3, 2012, http://www.measuredhs.com.

Kosek, M., C. Bern, and R. Guerrant. 2003. The global burden of diarrhoeal disease, as estimated from studies published between 1992 and 2000. *Bulletin of the World Health Organization*.

Lagarde, E., B. Auvert, M. Caraël, M. Laourou, B. Ferry, E. Akam, T. Sukwa, L. Morison, B. Maury, J. Chege, I. N'Doye, A. Buvé, and The Study Group on Heterogeneity of HIV Epidemics in African Cities. 2001. Concurrent sexual partnerships and HIV prevalence in five urban communities of sub-Saharan Africa. *AIDS* 15(7): 877–884.

Larson, B. A., M. P. Fox, M. Bii, S. Rosen, J. Rohr, D. Shaffer, F. Sawe, M. Wasunna, and J. L. Simon. 2012. Antiretroviral therapy, labor productivity, and gender: A longitudinal cohort study of tea pluckers in Kenya. *AIDS* 27(1): 115–123.

Larson, B. A., M. P. Fox, S. Rosen, M. Bii, C. Sigei, D. Shaffer, F. Sawe, K. McCoy, M. Wasunna, and J. L. Simon. 2009. Do the socioeconomic impacts of antiretroviral therapy vary by gender? A longitudinal study of Kenyan agricultural worker employment outcomes. *BMC Public Health* 9(1): 240.

Lawrence, M., J. Singh, F. Lawrence, and R.G. Whitehead. 1985. The energy cost of common daily activities in African women: increased expenditure in pregnancy? *The American Journal of Clinical Nutrition* 42(5): 753–763.

Liu, L., H. L. Johnson, S. Cousens, J. Perin, S. Scott, J. E. Lawn, I. Rudan, H. Campbell, R. Cibulskis, M. Li, C. Mathers, and R. E. Black. 2012. Global, regional, and national causes of child mortality: An updated systematic analysis for 2010 with time trends since 2000. *The Lancet* 379(9832): 2151–2161.

Lopez, A., C. Mathers, M. Ezzati, D. Jamison, and C. Murray. 2006. Global and regional burden of disease and risk factors, 2001: Systematic analysis of population health data. *The Lancet* 367(9524): 1747–1757.

Lunn, P. 2000. The impact of infection and nutrition on gut function and growth in childhood. *Proceedings of the Nutrition Society* 59: 147–154.

Luong, D. H., S. Tang, T. Zhang, and M. Whitehead. 2007. Vietnam during economic transition: A tracer study of health service access and affordability. *International Journal of Health Services* 37(3): 573–588.

Martorell, R., C. Yarbrough, S. Yarbrough, and R. E. Klein. 1980. The impact of ordinary illnesses on the dietary intakes of malnourished children. *The American Journal of Clinical Nutrition* 33(2): 345–350.

Mills, E. J., J. B. Nachega, I. Buchan, J. Orbinski, A. Attaran, S. Singh, B. Rachlis, P. Wu, C. Cooper, L. Thabane, K. Wilson, G. H. Guyatt, and D. R. Bangsberg. 2006. Adherence to antiretroviral therapy in sub-Saharan Africa and North America. *Journal of the American Medical Association* 296(6): 679–690.

Monforte, A. A., A. C. Lepri, G. Rezza, P. Pezzotti, A. Antinori, A. N. Phillips, G. Angarano, V. Colangeli, A. DeLuca, G. Ippolito, L. Caggese, F. Soscia, G. Filice, F. Gritti, P. Narciso, U. Tirelli, and M. Moroni. 2000. Insights into the reasons for discontinuation of the first highly active antiretroviral therapy (HAART) regimen in a cohort of antiretroviral naive patients. *AIDS* 14(5): 499–507.

Moriarty, P., J. Butterworth, and B. van Koppen, eds. 2004. Beyond domestic. Case studies on poverty and productive uses of water at the household level. IRC Technical Paper Series, no. 41. Delft: IRC International Water and Sanitation Centre.

Niehaus, M. D., S. R. Moore, P. D. Patrick, L. L. Derr, B. Lontz, A. A. Lima, and R. L. Guerrant. 2002. Early childhood diarrhea is associated with diminished cognitive function 4 to 7 years later in children in a northeast Brazilian shantytown. *The American Journal of Tropical Medicine and Hygiene* 66(5): 590–593.

Norgan, N. G., et al. 1974. The energy and nutrient intake and the energy expenditure of 204 New Guinean adults. *Philosophical Transactions of the Royal Society of London. B, Biological Sciences* 268(893): 309–348.

Omran, A. R. 1971. The epidemiologic transition: a theory of the epidemiology of population change. *The Milbank Memorial Fund Quarterly* 509–538.

Panter-Brick, C. 1992. The energy cost of common tasks in rural Nepal: Levels of energy expenditure compatible with sustained physical activity. *European Journal of Applied Physiology and Occupational Physiology* 64(5): 477–484.

Parashar, U. D., J. S. Bresee, and R. I. Glass. 2003. The global burden of diarrhoeal disease in children. *Bulletin of the World Health Organization* 81(4): 236.

Pickering, A., and J. Davis. 2012. Freshwater availability affects child health in sub-Saharan Africa. *Environmental Science & Technology*.

Pickering, A. J., J. Davis, S. P. Walters, H. M. Horak, D. P. Keymer, D. Mushi, R. Strickfaden, J. S Chynoweth, J. Liu, A. Blum, K. Rogers, and A. B. Boehm. 2010. Hands, water, and health: Fecal contamination in Tanzanian communities with improved, non-networked water supplies. *Environmental Science & Technology* 44(9): 3267–3272.

Pickering, A. J., T. R. Julian, S. J. Marks, M.C. Mattioli, A. B. Boehm, K. J. Schwab, and J. Davis. 2012. Fecal contamination and diarrheal pathogens on surfaces and in soils among Tanzanian households with and without improved sanitation. *Environmental Science & Technology* 46(11): 5736–5743.

Pronyk, P. 2012. Errors in a paper on the Millennium Villages project. *The Lancet* 379(9830): 1946.

Pronyk, P. M., M. Muniz, B. Nemser, M. Somers, L. McClellan, C. A. Palm, U. K. Huynh, Y. B. Amor, B. Begashaw, J. W. McArthur, A. Niang, S. E. Sachs, P. Singh, A. Teklehaimanot, J. D. Sachs. 2012. The effect of an integrated multisector model for achieving the Millennium Development Goals and improving child survival in rural sub-Saharan Africa: a non-randomised controlled assessment. *The Lancet* 379 (9832): 2179–2188.

Qamar, M. K. 2003. Facing the challenge of an HIV/AIDS epidemic: Agricultural extension services in sub-Saharan Africa. Rome: Food & Agricultural Organization.

Rao, S., M. Gokhale, and A. Kanade. 2008. Energy costs of daily activities for women in rural India. *Public Health Nutrition* 11(02): 142–150.

Rosen, S., and M. P. Fox. 2011. Retention in HIV care between testing and treatment in sub-Saharan Africa: A systematic review. *PLoS Medicine* 8(7): e1001056.

Rosen, S., M. Ketlhapile, I. Sanne, and M.B. DeSilva. 2007. Cost to patients of obtaining treatment for HIV/AIDS in South Africa. *South African Medical Journal* 97(7): 524–529.

Rosen, S., and J. R. Vincent. 1999. Household water resources and rural productivity in sub-Saharan Africa: A review of the evidence. Cambridge, MA: Harvard Institute for International Development, Harvard University.

Sachs, J., J. W. McArthur, G. Schmidt-Traub, M. Kruk, C. Bahadur, M. Faye, and G. McCord. 2004. Ending Africa's poverty trap. *Brookings Papers on Economic Activity* 2004(1): 117–240.

Smith, L., H. Alderman, and D. Aduayom. 2006. Food insecurity in sub-Saharan Africa: New estimates from household expenditure surveys. Research Report 146. Washington, DC: International Food Policy Research Institute.

Ulimwengu, J. 2009. Farmers' health and agricultural productivity in rural Ethiopia. *African Journal of Agricultural and Resource Economics* 3(2):83–100.

UN. 2010. The Millennium Development Goals Report. New York: United Nations Department of Economic and Social Affairs.

UNAIDS. 2001. HIV/AIDS, food security and rural development. Fact sheet for the United Nations special session on HIV/AIDS.

UNAIDS. 2012. Global report: UNAIDS report on the global AIDS epidemic.

United Nations Development Programme (UNDP). 2012. African Human Development Report 2012: Towards a food secure future. New York: UNDP Regional Bureau for Africa.

United Nations Human Settlements Program (UN-HABITAT). 2005. Kisumu City Development Strategies, 2004–2009.

Waddington, H., and B. Snilstveit. 2009. Effectiveness and sustainability of water, sanitation, and hygiene interventions in combating diarrhoea. *The Journal of Development Effectiveness* 1(3): 295–335.

Walker, C. L. F., Rudan, I., Liu, L., Nair, H., Theodoratou, E., Bhutta, Z. A., et al. 2013. Global burden of childhood pneumonia and diarrhoea. *Lancet* 381(9875): 1405–1416.

Weiser, S. D., K. A. Fernandes, E. K. Brandson, V. D. Lima, A. Anema, D. R. Bansberg, J. S. Montaner, and R. S. Hogg. 2009. The association between food insecurity and mortality among HIV-infected individuals on HAART. *AIDS Journal of Acquired Immune Deficiency Syndrome* 52(3): 342.

Weiser, S. D., M. Heisler, K. Leiter, F. Percy-de-Korte, S. Tlou, S. DeMonner, N. Phaladze, D. R. Bansberg, and V. Iacopino. 2006. Routine HIV testing in Botswana: a population-based study on attitudes, practices, and human rights concerns. *PLOS Med* 3(7): e261.

White, G. F., D. J. Bradley, A. U. White. 1972. *Drawers of water*. University of Chicago Press.

World Health Organization. 2010a. (The) Right to Water, Fact Sheet No. 35. United Nations, OHCHR, UN-HABITAT.

———. 2010b. Antiretroviral therapy for HIV infection in adults and adolescents: Recommendations for a public health approach: 2010 revision. Geneva: World Health Organization, 1–359.

8

Land Institutions and Food Security in Sub-Saharan Africa

Whitney L. Smith and Rosamond L. Naylor

In sub-Saharan Africa (SSA), food insecurity is often associated with low productivity in staple crop systems dominated by smallholder farmers. Within this agricultural context, the region's food insecurity is as varied as the individual histories of the SSA countries. It differs substantially from that in Latin America and Asia, two regions that experienced more substantial investments in infrastructure, irrigation, high-yielding seed varieties, and synthetic fertilizers, and more use of Green Revolution technologies since the early 1970s (see discussions by Falcon in Chapter 2, Burney in Chapter 6, and Vitousek and Matson in Chapter 10). Sustained food insecurity in SSA can be explained in part by the region's heterogeneous agro-ecosystems and fractured groundwater resources, but poor governance and a history of violent conflict stemming from SSA's colonial legacy are also at fault. Increasing the amount and quality of food produced per hectare, in environmentally and economically sustainable ways, is a necessary precondition for improving food security and human health in this uniquely insecure region.

Why is history so important for understanding Africa's food insecurity? The continent's colonial heritage has left many now-independent nations with a complex set of land institutions governing land tenure and property rights that affect agricultural productivity, economic growth, and rural income distribution. These institutions include both customary (derived from traditional tribal customs) and statutory (derived from national legislation) laws that are rarely compatible (Bassett and Crummey 1993; Binswanger et al. 1995). Throughout most of SSA, the two sets of laws do not recognize each other and often permit dissimilar and conflicting rights to land.

As a result of these conflicting systems of land governance, two major paradoxes have emerged in the context of sub-Saharan Africa's rural development and food security. First, the region is recognized as having the world's largest available and under-utilized land area for agricultural expansion and improvement—yet a growing share of farmers live in densely settled areas where land area per capita is declining and where farm productivity is not sufficient to stave off hunger (Jayne et al. 2014). Much of the under-utilized land remains under government control and is not accessible to smallholders, or it lacks the necessary water or transportation infrastructure for profitable expansion by smallholders. The second paradox concerns the purchase of available land in large tracts by both domestic and foreign investors, a practice that is increasing. This escalating demand for land since 2005, as determined by reported land acquisitions, signed contracts, and projects in production, has been significantly higher in countries where land institutions are weak (Arezki et al. 2011; Anseeuw et al. 2012b).

The distribution of under-utilized land in SSA will impact long-term agricultural productivity, supply-chain development, rural employment, and smallholder farm incomes. How land is distributed and used will also strongly influence food security, and pre-conditions for conflict, throughout the region. The greater the inequality in assets, such as land, the more difficult it is for the poor to participate in the economic growth process (as reviewed by FAO 2013).

This chapter explores the issues of land rights, land distribution, and food security in sub-Saharan Africa through a legal and economic lens. Broadly speaking, land law is at the heart of land tenure issues. Legal institutions such as property rights, land statutes, customary tenure principles, and land management systems determine people's ability to acquire and benefit from land. Legal analysis is thus helpful for understanding the institutional weaknesses that create tenure insecurity, and also for assessing whether the applicable land law, that creates and reinforces these weaknesses, might also be used to attempt to remedy them.

We begin by discussing the potential for improving rural incomes and food security in SSA by developing large tracts of high productivity land. Then we turn to the topic of large land acquisitions—commonly dubbed "land grabs" by the media—by foreign and national interests. A framework for understanding these emerging trends is then presented, first with a wide-angle view of the legal context in which land institutions operate, and then with a more detailed review of land legislation in two countries. Mozambique and Zambia illustrate two different approaches to and outcomes from institutional reform in land markets. The main objectives of our analysis are to examine the nature of land laws that form a shaky foundation for current land acquisitions, and to assess, from a legal point of view, the success or failure of attempts to remedy those weaknesses.

Farmland Distribution and Food Security

Two trends currently dominant sub-Saharan Africa's land markets: (1) a decline in per capita land holdings among smallholder communities under the jurisdiction of customary land laws; and (2) a rising number of large-scale land acquisitions approved by governments via statutory law. Most SSA countries have experienced a steady decline in average farm size since the 1960s, with arable land per person in agriculture falling as low as 2–4 hectares in Ethiopia, Zambia, Kenya, Mozambique, Malawi, and Uganda (Jayne et al. 2014). High fertility rates within rural communities, and the division of fixed land holdings among family members as generations proceed, have been largely responsible for this trend. Although the mean farm size in SSA tends to be larger than in Asia, yields of staple crops such as maize are substantially lower. Moreover, significant disparities in landholding size persist even within the smallholder sector. In southern and eastern SSA, most households now operate on less than one hectare of land and do not produce enough food to meet their nutritional needs. At least half of the households on the smaller plots are net purchasers of grain, and family members often suffer from chronic hunger during much of the year (as described by Burney in Chapter 6). Meanwhile the top 5 percent of smallholders, those with access to relatively larger and more fertile plots, can grow a surplus and contributing to over half of the marketed maize surplus throughout the region (Jayne et al. 2010). Even those smallholders who have sufficient income to enlarge their farms often fail to do so. There is a widespread perception that additional land cannot be purchased under customary law, even when vacant arable land is in sight (as reviewed in Jayne et al. 2014).

Conflicts between governments and traditional authorities over rights to customary land have occurred throughout SSA since independence. These conflicts, while not new, have escalated since 2005 as global food prices and land values have risen (Binswanger et al. 1995; Herbst 2000; Deininger and Byerlee 2011; Kugelman and Levenstein 2012). In recent years, large tracts of land have been allocated to national elites who have no customary claims, and foreign entities have acquired long-term leases from governments for land that was previously under customary control. Many governments have justified these land transactions by claiming that new entrepreneurs, emergent farmers, and large-scale farming systems are needed to improve agricultural productivity and food security.

There are two sides to this story. The positive view of large-scale agriculture focuses on the vast areas of under-utilized and potentially high-productivity land, including land within the Guinea Savannah (Figure 8.1) (World Bank 2009). This agro-ecosystem is comparable to the Brazilian *cerrado* (native grasslands), where agriculture has expanded dramatically in recent decades through large-scale, mechanized, and high-input practices. It is also comparable to the northeast region of Thailand, where erratic and unreliable, though abundant, rainfall and poor soils characterize the physical landscape. Nonetheless, agricultural production has also expanded there in the past 30 years, but in that case through smallholder systems.

The critical policy question for sub-Saharan Africa is whether the Guinea Savannah can be developed in a way that both enhances global food security

FIGURE 8.1 Map of the Guinea Savannah (shaded in light grey)

by increasing aggregate food production (availability), and improves regional food security by boosting rural employment and incomes (access). By attracting large-scale investors into the Guinea Savannah and other potentially productive regions of SSA, governments have the opportunity to reverse the long-term decline in agricultural investments that has thwarted rural economic growth. The vision of some development experts is that commercial enterprises will be able to develop supply chains, employ unskilled labor, and create technological spillovers in regions where small-scale farming has historically failed to provide food security (Collier and Dercon 2013).

The negative side of the story is that such investments may take land away from smallholders, widen income disparities, and leave many households more destitute than they were prior to the new land deals (von Braun and Meinzen-Dick 2009; Anseeuw et al. 2012a; Oxfam 2012). Weak land institutions and a lack of tenure security in many SSA countries make it easy for governments to appropriate land for which there is no formal legal title, even if it is already occupied. As such, governments that are eager to attract foreign investment may dispossess local communities of their traditionally held lands; lacking formally recognized land rights, such communities are without legal recourse to substantiate and enforce their claims (Anseeuw et al. 2012a). This outcome can lead to greater rural poverty, unless investors or government agencies compensate communities for the land or the investment generates significant alternative employment. Without access to land for growing food—or without use of communal lands for hunting, gathering, and grazing—household, local, and even national food security may be severely jeopardized by land acquisitions.

The Extent and Location of Large-scale Land Acquisitions

Since the turn of the twenty-first century, the number of large-scale farmland acquisitions has risen dramatically across the globe. The Land Matrix Partnership,[1] a major research and data-gathering effort to verify and analyze data on land deals internationally, defines "large-scale" as any transaction that involves at least 200 hectares (ha) of land (Anseeuw et al. 2012b; Land Matrix 2013). According to their database, 1,071 potential or actual large-scale agricultural land deals were reported from 2000 through June 2013, covering almost 49 million ha of land in developing countries.[2] This figure includes 755 concluded deals (written or oral agreement), 145 intended deals (interest expressed or under negotiation), 50 failed deals (either contract cancelled or negotiations terminated), and 121 additional reported deals for which no reliable information is available (Land Matrix 2013). Yearly data on large-scale land acquisitions are difficult to obtain. However, it appears that growth in land acquisitions was generally low until 2005, and then began to surge, reaching its peak in 2009 and then decreasing again between 2010 and 2012 (Deininger and Byerlee 2011; Anseeuw et al. 2012a and 2012b).

The majority of nations hosting land acquisitions—also called *target countries*—are developing countries in sub-Saharan Africa, Latin America, Central Asia, Eastern Europe, and Southeast Asia with large tracts of land perceived

---

[1] The Land Matrix is an online public database of large-scale land deals facilitated by a partnership among the International Land Coalition (ILC), The Centre for Development and Environment (CDE) at the University of Bern, the Centre de Coopération Internationale en Recherche Agronomique pour le Développement (CIRAD), the German Institute of Global and Area Studies (GIGA), and the Gesellschaft für Internationale Zusammenarbeit (GIZ). The aim of the Land Matrix project is to promote transparency and open data in decision making over land and investment. For information about the project and access to the Land Matrix database, see http://landmatrix.org/ (accessed June 14, 2013).

[2] Data accuracy and reliability have been major issues in media reports of "land grabbing." Despite its efforts to address the lack of reliable data on large-scale land acquisitions, the Land Matrix itself has experienced widespread criticism of its accuracy and reliability. The Land Matrix cautions that its data are provisional and dynamic, as the group depends on reporting and feedback from the larger academic and civil service communities. For a discussion of the development of the Land Matrix and its challenges, see Anseeuw et al. (2013).

to be under-utilized and/or available for agricultural development (Deininger and Byerlee 2011). Sub-Saharan Africa is the prime target for land deals globally, with almost half (48 percent) of all 950 large-scale agricultural land transactions reported in the Land Matrix Database for which reliable information exists (Land Matrix 2013). Asia accounts for 37 percent, followed by Latin America at about 8 percent. The concentrated activity in sub-Saharan Africa may be a function of media and organizational interest, and therefore more reporting, in this region. It may also reflect investors' perception—often borne out—that large land areas are "available" and can be acquired at very low cost from the governments (Schoneveld 2011; Anseeuw et al. 2012b).

The investors, or parties acquiring the land, include governments, private sector entities such as agribusinesses or investment funds, and public-private partnerships, any of which may be foreign or domestic (Deininger and Byerlee 2011). The nations where these entities are based are referred to as *origin countries*. Although the common perception is that wealthy foreign governments or enterprises from the United States, EU, Gulf States, or East Asia have been racing against each other to buy up farmland in low-income countries, the reality is that the target countries are also involved. Intra-regional transactions within SSA are on the rise, and local elites from African countries are increasingly involved in large-scale land acquisitions (Kugelman 2012; Anseeuw et al. 2012a).[3]

A major share of large-scale acquisitions has occurred in countries where governance is poor and where land institutions are weak, suggesting that the demand for land is significantly higher in settings where local land rights are not well protected. A 2011 report by the International Monetary Fund (IMF) on the determinants of global land acquisitions confirms this pattern; in particular, it demonstrates a highly significant and negative correlation between the number of land deals and target countries' quality of land governance and recognition of local rights (Arezki et al. 2011). Table 8.1 highlights several interesting results from this report. First, of the 84 target countries in the study, only half (43 countries) had projects actually implemented on the acquired land. In the remaining countries, the acquired land has remained idle. Second, 56 of the 84 target countries were also origin countries, meaning that they were a source of investment capital as well as a host country for investment. Only 23 nations in the study were *origin-only countries*, meaning that they acquired land in other countries but were not target countries themselves for land acquisitions. Finally, nations that were targets for land deals had significantly lower land tenure security than countries in the origin-only grouping.[4]

---

[3] In nations where domestic investments in large-scale land deals occur, the origin country is also a target country.

[4] Arezki et al. (2011) developed their land tenure security index using land governance data compiled by the French Development Agency (AFD) and the French Ministry for the Economy, Industry and Employment (MINEIE). Key variables of the land tenure security index include: the recognition

What is especially interesting about the 2011 IMF study is that the rankings of land tenure security were strongly correlated with governance rankings. The quality of a country's governance can be measured in a number of ways, including: regulatory quality, rule of law, control of corruption, political stability and absence of violence, and voice and accountability. These measures are typically expressed in percentage rankings within a larger study population.[5] In target countries, the governance indicators were generally weak (34 to 40 percent), whereas in origin-only countries, the rankings were much higher (64 to 80 percent) (Table 8.1).

From a food security perspective, the implications of land acquisitions under poor governance and land tenure structures can be serious when smallholder communities or pastoralists are displaced from their land without adequate compensation or long-term employment alternatives. Data on former land use are scarce and difficult to obtain. For the relatively small number of cases that have been reported, however, it appears that a significant share of the land traded in large-scale acquisitions (50 percent of reported land deals and 40 percent of land area) was formerly in the hands of smallholder farmers and pastoralists (Table 8.2) (Land Matrix Newsletter 2013).

Why Are There so Many Land Deals?

Given SSA's long and complicated history of land institutions and governance, what can account for the sudden surge in large-scale land transactions? And why might foreign entities, in particular, be willing to take large risks of acquiring or leasing agricultural land within a weak institutional setting? There are several possible explanations, but the main reason is that, while high risk investments, the potential payoffs to agricultural land investments have escalated since the mid-2000s, while the returns on many financial instruments have declined since the global financial crisis in 2008. Increased global demands for

---

of local land rights (even if not formalized); the existence of a land policy; and levels of land-related conflict. Principal component analysis was used to create the index, with the first component used as an indicator of overall tenure security. The index was created on the basis of data for 215 total countries, and low values reflect high tenure insecurity. The AFD database is published as the 2009 Institutional Profiles Database, available at http://www.cepii.com/anglaisgraph/bdd/institutions.htm (accessed June 20, 2013).

[5] Governance rankings on Regulatory Quality, Rule of Law, Control of Corruption, Political Stability and Absence of Violence, and Voice and Accountability are from the World Bank's Worldwide Governance Indicators database, available at http://info.worldbank.org/governance/wgi/index.asp. The Worldwide Governance Indicators report on these five broad dimensions of governance for more than 200 countries during the period 1996–2011. For example, rankings for the United States span from 64 percent (Political Stability/Absence of Violence) to 92 percent (Regulatory Quality). At the other end of the spectrum, rankings for Zimbabwe range from a low of 0.9 percent (Rule of Law) to a high of 16 percent (Political Stability and Absence of Violence).

TABLE 8.1

Institutional Indicators for Countries with Reported Land Deals[a]

Variable	Origin Countries[b]	Origin Only Countries[c]	Target Countries[d]	Target w/ Implemented Projects[e]
No. Observations (countries)	56	23	84	43
Land Tenure Security Index[f]	.61	2.15	−.98	−.95
Governance Rankings[g]	50%–63%	64%–80%	34%–39%	35%–40%

[a] This table is adapted from a study by the International Monetary Fund focusing on a set of countries with reported land deals throughout the world (Arezki et al. 2011).

[b] *Origin countries* represent the subset of countries that acquire land in other countries. They may also be targets of land acquisition.

[c] *Origin-only countries* are the subset of countries from which land deals originate from resident investors, but which are not targets for land acquisitions themselves.

[d] *Target countries* are those countries that are reported to be the hosts of investment in land.

[e] *Target countries with implemented projects* are those target countries that are hosts of the investment and have at least one project actually in production, as opposed to being only the target of a reported investment.

[f] See footnote 4.

[g] See footnote 5.

TABLE 8.2

Former Land Use in 755 Concluded Deals (2000–June 2013)

Former Land Use	No. of Concluded Deals	Size ('000 ha)
Commercial (large-scale) Ag	42	1,180
Smallholder Agriculture	56	1,041
Pastoralists	6	397
Forestry	15	746
Conservation	5	152
No information for the deal	631	29,034
Total	755	31,370

Source: Author's creation using Land Matrix Database 2013; Land Matrix Newsletter 2013.

food and animal feed have continued to rise with population growth, income growth, and the associated transition toward more diversified (meat-based) diets. In addition, widespread attention to energy security and associated policies encouraging the use of crop-based biofuels have created a much higher demand for agricultural products that serve as feedstocks for the industry, such as maize (corn), sugar, rapeseed (canola), and soybeans (as discussed by Lobell and colleagues in Chapter 9). Despite continued (but variable) growth in supply, real, inflation-adjusted prices for staple foods, livestock feeds, and biofuel feedstocks rose sharply in real terms between 2006 and mid-2008, and have since fluctuated at historically high levels (Naylor and Falcon 2010, 2011).

With such patterns of demand dominating the world food system, expected returns to agricultural land, and hence land values, have risen to all-time highs. Moreover, the dramatic increase in agricultural prices between 2006 and mid-2008 has underscored the vulnerability of many import-dependent countries to price and supply shocks (Naylor and Falcon 2008). Some of these nations began acquiring more land abroad for agricultural use with the intention of exporting the crops produced for consumption at home; in so doing, they sought to escape the economic impacts of volatile world markets (von Braun and Meinzen-Dick 2009; Kugelman 2012).

Global financial groups have also turned toward land and land-based instruments as a relatively high-yielding, long-term investment. Weak financial markets worldwide have caused private sector financiers to look beyond the bond and equity markets for lucrative investments. Agricultural land has emerged as an object of speculation, especially given the expectation of rising prices that will result from rising demand for land (Kugelman 2012; Anseeuw et al. 2012a). In addition to these factors, large tracts of land have also been purchased in SSA for timber and other raw materials, for access to freshwater, and for purposes of carbon sequestration in compliance with Reducing Emissions from Deforestation and Forest Degradation policies and programs. How these multiple demands for land play out in the market—and what they mean for rural incomes, food security, and regional conflict in SSA—depends critically on the structure and functioning of land institutions in individual countries.

Conceptual Framework

A conceptual framework for understanding land institutions in sub-Saharan Africa is shown in Figure 8.2. At the heart of land acquisitions and land

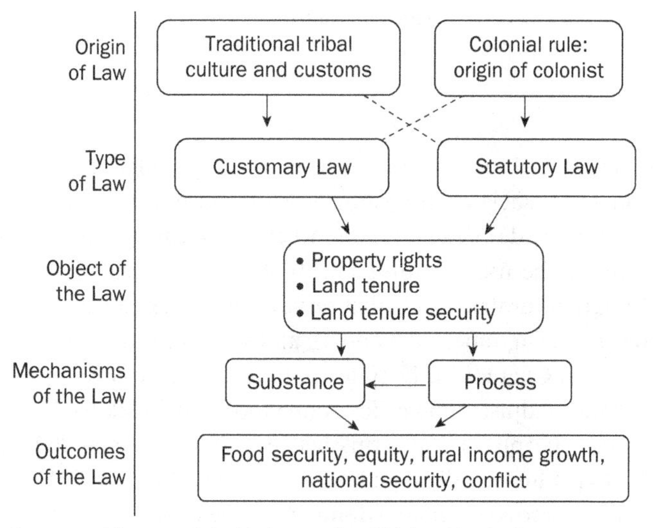

FIGURE 8.2 Conceptual Framework for Understanding SSA Land Institutions

Source: Author.

conflicts in SSA lies the issue of legal pluralism, defined as the co-existence of parallel, but separate, legal systems governing people's lives. In much of sub-Saharan Africa, statutory law and customary law have evolved together to create two different land tenure systems that often overlap and conflict with each other. Both systems define and circumscribe property rights in land, land tenure regimes, and land tenure security (Box 8.1). A key issue for food security

BOX 8.1

Legal Definitions of Property Rights,[6] Land Tenure, and Land Institutions

Legal scholars and practitioners typically rely on technical definitions and jargon that are confusing to many readers outside of the field. Some basic concepts and definitions are therefore helpful for understanding land conflicts and potential resolutions in a sub-Saharan African context.

Property rights are a social and legal construct, defined as "the interest, claim, or ownership that one has in tangible or intangible property" (Garner 1999). Commonly referred to as a "sticks in a bundle," each piece of property has its own specific bundle, or set of associated rights, such as the rights to possess, use, or exclude others from the property. Importantly, separate sticks in the same bundle may be held by different people, acquired in different ways, and held for different periods (FAO 2002). Property rights to land, or "land rights," can include the freedom to: occupy, build upon, or enjoy the land; bequeath, sell, lease, or grant the land or the rights to use the land to others; restrict others' access to the land; and use or extract natural resources located on or below the land (Knight 2010).

Land tenure refers generally to the system of rights and institutions that governs access to and use of land and other resources (Maxwell and Wiebe 1998). Land tenure incorporates property rights to land, but also captures the bodies of rules and institutions that govern how land is held, managed, used, and transacted (Cotula 2007). In concrete terms, the system of land tenure determines who can come onto land, what can be done on the land, and with what spatial and temporal limitations.

Land tenure security connotes the degree of certainty that an individual or group's bundle of rights to land will be recognized by others, and protected in case the rights are challenged, for example in litigation (FAO 2002). The legal context in which the land tenure system operates creates tenure security; if the legal system consistently and effectively recognizes and enforces land rights, tenure security follows because people can be confident that their rights to land will be protected and not arbitrarily taken away (Knight 2010).

Land institutions encompass land rights, land laws, and customary norms, as well as the management regimes that administer them. Institutions are defined broadly, reaching beyond the traditional notion of organizations or agencies formed to achieve a common objective (Cousins 2000). Institutions may be thought of as the "rules of the game in a society. . . [and] perfectly analogous to the rules of the game in a competitive sport" (North 1990; see also Blocher 2006). Therefore, land institutions include groups such as government agencies, private sector entities, and educational centers concerned with land and agriculture. They also include the rules and practices, including laws and customary norms, which define and dictate land tenure principles.

[6] We use Black's Law Dictionary for definitions of basic legal concepts, for ease of reference.

in SSA is whether an integrated land tenure regime can be implemented with adequate substantive and procedural mechanisms to ensure proper function—and ultimately, to ensure fair and consistent rules to guide land access for the rural population, including smallholder farmers.

The Challenge of Legal Pluralism

The duality of customary and statutory laws extends far beyond rules governing land tenure and property rights. But the issue as it applies to land institutions is particularly important for rural income levels and distribution, agricultural productivity, and food security in sub-Saharan Africa. Legal duality can create uncertainty over which set of rules applies in a given circumstance, and, in the event of conflict, whose rights to land are valid and enforceable. It thus fosters significant unpredictability and insecurity with respect to land tenure (Knight 2010).

Statutory law is "the body of law derived from statutes," and statutes are "laws passed by a legislative body" (Black's Law Dictionary 1999). Accordingly, a statutory system of land tenure refers to the set of land-related rules contained within written, legislated laws enacted by a national government. Customary law, by comparison, is typically an unwritten "body of rules founding its legitimacy in tradition" (Cotula 2007). In the land tenure context, it is the locally defined system under which land and natural resources are held, managed, used, and transacted (Box 8.2). Customary rules and norms govern the actions of members of a community and are administered by customary authorities such as chiefs or headmen. Like other forms of law, these rules exist by virtue of their legitimacy in the eyes of community members (Knight 2010). Communities can be delineated in any number of ways, for example as a village, a clan, a tribe, or a lineage. Customary law is thus diverse, reflecting a wide range of cultural, religious, economic and social norms (Cotula 2007).

Statutory law is often referred to as the "formal" legal system, while customary law is referred to as "informal." This distinction is misleading, however, because customary law, though unwritten and highly variable over space and time, is the operative legal system used to adjudicate and resolve most land conflicts in rural sub-Saharan Africa. Indeed, it is often the only system recognized by much of the rural population (Alden 2008; Knight 2010). Moreover, the dichotomy between statutory law and customary law in SSA is not clean. Customary systems have changed over time under the influence of colonial and post-independence interference by courts and government. Likewise, government officials applying statutory law often use customary principles to settle disputes through mediation (Cotula 2007). Customary land institutions are also continuously evolving as they interact with changing environmental conditions, political-economy dynamics, and globalization (Cotula 2007; Knight

BOX 8.2

Characteristics of SSA Customary Land Tenure Systems

While customary land tenure systems vary widely across sub-Saharan Africa, several significant characteristics are shared across most systems. Examples include:

- Multiple sets of rights to the same land or resources may exist and overlap in time and space. For instance, herders and farmers may use the same land in different seasons, or men and women may cultivate the same plot with different crops during the same season (Lavigne Delville 2000).
- Rights may be individual and allocated to a specific bounded area, or they may be communal and shared among a particular group; customary systems provide for both forms of land tenure, therefore "customary" should not be equated to "communal" or "common property" in all cases (Alden 2008).
- While customary land rights are usually inheritable and may be given as gifts, many customary land tenure systems prohibit sales, especially to outsiders (Cotula 2007).
- Customary land tenure systems may be inequitable, disadvantaging women or otherwise discriminating based on age, social status, or type of land use (Cotula 2007). Moreover, the customary system is just as vulnerable as any to capture by a few powerful interests, and thus must not be idealized as a necessarily more "fair" system.

2010). In this sense, customary law is analogous to the common law of the United States or Great Britain, which changes as judges reinterpret sets of facts in light of the current social or political context (Knight 2010).

HISTORY AND CONTEXT

Legal pluralism has its roots in the history of colonization on the African continent. Most colonial governments brought property rights systems and land laws from their native countries to their newly acquired African territories; in the process, they often marginalized or replaced customary land tenure systems that were in place at the time (Toulmin and Quan 2000). This denial of customary rights to land, along with the imposition of European property rights and land laws, expedited the expropriation of land for European settlers.

Colonial powers largely ignored customary land tenure systems that were inconvenient to their settlement or resource extraction goals. However, they also often co-opted for their own use other aspects of customary law that furthered their interests (McAuslan 2000). For example, colonizers chose to overlook evidence of strong individual and family claims to land, and instead promoted the idea that all land was held communally by the state, and was thus potentially available for settlement (Knight 2010). Colonial officials relied on traditional chiefs to administer and adjudicate land policy and law, often manipulating the distribution of land to the colonialists' favor. Over time, customary land tenure principles were redefined through European legal constructs, and permanently altered as a result. As summed up by the African historian McAuslan (2000): "Customary law in fact long ago ceased to be part of traditional society—a bulwark against the colonial authorities—and became, instead, part of the colonial apparatus of rule."

The historical clash of legal systems and the subjugation of customary law eventually led many sub-Saharan African governments to nationalize all land within their borders upon gaining independence. Nationalization made land the official property of the state, to be held in trusteeship on behalf of all citizens (Toulmin and Quan 2000). Nascent governments in sub-Saharan Africa nationalized land in order to secure absolute authority over this most fundamental of resources, and its allocation and development—recognizing, as the European colonial powers did, that land was a vital source of political power and control (Knight 2010). Thus in one sense the nationalization of land mirrored the colonial patterns and entrenched the existence of two parallel land tenure systems—one centralized in the national government and the other reflecting and guiding the reality of land tenure and administration at a local level.

As trustees of the land, sub-Saharan African governments became obligated to ensure access and to promote agricultural development for their rural populations. They have not always proven to be trustworthy guardians. Many governments have struggled to administer effective land tenure policies and have ignored the interests of smallholder farmers who have little political power. With encouragement from international development organizations such as the World Bank, many sub-Saharan African governments embarked in the late twentieth century on efforts to create "modern" national land legislation intended to replace customary land tenure systems entirely (Cotula 2007).[7] The predominant view was that titling and registration of land rights were essential ingredients for tenure security, and that this formalization of individual property rights was a prerequisite incentive for increased agricultural productivity, efficiency, and rural development (e.g., DeSoto 2000). However, the procedures for ensuring formal recognition of land, and for adjudication of land disputes, remain cumbersome and expensive. They are ultimately inaccessible to most poor smallholders (Deininger 2003).

The presumed benefits of formal land titles have not materialized in most parts of SSA. And history has shown that customary tenure systems have not necessarily constrained agricultural productivity (Migot-Adholla et al. 1991;

---

[7] These efforts were largely in response to academic and donor organization focus on titling and registration programs designed to mimic Western property rights systems. See, for example, Deininger (2003), explaining that the World Bank's policy as of 1975 was that formal land titling was a prerequisite for development; and Platteau (1996), citing the major property rights theorists that advocated establishment of private ownership rights.

Bruce and Migot-Adholla 1994; Platteau 1996). On the ground, statutory land law remains irrelevant to the lives of many rural Africans who continue to follow customary rules for accessing land (Cotula 2007). National land statutes have not prevailed for a variety of reasons, including mismanagement and corruption at the highest levels of government, and inadequate administrative capacity and budgetary resources at lower levels of government. More generally, most rural communities lack knowledge of statutory laws and remain wary of the intentions of central governments. As a result, legal pluralism persists throughout sub-Saharan Africa today, despite earlier efforts to transition towards a unified, Western-style system of private property rights.

PROBLEMS STEMMING FROM LEGAL PLURALISM

The existence of two legal systems that function in parallel is not necessarily a problem, and has arisen in history any time two political regimes have overlapped (Lavigne Delville 2000; Pimentel 2011). For example, the U.S. legal system is a pluralistic system that functions, albeit imperfectly, with state and federal law operating alongside each other under the principle of federalism. The Tenth Amendment to the United States Constitution speaks to the issue,[8] and over the course of two centuries, American courts and legislatures have worked out rules for which law applies in what contexts, as well as processes for conflict resolution.

Legal pluralism is only problematic when the two separate legal systems contain conflicting rules but do not recognize each other. In the case of land tenure systems in sub-Saharan Africa, uncertainty exists over land rights because statutory law defines those rights differently than customary law in most locations. As far as rural residents are concerned, there may be no ambiguity over which law determines their access to and use of land; they simply follow customary norms and practices. However, tenure insecurity arises when customary rules are challenged or trumped outright by statutory land law and state authorities. Local smallholders risk violating statutory law without knowing it by virtue of following customary principles, and land to which they traditionally had access may be allocated to others under statutory procedures.

This ambiguity creates room for exploitation by opportunistic local elites or foreign investors interested in obtaining rights to land and water resources for speculative or investment purposes. Furthermore, the existence of multiple arbitration or adjudication authorities leads to inconsistency, unpredictability, and the perception that no decision is ever final. Parties may "forum shop"—meaning they pick and choose the aspects of customary or statutory law that

---

[8] The text of the Tenth Amendment reads: "The powers not delegated to the United States by the Constitution, nor prohibited by it to the States, are reserved to the States respectively, or to the people."

best suit their needs—and select a dispute resolution authority that will best support their claims and serve their ends. Where there are no clear rules dictating which law applies and which forum is the final dispute resolution authority, legal pluralism weakens the integrity of both legal systems, and ultimately compromises the rule of law.

INTEGRATING STATUTORY AND CUSTOMARY LAW

In order to avoid these outcomes, statutory and customary land tenure systems must recognize each other and operate together in an integrated fashion (Fitzpatrick 2005; Blocher 2006; Knight 2010; Pimentel 2011). It is critical that formal statutory law recognize and give equal weight to the land rights established, prescribed, and adjudicated by customary law, so long as the customary law meets conditions of fairness and equity (Deininger 2003; Cotula 2007; Knight 2010). Customary land tenure systems are familiar to rural smallholders, and widespread reliance on customary systems needs to be weighed against the enormous costs and risks associated with trying to replace them (Fitzpatrick 2005). At the same time, an integrated land tenure system should not make customary tenure principles inflexible by crystallizing them at a single point in time, nor should it legitimize or codify aspects of customary tenure that are inequitable or dysfunctional (Pimentel 2011). When land rights designated under customary law are recognized as legitimate, the rural poor can be protected more thoroughly from exploitation by national elites and foreign investors (Knight 2010).

However, there remain challenging questions of how best to integrate statutory and customary land laws and what exactly a hybrid model of tenure should look like (Fitzpatrick 2005; Knight 2010; Pimentel 2011). Many current African land laws already attempt to reconcile the two systems by recognizing and incorporating customary land tenure principles. The enactment of these statutes signals progress toward an integrated, hybrid approach to land tenure—and therefore a reliable, sound set of land institutions. The inherent challenges of integration are exacerbated by a number of problematic features of current land laws themselves, as well as by intractable issues associated with implementation and enforcement.

Land statutes vary throughout SSA in the degree to which they recognize, incorporate, and enforce customary land law. More progressive statutes, such as Mozambique's 1997 Land Law, actually embed customary tenure principles into the language of the law by stating that a formal right to use land may be established by continuous ownership under customary principles. Other land statutes, such as Zambia's, leave open to interpretation the legitimacy of customary land rights in the eyes of the government. In order to understand how well a hybrid system of land tenure operates, two levels of analysis are required. First, one must examine whether customary land rights are protected and enforced under the law, and if so, by what mechanisms. Second, one must assess the rules for conflict resolution. These rules encompass, for example, jurisdiction over land disputes and safeguards against the broad powers of the national government, which remains the actual owner of the majority of land in most SSA countries. A comparative analysis of land laws in Mozambique and Zambia demonstrates the successes and failures of hybrid land tenure and property rights systems.

Land Laws of Mozambique and Zambia

Mozambique and Zambia fall broadly within the Guinea Savannah and share a 419-kilometer border (see Figure 8.1). Mozambique is narrow, stretching north-south along the eastern edge of Africa with a coastline along the Indian Ocean of almost 2,500 kilometers (California's is 1,350 km, for comparison). Zambia is landlocked and shares borders not only with Mozambique, but also Malawi, Zimbabwe, Botswana, Namibia, Angola, the Democratic Republic of the Congo, and Tanzania. Mozambique was a Portuguese colony while Zambia was British, and both countries were among the last to gain independence from colonial rule, with Zambia in 1964 and Mozambique not until 1975.

According to the World Bank, more than half of the global land base that could be used to expand cultivated area is in just 10 countries, 6 of which are in sub-Saharan Africa. Mozambique and Zambia are among those six, and have low ratios of cultivated to total suitable land area (Deininger and Byerlee 2011).[9] In addition, they are characterized as having high yield gaps (difference between experimental yields and yields in farmers' fields) due to deficiencies in technology, capital markets, infrastructure, and institutions including land tenure and property rights (Deininger and Byerlee 2011). As a result, they are prime targets for agricultural investment and have experienced relatively large numbers of land deals since 2000 (Land Matrix 2013).

There have been dramatic changes in land tenure laws in both countries in recent decades, with new land legislation in Zambia in 1995 and Mozambique in 1997. In both cases, customary land tenure is at least recognized as a legitimate source of rights to land. However, customary rights are not necessarily respected by the central government to the same extent as statutory rights, leaving land rights of rural populations vulnerable to exploitation or dispossession. Increased attention from national and international investors has accelerated the need to understand and remedy the shortcomings of these land laws.

---

[9] Not all unused land is suitable for cultivation. Agro-ecological constraints, as well as economic and political limitations render a large share of uncultivated land "off base" for future development, as discussed by Rueda and Lambin in Chapter 12 of this volume.

MOZAMBICAN LAND LAW

Mozambique's 1997 Land Law is considered one of the most progressive in all of Africa (USAID 2007; German et al. 2011; Nielsen et al. 2011). The law emerged from extensive collaboration among governments, academics, and donor agencies to address Mozambique's land conflicts and challenges. One of the express purposes of the law was to recognize the customary tenure regimes that continued to determine land access and rights across most of the country (Tanner 2002; Knight 2010). Mozambique's customary law reflects a wide range of cultures, tribes, and languages, without a dominant set of rules, practices, or structures governing land tenure (Knight 2010).

After Mozambique achieved its independence from Portugal in 1975, the new Front for Liberation of Mozambique (FRELIMO) government retained state ownership of all land. FRELIMO opted for a socialist land reform model, creating an extensive collective production system of state farms. Part of the FRELIMO campaign was to dissolve traditional leadership, stripping chiefs of their powers because they were perceived to be puppets of former colonial authorities. Popular resistance to these socialist policies led Mozambique into a violent, 16-year civil war, which devastated vast areas of countryside as well as infrastructure, and displaced upwards of six million people (USAID 2007; Knight 2010; Nielsen et al. 2011). After the war ended in 1992, the Mozambican government moved quickly to address land conflicts arising from competing claims by returning peasants, former colonial landholders, and new investors all asserting rights to the same land (Tanner 2002).

Eager to resolve these conflicts and to spur investment in agriculture, the government formed a Land Commission to develop and implement a new lands policy framework.[10] At that time, at least 25 different cultural and ethnically based regimes had survived civil conflict and were governing an estimated 90 percent of the country's land use. The Land Commission recognized that customary land tenure principles represented an already-established, accepted, and dominant land tenure regime, and that these customary systems provided an important service at very low cost to the state. As a result, it gave customary systems full legitimacy under Mozambique's Land Law (Tanner 2002).

Mozambique's National Land Policy was approved in 1995, and serves as the policy framework within which the 1997 Land Law operates. The stated goal of the Policy was "to safeguard the diverse rights of the Mozambican people over the land and other natural resources, while promoting new investment and the sustainable and equitable use of these resources" (Government of Mozambique National Land Policy 1995, as cited in Tanner 2002). The

---

[10] For a fascinating discussion of the Land Commission and the process of formulating and drafting Mozambique's 1997 Land Law, see Christopher Tanner's "Law-Making in an African Context: the 1997 Mozambican Land Law" (Tanner 2002). Tanner was the Chief Technical Assistant with the FAO Technical Cooperation program that worked with the Mozambican Land Commission in 1995–1996 and played a central role in drafting the 1997 Land Law. The insights he captures in his discussion are highly valuable lessons for any new or ongoing legislative efforts in sub-Saharan African countries.

Policy aligned the government's approach to land rights with principles of the 1990 Constitution[11], and served as the platform upon which the Land Law was drafted. The Land Law, enacted in 1997, is short—35 Articles in 12 pages—and its major provisions are outlined in Box 8.3.

After the passages of the 1997 Land Law, regulations were established to govern the market for long-term leases (DUATs), as well as the procedures for delineating rights obtained by statutory or customary means (USAID 2011). However, these regulations fall short of mapping out procedures for dispute resolution, identifying oversight mechanisms for customary land tenure, or establishing rules for enforcement and penalties for violations of the law

BOX 8.3

1997 Land Law of Mozambique

- The state is the owner of all land in Mozambique (Land Law Article 3; Constitution Article 109).
- Individuals, communities or entities may obtain long-term rights to use and benefit from the land (Land Law Article 10.1). These rights are called DUATs, or *Direitos de Uso e Aproveitamento dos Terras* (Portuguese).
- Land cannot be sold, mortgaged, or otherwise encumbered or alienated (transferred) (Land Law Article 3; Constitution Article 109).
- The right to use land may be transferred by inheritance (Land Law Article 16.1; Constitution Article 111).
- Improvements such as buildings made on the land, as opposed to the land itself, are private property, and can be bought, sold, or mortgaged (Land Law Article 16.2).
- The customary rights of local communities to their lands are intact and valid; rights obtained under customary norms, or by virtue of good-faith occupancy for at least 10 years, are equivalent to rights granted by the government (Land Law Articles 12.a, 12.b, 13.2 and 14.2).
- The government may authorize an application by an individual or corporation for a 50-year leasehold, after consultation and approval by the community within which the land requested is located (Land Law Articles 12.c, 13.3, 17.1). This application process is the only means by which foreigners or companies may obtain rights to land.
- The rights of lands to "nationals" are unrestricted; foreign individuals and entities must have local residence and investment plans in order to obtain rights to land (Land Law Articles 10 and 11).
- Women's rights to land are equivalent to those of men (Land Law Article 10; Constitution Article 36).
- Rights to use land held by a community are legally equivalent to rights held by individuals or entities (Land Law Article 10.1).
- Local communities are active participants in the management of natural resources, in the resolution of conflicts, and in the titling process (Land Law Articles 13.3 and 24).

Sources: Government of Mozambique 1997, 1998, 2004.

[11] The Constitution was amended in 2004, with all provisions relevant here remaining the same. We cite the 2004 version.

(Knight 2010). Lack of these procedures leaves the statute without adequate "teeth" to establish a functioning, integrated land tenure regime.

ZAMBIAN LAND LAW

Zambia's Land Law was enacted in 1995. Although the law recognizes customary rights to land, it is not as sophisticated or progressive as Mozambique's 1997 Land Law. Like most sub-Saharan African countries, Zambia's land tenure system is a product of its colonial history. The British colonists designated areas of "crown land" in what was then Northern Rhodesia for European settlement. Crown lands generally bordered the north-south rail line and comprised only 6 percent of the colonial territory (Adams 2003), but they contained all known mineral resources. The remaining land was considered "customary land" and was categorized as either native reserves or native trusts. The British rulers recognized rights of occupancy, use and common benefit in the reserve and trust areas; under colonial law these lands could not be converted to crown land or permanently alienated (sold or transferred) to Europeans or non-indigenous people. The colonial regime assigned administrative authority over customary lands to the various chiefs, granting them a great deal of control over use and allocation of land (Brown 2005).

Upon independence in 1964, the new Zambian government renamed all crown lands as state lands, and vested ownership of all lands, state and otherwise, in the president, following the socialist models in other sub-Saharan African countries after independence. Private ownership of land was abolished, and only leasehold tenure was allowed on state lands. The new government otherwise maintained most of the colonial land tenure regime, including restrictions on transfers to foreigners.

In the 1990s, the Movement for Multiparty Democracy (MMD) government sought to introduce market reforms in the land law, and to eliminate restrictions on sales and transfers that impeded foreign agricultural investment. With the technical and financial assistance of international organizations (World Bank, IMF, and USAID), the Zambian government drafted the 1995 Lands Act amid much controversy and national debate (Adams 2003; Brown 2005).

Zambia's Lands Act also is short—32 articles in 11 pages—and focuses predominantly on the powers of the president to transfer Zambian land. Although existing customary rights are recognized, the Act does not attempt to integrate customary land law into statutory law in any explicit way. It recognizes customary tenure, but does little to elevate it to the level of statutory title (Metcalfe and Kepe 2008). The Act is essentially a reaction to the earlier socialist land policies following independence, wherein the major reform is to permit leasehold land to be titled, bought, and sold. In contrast to Mozambique's 1997 Land Law, Zambia's Lands Act does not reflect any progressive thinking regarding customary land institutions. Instead, the law's design facilitates

BOX 8.4

1995 Lands Act of Zambia

- All land in Zambia is vested in the President and held on behalf of all Zambians (Lands Act Article 3.1).
- The President is authorized to alienate, or transfer leasehold rights to, land to any Zambian, or to any non-Zambian under certain circumstances, including to an investor (Lands Act Articles 3.2–3.3).
- However, the President cannot alienate customary land to a Zambian or non-Zambian unless he takes into account the local customary law, consults with the chief and the local authority of the area in which the land to be alienated is situated, and consults with anyone else whose interest might be affected by the grant (Lands Act Article 3.4).
- The maximum leasehold term the President may grant is 99 years, unless the President determines it is in the best interests of the country or a two-thirds majority of the National Assembly approves a longer lease term (Lands Act Article 3.6).
- The Lands Act recognizes customary tenure over land and allows for the continuation of customary holdings (Lands Act Article 7).
- Customary tenure can be converted to private leasehold tenure over statutory land at the election of the customary tenure holder (Article 8.1).
- Occupation of vacant [titled] land is unlawful and violators are subject to eviction (Lands Act Article 9).
- A national Lands Tribunal is established, and its jurisdiction extends to resolving any disputes relating to land under the Act (Lands Act Part IV, Articles 20–29).

Sources: Government of Zambia 1991, 1995, 2012.

the permanent conversion of customary lands to state leasehold lands (Brown 2005; Honig 2012). The major features and provisions of the 1995 Lands Act are described in Box 8.4.

Importantly, Zambian chiefs retain a great deal of authority over administration and allocation of customary land under the 1995 Lands Act (Metcalfe and Kepe 2008; Honig 2012). Chiefs are the primary point of contact between their rural communities, the government, and investors or other foreigners. They regulate access to land and common pool resources, and adjudicate land-related disputes—despite the existence of local courts or the national Land Tribunal established under the 1995 Lands Act. The Act thus preserves and enhances the traditional power of chiefs, in the short term. In so doing, it also provides an avenue for corruption via chiefs' authority over transferring customary land to state leasehold and private ownership. At the same time, the Act threatens the power of Zambian chiefs in the long term by encouraging the permanent conversion of customary tenure to state leasehold tenure. When that happens, the influence and presence of the state in any particular region increases (Brown 2005; Honig 2012). Once land is transferred out of the chiefs' power by conversion to state land, that customary power is extinguished forever. In fact, the law has tended to erode customary land tenure structures over time, and has diminished the extent of customary land (Honig 2012).[12]

Zambian land institutions thus remain in a state of tension between customary and statutory authority. The government of Zambia has enacted few implementing regulations to flesh out the sparse Lands Act, leaving the Act inadequate to resolve any of this tension. Although many policymakers recognize the weaknesses in the land law, the political will to find solutions is lacking, primarily because land issues are so controversial (Oakland Institute 2011b). In 2007, the government of Zambia set up a National Constitutional Conference to revise the constitution specifically to address land issues. The proposed constitutional revisions included rules to prevent land speculation, specific rules covering the process of converting land from customary to statutory status, and the establishment of size limits on holdings of arable land by any single entity (USAID 2010). However, the proposed constitutional revisions maintained the customary vs. state dual tenure system of the 1995 Lands Act, with all land remaining vested in the president as before. Even so, the draft constitution was defeated in March 2011, failing to attract the necessary two-thirds vote in Parliament (Oakland Institute 2011b). The Zambian government still has yet to finalize a draft of a new constitution, after years of effort.

Moreover, there has never been a National Land Policy in place in Zambia. There is a draft Land Policy, but its approval has stalled due to the aforementioned constitution-making process (Government of Zambia 2013). In a report to the National Assembly, the government Committee on Lands, Environment, and Tourism strongly urged the adoption of such a Land Policy to "guide the overall administration of land in Zambia. . . [setting] out goals and direction for the present and future" (Government of Zambia 2013).

COMPARATIVE ANALYSIS OF THE TWO LAND LAWS

It is clear that legal pluralism and its negative effects persist in spite of both the Mozambican and Zambian land laws. The central question of this chapter is whether or not the integration of customary and statutory law is sufficient to guarantee the rights of smallholder farmers, in order to improve rural incomes and food security. Determining whether or not integration of customary practice is sufficient from a legal perspective requires an examination of both the substantive and the procedural mechanisms of existing legislation. Table 8.3

[12] The official government records still show the breakdown of customary and state land remains at 94 percent and 6 percent, respectively; however, current estimates suggest that the area for customary lands is much lower. Estimates of customary holdings vary from 61 to 84 percent, and one report states that only 37 percent of land is effectively controlled by customary authorities (USAID 2010; German et al. 2011; Honig 2012). The government of Zambia recognizes this lack of accurate data as a major land administration challenge and has called for a countrywide land audit to quantify the area of various categories of land (Government of Zambia 2013).

TABLE 8.3

Parameters for Assessing the Successful Integration of Statutory and Customary Law

SUBSTANCE: Recognition of customary rights and principles found within land statute	Does the language of the statute make customary land rights legally equivalent to those obtained from the government under the terms of the statute? The best answer to this question is "yes." Are any affirmative steps, such as registration or titling of land, required to designate or otherwise sanction customary land rights? The best answer to this question is "no."
PROCESS: Integrated function of land tenure systems	Are there transparent, accessible, and predictable procedures for conflict resolution? The best answer to this question is "yes." Does the statutory law provide safeguards against the abuse of power and discretion on the part of the state? The best answer to this question is "yes."

provides a list of useful questions for assessing substance and process. This list is not exhaustive, but it serves as a useful starting point for analysis.

Substantive law is law that creates, defines, and regulates the rights, duties, and powers of parties (Black's Law Dictionary 1999). A relevant example is Article 10.1 of the 1997 Land Law of Mozambique, which entitles individuals, communities, and corporations to obtain a right of use and benefit from the land. Procedural law, on the other hand, refers to the rules that prescribe the steps for having a right or duty judicially enforced (Black's Law Dictionary 1999).

The substantive questions of whether the land statutes in Mozambique and Zambia fully respect customary land tenure, and whether such recognition requires affirmative steps on the part of the rights-holder, are assessed in Table 8.4. In general, the 1997 Land Law of Mozambique reflects a strong substantive recognition of customary rights to land. The law explicitly recognizes those rights and does not require that customary rights-holders take any affirmative steps to dignify their rights. In short, customary rights to land are automatically existent and enforceable under the law.

On the other hand, while the Lands Act of Zambia recognizes existing rights to land in customary areas, it does not equate those rights to leasehold rights to state land. It subjects customary rights-holders to the possibility of eviction from titled lands, even where title was unknown or obtained in violation of preexisting customary rights. Furthermore, customary land rights in Zambia are not automatically enforceable, and require some action on the part of the chief to defend customary claims in the event of a conflict. Although the law does not require any registration or documentation of customary tenure, it does not explicitly state that such registration is not required to enforce customary rights, leaving open to interpretation the strength of an oral claim to land based on good-faith occupancy.

There are signs of recent progress in Zambia, however. In its June 2013 report to the legislature, the Committee on Lands, Environment, and Tourism included

TABLE 8.4

Assessment of Land Statutes in Mozambique and Zambia

SUBSTANCE	Mozambique	Zambia
Legal equivalence between statutory and customary rights to land?	**Yes.** Under Articles 12(a) and (b), of the 1997 Land Law, land use rights may be obtained by occupancy "in accordance with customary norms and practices" or by good faith occupancy for 10 years. Rights obtained in these ways are no different from rights obtained by allocation from the government under Article 12(c). While the text of the statute does not specifically state that the two are equivalent, the implication is clear from the construction of the statute because the land right is the same regardless of how it is obtained.	**Yes and No.** Under Article 7, land held by a person under customary tenure, in a "customary area," shall continue to be so held and recognized after the commencement of the Act. The law recognizes customary tenure and distinguishes between customary and state leasehold tenure without explicitly equating the two forms of tenure. However, recognition of customary tenure under Article 7 is subject to Article 9 of the statute, which criminalizes the occupation of vacant titled land. Therefore in the event of conflicting claims, it is likely that claims under statutory law will prevail over those of customary law.
Automatic recognition of customary rights?	**Yes.** No affirmative steps such as titling or registration are required to establish rights to land under the 1997 Land Act. Under Article 14.2, "[t]he absence of registration does not prejudice the right of land use and benefit" acquired under customary principles per Articles 12(a) and (b).	**Not always.** Article 7 of the 1995 Lands Act does not require that any affirmative steps, such as titling or registration, be taken in order to secure customary rights to land. However, the Act does not explicitly state that registration is **not** required. Issues may arise when individuals (either local or foreign) wish to obtain leasehold tenure and convert customary land to statutory land, under Articles 8 and 3, because action by the chief is required to defend and protect customary rights.

in its findings that unlike other countries such as Tanzania, certificates of title could only be obtained for state land but not customary land (Government of Zambia 2013). In order to obtain a certificate for customary land, one would need consent from the chief to convert the land to state land. According to the Committee, the President of the Republic of Zambia has requested legislation to give greater security of tenure to customary land, and the Committee recommends such legislation to provide for universally recognized and enforceable customary land certificates. In the words of the Committee, "[e]nsuring tenure security for customary land rights in [Zambia] is an essential element for sustainable development, given the preponderance of customary tenure."

The procedural question of whether the land laws include a mechanism for fair conflict resolution is assessed in Table 8.5. Procedures that prescribe the appropriate forum and adequate rules for a dispute resolution require transparency, accountability, and predictability. In addition, the existence of safeguards against abuse of discretion by the government, whose powers over land are broad in both Mozambique and Zambia, is an important component of procedural law.

TABLE 8.5
Assessment of Procedures of Mozambique and Zambia Land Laws

PROCEDURE	Mozambique	Zambia
Fair conflict resolution procedures?	**No.** The 1997 Land Law states only that: "Conflicts over land shall be resolved in a Mozambican forum" (Article 32.2). Regulations provide that holders of land rights have the right to defend their rights from encroachment and appeal any decisions, as stated in Articles 13.1 and 40. However, the Land Law sets forth no procedures or rules for conflict resolution, leaving resolution of conflicts to either customary systems with no oversight or a judicial system that is inaccessible and ineffective.	**Yes and No.** Within the 1995 Lands Act, Part IV (Article 22a) created a Lands Tribunal with jurisdiction to "inquire into and make awards and decision in any dispute relating to land under this Act." The Act includes provisions for the make up of the Tribunals as well as their applicable rules. However, experts agree that the Tribunals are ineffective, inaccessible, and too expensive, and have been a wholesale failure where they are even known to exist.
Adequate safeguards against abuse of discretion by the State?	**No.** Despite its progressive approach to protecting customary forms of tenure, the 1997 Land Law is vague and provides little structure for enforcement of those rights against state actors. The state remains the owner of all land in Mozambique, by constitution and statute, and there is little accountability on the part of the state.	**No.** Although the 1995 Lands Act liberalizes the transfer of land and recognizes customary tenure, it contains no checks and balances to restrain or otherwise counter the president's broad powers as trustee under the Act to transfer land to any Zambian, any foreigner who is a permanent resident, or any entity that holds an investment certificate or has obtained the President's consent (i.e., any individual or group he likes).

Neither the 1997 Land Law of Mozambique nor the 1995 Lands Act of Zambia goes far enough to ensure adequate conflict resolution procedures or restrain the power of the government over land. The mere recognition of customary rights to land in each case is insufficient to guarantee the ability to enforce and defend those rights, which leads to persistent tenure insecurity.

In the case of Mozambique, the lack of a requirement to register customary rights means there is a corresponding lack of will on the part of the public sector to record those rights (Tanner 2010). Even though customarily derived DUATs exist all over Mozambique, very few have been formally mapped and registered. As of 2010, only 292 communities had their DUATs delimited and registered, compared with more than 20,000 delimitations for statutorily derived (and legally mandatory) DUATs (Tanner 2010). Even though these customary rights are legally recognizable and enforceable, they are invisible to anyone other than the local smallholders. Any official map of land ownership does not show these customary DUATs. Therefore, customary land remains vulnerable to expropriation, and without means to defend their rights, local smallholders may still lose them to outside interests.

Claimants of customary land tenure rights are left to navigate inaccessible and often corrupt judicial systems, and often lack the time, money, or savvy to defend their claims to land adequately. In both cases, the governments need to introduce amendments to the land law or regulations that set forth the procedural rules for conflict resolution.

As far as checks and balances on the power of the state, both the Mozambican Land Law and the Zambian Lands Act require the government to consult with and seek approval of local communities before transferring customarily held land. These provisions are not always effective or enforced, however, and they ultimately fail to create adequate safeguards against the national government's broad powers as landowner and trustee of all land. Two specific cases—one for each country—are described here to illustrate the relative weaknesses in these land tenure systems and the tension that persists between customary and statutory laws.

CASE 1: THE HOYO HOYO PROJECT IN ZAMBÉZIA, MOZAMBIQUE

Zambézia is one of Mozambique's most promising provinces in terms of agricultural potential, but it is also one of the country's poorest regions. There have been several foreign agricultural investment projects in Zambézia; one, in particular—the Hoyo Hoyo project—highlights the country's land tenure tensions and overlapping land institutions.[13] The Hoyo Hoyo project comprises 10,000 ha in the area of Lioma, Zambézia. Lioma was a Portuguese colonial settlement, and became a government-owned state farm after independence.

The farm was abandoned in the 1980s during the Civil War, but peasants and former state farm workers returned to the area after the war ended. The farmers began to occupy the land, clearing it, preparing it, and eventually cultivating it, with full recognition of local authorities. The Cooperative League of the USA (CLUSA), an American non-profit organization focused on smallholder agricultural development, began a relatively modest project in Lioma in 2003. CLUSA introduced soybean cultivation and promoted farmer associations, and the project was highly successful in raising local incomes, as well as attracting additional donor investment. In spite of this success, in December 2009, the government of Mozambique awarded a 10,000 ha concession of land, formerly part of the old state farm, to the Portuguese company Quifel. Quifel's project Hoyo Hoyo included 490 ha of land occupied by 244 farmers growing soybeans with the assistance of CLUSA. These farmers had rights to use the land under the 1997 Land Law, as they had occupied the land in good faith for more than 10 years. However, the Mozambican government ignored these rights in granting the land to Quifel. The government further ignored Land Law regulations requiring the company to demarcate its concession within one year and carry out a substantial part of its project within two years. Conflicts arose quickly between the company and the farmers,

[13] This case study is based on the work of Norfolk and Hanlon (2012).

especially when Quifel plowed 500 ha outside their allocated area—land that had already been cleared by the local farmers—in December 2010.

In the end, Quifel planted only 100 ha of the 500 acres of land outside their allocated area in 2010, and cleared and planted very little additional land in the 2011–12 growing season. Under the 1997 Land Law, the company is required to have its concession demarcated and a map created showing its boundaries within a year, as well as to begin substantial operations under the project plan within two years. Neither had happened by the end of 2011, and under the 1997 Land Law, Quifel should have lost its land lease on this basis.

The Hoyo Hoyo project highlights the fundamental flaws in Mozambique's 1997 Land Law, which create a vacuum where implementation and enforcement should be. First, as mentioned, evidence presented by Norfolk and Hanlon (2013) shows that Quifel failed to meet its obligations to survey, demarcate, and develop the land within the time period prescribed by the Land Law. Second, it appears that the consultation requirements under the Land Law were not met. Further, promised out-grower schemes, which would have provided local farmers with suitable alternative land and technical assistance, never materialized.

As of mid-2012, Quifel had hired and trained some local workers, and begun construction of its initial project building (Hanlon and Smart 2012). However, Quifel had still not done its required survey and demarcation, nor started actual production. In July 2012, the government finally forced Quifel to survey 3,500 ha that it intended to plant in late 2012 for the 2012–13 growing season, with the participation of the community. The survey identified 836 farmers using 1,945 ha within the 3,500 ha; many of these farmers were prepared to move to nearby land that had been allocated for resettlement. However, community members requested new land that is already cleared and plowed, and while Quifel agreed to these demands, clearing of the land had only barely begun as of September 2012. According to Hanlon and Smart (2012), the district administrator has backed the community, not requiring them to move until alternative plots with cleared land were ready (unlikely in time for the mid-December 2012 planting deadline).

The Hoyo Hoyo story bears an important lesson for governments, investors, and local communities regarding the coexistence of statutory and customary rights to land. In reality, even where the land law is "progressive" in its support of customary rights, the rural poor remain vulnerable to more powerful government and investment interests when they are left without recourse to protect and enforce their rights to land. Thus while the Mozambican Land Law's recognition of customary rights is a good start, much more is needed in the way of enforcement mechanisms and oversight to create an institutional framework in which rural smallholder farmers and pastoralists may defend their rights to land. The case also shows that investors who disregard existing customary rights to land are likely to face challenging situations, especially where the investors fail to conduct adequate consultations with local land users.

CASE 2: THE FARM BLOCK DEVELOPMENT PROGRAM IN ZAMBIA

Zambia presents an interesting, and contrasting, illustration of the tension between customary and statutory land rights in sub-Saharan Africa. In this case, the government of Zambia has been directly involved in many land acquisitions through the Farm Block Development Plan. Farm blocks, defined as large tracts of land set aside for commercial farming, have been used both pre- and post-independence as incentives to attract settlers to the land. Many of the pre-independence farm blocks were awarded to urban dwellers, professionals, and bureaucrats, with some larger plots dedicated to tobacco and farmed by South Africans or Zimbabweans (Oakland Institute 2011b; Chu 2013). As of 2005, the government of Zambia has revitalized its agricultural development strategy with a new Farm Block Development Plan, wherein a farm block of approximately 100,000 ha is to be developed in each of Zambia's nine provinces (Government of Zambia 2005; Chu 2013). The government's development strategy emphasizes the role of foreign and domestic investment.

With the Farm Block Development Plan, the government is actively acquiring lease rights to large areas of land, much of which lies within customary chiefdoms, and plans to develop this land into integrated commercial farming estates with publicly financed infrastructure (German et al. 2011; Honig 2012). Some of the designated land may already be titled and held by the state; however, much of the land is held and/or occupied customarily, administered by a customary chief. In the latter case, the land must be converted from customary to statutory tenure status before implementation of the Farm Block Development Plan (Honig 2012). Each farm block is organized around a central "core venture" funded by private investment, with several smaller surrounding commercial farms whose production is managed by and feeds into the core venture. In addition, a number of smaller out-grower plots are included in each farm block intended for currently resident smallholders and other interested participants. Within each farm block, and consistent across the blocks, the central government provides critical infrastructure, including roads, power, communication facilities, dams, and irrigation systems (Government of Zambia 2005).

The Government of Zambia acquires customary lands for farm block projects through negotiations and agreements with the chiefs who control the land. The decision of a chief to cede customary land to the state is irreversible; if the investment fails, the land remains state land and can be otherwise leased by the state. Thus, many observers worry that the Farm Block Development Plan really amounts to a state land grab of customary land (Honig 2012).

As of 2012, only one farm block, the Nasanga farm block development in the Serenge District, had been completed (Oakland Institute 2011b; Honig 2012). However, government officials have at least negotiated with chiefs in each of the other provinces for 100,000 ha parcels of land; in some cases the parcel will comprise tracts of land from contiguous chiefdoms (Honig 2012).

Convincing chiefs to go along with the Farm Block Development Plan and cede their lands has largely been an exercise of the government advertising the potential benefits from infrastructure and employment, and the chiefs in turn garnering political favor in return for their willingness to get on board.[14] There is some evidence to support the notion that the potential benefits of farm blocks will materialize: the completed Nasanga farm block does include improved roads and electricity supply.

A chief's decision whether to transfer customary land under his control to the state and convert the land to statutory title is highly political, and consequential. There is evidence to suggest that where chiefs are secure in their positions, face no challenges or conflicts within their territory, and have other sources of monetary income, they are financially and politically independent from the state (Honig 2012). As a result, they are not highly motivated to cede their lands to the state for farm block developments. On the other hand, when a chief's power is in question, or where there are border disputes, the chief is more likely to depend on the state to support his existence. This is particularly true when the chief in question is lured by the promise of personal financial gain. Weaker chiefs are thus more likely to cede their lands, opting for the short-term potential of farm block development over the long-term security of retaining customary tenure. It is also possible that these chiefs are resigned to a perceived inevitability of state control over their lands, and therefore consent to transfer of their lands to state control to obtain immediate gains (Honig 2012).

The dual system of land tenure in Zambia and the operation of the 1995 Land Act create competition for control over land between state and customary authorities. As illustrated here, implementation of the Farm Block Development Program relies on the opportunities created by this dual system of tenure, and accelerates the conversion of customary land to statutory. In the middle of this imbalanced process are the customary chiefs, under whose authority smallholder farmers stand to gain or lose much from the decision to convert customary land to farm block development. The chief may be highly susceptible to political pressure in favor of conversion, while local farmers have little or no say in the eventual outcome. Most smallholders cannot afford to convert their own plots of land to individual leasehold title over state land, and their land tenure remains vulnerable to the larger political and legislative forces at play.

---

[14] For an enlightening discussion on the dynamics and politics of the chieftaincy in Zambia as it relates to land and transfers of land to the state for farm block development, see Lauren Honig's paper "State Land Transfers and Local Authorities in Zambia," presented at the Land Deal Politics Initiative Global Land Grabbing II Conference at Cornell University, October 17–19, 2012.

The Future Course of Land Institutions and Food Security

The examples from Mozambique and Zambia underscore the conclusion that although the integration of customary law and statutory law is *necessary* to support smallholder land rights, it is not *sufficient*. In Mozambique, the substance of customary rights over land is well articulated in the law but poorly enforced. The rural poor, in particular, have little political power and virtually no legal recourse to secure their claims over land. In Zambia, both the substance and the process of recognizing customary law are weak within the country's land institutions. In both cases, the details of land laws often end up determining outcomes more than the intent.

As instructive as the cases of Mozambique and Zambia are, the diversity of legal, traditional, and agro-ecological conditions throughout SSA is great. Few sweeping generalizations are possible about how land tenure and land rights law can be used to ensure improved food security among smallholder farmers. How best to design—or reconstruct—a legal framework for land institutions in any given country depends on its colonial history, its political and social dynamics, and its rule of law more generally.

The agro-ecological and biophysical setting within which rural development takes place is also important. Along these lines, significant differences exist between sub-Saharan Africa and Asia. Whereas many Asian agricultural systems have historically been developed around irrigation (both low- and high-technology systems), over 95 percent of SSA agriculture remains rainfed (as discussed by Burney in Chapter 6). Sub-Saharan Africa has been characterized by shifting agriculture, long fallow seasons, and open grazing systems. As a result, farming has not developed in settled plots within fixed land regimes as in much of Asia. The linkages between land institutions and water institutions (see Thompson, Chapter 11) also are critical for rural development, and warrant much greater study in the sub-Saharan Africa context.

A new field of study is rapidly emerging around land tenure, property rights, and land institutions in the developing world. Even in emerging economies such as Brazil—now considered a major powerhouse in global agricultural production—land rights and land tenure for farmers remain unclear. This is especially true in key agricultural states where land expansion has been occurring in recent decades, such as Mato Grosso and Pará (Garrett et al. 2012). In sub-Saharan Africa, the combination of weak land institutions and poor governance in many countries opens the door for corruption and questionable land deals. Due to the large number of land acquisitions by foreign entities and national elites since the mid-2000s—and the potential consequences of such transactions for the rural poor—there is urgency to improving land institutions and securing land rights for smallholders.

There is no easy "low hanging fruit" to grab in the effort to achieve strong and effective land institutions. The typical approach of promoting

Western-style property rights is much too simplistic and is fraught with potential conflict. In situations where corruption and vested interests are in play, little can be done to solve land tenure issues short of eliminating the large capacity constraints on good governance. However, in instances where problems of legal pluralism persist mainly due to confusion and limitations on administrative capacity, the use of an independent review board or independent land commission offers one possibility for providing the due diligence necessary for the protection of customary property rights. Most important, the use of independent institutions capable of navigating the complex terrain of legal pluralism in land tenure situations would help to distinguish productive land investments from land deals that are disrupting and disenfranchising the lives of smallholders with traditional claims. Advancing rural development and food security in most parts of sub-Saharan Africa will be possible only when this distinction is made.

Acknowledgments

The authors gratefully acknowledge the comments and contributions from Walter Falcon, Derek Byerlee, Bill Burke, Erik Jensen, and Chris Fedor.

References

Adams, M. 2003. *Land Tenure Policy and Practice in Zambia: Issues relating to the Development of Agricultural Sector*. Draft. Makoro Ltd. for DFID, retrieved from http://www.aec.msu.edu/fs2/zambia/resources/Land1.pdf.

Alden W. L. 2008. Custom and commonage in Africa rethinking the orthodoxies. *Land Use Policy* 25: 43–52.

Anseeuw, W., M. Giger, C. Althoff, P. Messserli, K. Nolte, M. Taylor, and A. Seelaff. 2013. Creating a public tool to assess and promote transparency in global land deals: The experience of the Land Matrix. Paper presented at the World Bank Conference on Land and Poverty, April 8–11, 2013. Washington, DC: World Bank.

Anseeuw, W., L. Alden Wily, L. Cotula, and M. Taylor. 2012a. *Land Rights and the Rush for Land: Findings of the Global Commercial Pressures on Land Research Project*. Rome: International Land Coalition, retrieved from http://www.landcoalition.org/sites/default/files/publication/1205/ILC%20GSR%20report_ENG.pdf.

Anseeuw, W., M. Boche, T. Breu, M. Giger, J. Lay, P. Messerli, and K. Nolte. 2012b. *Transnational Land Deal for Agriculture in the Global South: Analytical Report Based on the Land Matrix Database*, retrieved from http://landportal.info/landmatrix/media/img/analytical-report.pdf.

Arezki, R., K. Deininger. and H. Selod. 2011. "What Drives the Global Land Rush?" IMF Working Paper No. 251. Washington, DC: International Monetary Fund, retrieved from http://www.imf.org/external/pubs/ft/wp/2011/wp11251.pdf.

Bassett, T., and D. Crummey, eds. 1993. *Land in African agrarian systems*. Madison, WI: The University of Wisconsin Press.

Binswanger, H., K, Deininger, and G. Feder. 1995. Power, distortions, revolt, and reform in agricultural land relations. In *Handbook of Development Economics*, Vol. III, Part B, ed. T. Behrman and T. N. Srinivasan, 2659–2772. Amsterdam: Elsevier.

Blocher, J. 2006. Building on custom: Land tenure policy and economic development in Ghana. *Yale Human Rights and Development Law Journal* 9: 166–202.

Brown, T. 2005. Contestation, confusion and corruption: Market-based land reform in Zambia. In *Competing Jurisdictions: Settling Land Claims in Africa*, ed. S. Evers, M. Spierenbug, and H. Wels, 79–108. Boston: Brill.

Bruce, J., and S. Migot-Adholla, eds. 1994. *Searching for tenure security in Africa.* Washington, DC: World Bank.

Chu, J. 2013. "Creating a Zambian breadbasket: 'Land grabs' and foreign investments in agriculture in Mkushi District, Zambia." Working Paper 33, Land Deal Politics Initiative, available at: http://www.iss.nl/fileadmin/ASSETS/iss/Research_and_projects/Research_networks/LDPI/LDPI_WP_33.pdf.

Collier, P., and S. Dercon. 2013. African agriculture in 50 years: Smallholders in a rapidly changing world? Paper presented at the Food and Agriculture Organization of the United Nations Expert Meeting on How to Feed the World in 2050, Rome, June 24–26.

Cotula, L. 2007. Introduction. In *Changes in "Customary" Land Tenure Systems in Africa*, ed. L. Cotula, 5–13. International Institute for Environment and Development, retrieved from http://pubs.iied.org/12537IIED.html.

Cousins, B. 2000. Tenure and common property resources in Africa. In *Evolving Land Rights, Policy and Tenure in Africa*, ed. C. Toulmin, and J. Quan, 151–179. London: DFID, IIED, NRI.

Deininger, K. 2003. *Land policies for growth and poverty reduction.* Washington, DC: The World Bank.

Deininger, K., and D. Byerlee. 2011. *Rising global interest in farmland: Can it yield sustainable and equitable benefits?* Washington, DC: The World Bank.

DeSoto, H. 2000. *The mystery of capital.* New York: Basic Books.

Fitzpatrick, D. 2005. 'Best practice' options for the legal recognition of customary tenure. *Development and Change* 36(3): 449–475.

Food and Agriculture Organization of the United Nations (FAO). 2013. "Food Balance Sheets." Accessed August 12, 2013, http://faostat.fao.org/site/354/default.aspx.

Food and Agricultural Organization of the United Nations (FAO). 2002. *Land tenure and rural development.* FAO Land Tenure Studies #3. Rome: FAO.

Garner, B. A., ed. 1999. *Black's Law Dictionary*, 7th ed. St. Paul, MN: West Group.

Garrett, R. D., Lambin, E. F., and R. L. Naylor. 2012. Land institutions and supply chain configurations as determinants of soybean planted area and yields in Brazil. *Land Use Policy* (31): 385–396.

German, L., G. Schoneveld, and E. Mwangi. 2011. *Contemporary processes of large-scale land acquisition by investors: Case studies from sub-Saharan Africa.* Occasional Paper 68. Bogor, Indonesia: Center for International Forestry Research, retrieved from http://www.cifor.org/publications/pdf_files/OccPapers/OP-68.pdf.

Government of Mozambique. 1997. Land Law (Law no. 19/97 of 1 October). Maputo: MOZLegal Lda.

———. 1998. Land Law Regulations (Decree no. 66/98 of December 8). Maputo: MOZLegal Lda.

———. 2004. Constitution of the Republic of Mozambique, retrieved from http://www.unhcr.org/refworld/pdfid/4a1e597b2.pdf.

Government of Zambia. 2013. Report of the Committee on Lands, Environment, and Tourism for the Second Session of the Eleventh National Assembly Appointed on September 27, 2012, available at: http://www.parliament.gov.zm/index.php?option=com_docman&task=cat_view&gid=182&Itemid=112&limit=5&limitstart=0&order=hits&dir=DESC.

Government of Zambia. 2012. First Draft Constitution. Technical Committee on Drafting the Zambian Constitution, retrieved from http://zambianconstitution.org/downloads/First%20Draft%20Constitution.pdf.

Government of Zambia. 2005. Farm Block Development Plan 2005–2007. Ministry of Finance and National Planning, available at: http://www.oaklandinstitute.org/zambia-farm-block-development.

Government of Zambia. 1995. Lands Act, retrieved from http://www.parliament.gov.zm/downloads/VOLUME%2012.pdf.

Government of Zambia. 1991. Constitution of the Republic of Zambia, retrieved from http://unpan1.un.org/intradoc/groups/public/documents/cafrad/unpan004847.pdf.

Hanlon, J., and T. Smart. 2012. Small farmers or big investors? The Choice for Mozambique. Research Report 1, available online at http://www.open.ac.uk/technology/mozambique/sites/www.open.ac.uk.technology.mozambique/files/files/Soya_boom_in_Gurue_Hanlon-Smart.pdf.

Herbst, J. 2000. *States and Power in Africa: Comparative Lessons in Authority and Control.* Princeton, NJ: Princeton University Press.

Honig, L. 2012. State Land Transfers and Local Authorities in Zambia. Draft Paper prepared for the Land Deal Politics Initiative: Global Land Grabbing II Conference, October 17–19, 2012, available at http://www.cornell-landproject.org/download/landgrab2012papers/honig.pdf.

Jayne, T.S., J. Chamberlin, and M. Muyanga. 2014. Emerging land issues in African agriculture: implications for food security and poverty reduction strategies. In *Falcon, W. P. and R. L. Naylor (eds.), Frontiers in Food Policy: Perspectives on sub-Saharan Africa*, 265–308. Stanford, CA: Printed by CreateSpace.

Jayne, T. S., D. Mather, and E. Mghenyi. 2010. Principal challenges confronting smallholder agriculture in sub-Saharan Africa. *World Development* 38(10): 1384–1398.

Knight, R. S. 2010. *Statutory Recognition of Customary Land Rights in Africa: An Investigation into Best Practices for Lawmaking and Implementation.* FAO Legislative Study No. 105. Rome: FAO.

Kugelman, M. 2012. Introduction. In *The Global Farms Race: Land Grabs, Agricultural Investment, and the Scramble for Food Security*, ed. M. Kugelman and S. L. Levenstein, 1–20. Washington, DC: Island Press.

Kugelman, M., and S. L. Levenstein, eds. 2012. *The Global Farms Race: Land Grabs, Agricultural Investment, and the Scramble for Food Security*. Washington, DC: Island Press.

Land Matrix Database. 2013. Online public database, available at: http://landportal.info/landmatrix.

Land Matrix Newsletter. 2013. Available at http://landmatrix.org/media/filer_public/2013/06/10/lm_newsletter_june_2013.pdf. June 11, 2013.

Lavigne Delville, P. 2000. Harmonizing formal law and customary land rights in French-speaking West Africa. In *Evolving Land Rights, Policy and Tenure in Africa*, ed. C. Toulmin and J. Quan, 97–121. London: DFID, IIED, NRI.

Maxwell, D. and K. Wiebe. 1998. Land Tenure and Food Security, A Review of Concepts, Evidence, and Methods. Research Paper No. 129, Wisconsin: Land Tenure Center.

McAuslan, P. 2000. Only the name of the country changes: The diaspora of "European" land law in Commonwealth Africa. In *Evolving Land Rights, Policy and Tenure in Africa*, eds. Toulmin, C. and J. Quan, 75–96. London: DFID, IIED, NRI.

Metcalfe, S. and T. Kepe. 2008. Dealing land in the midst of poverty: Commercial access to communal land in Zambia. *African and Asian Case Studies* 7: 235–257.

Migot-Adholla, S., Hazell, P., Blarel, B. and F. Place. 1991. Indigenous land rights systems in Africa: A constraint on productivity? *World Bank Economic Review* 5(1): 155–175.

Naylor, R. and W. Falcon. 2011. The global costs of American ethanol. *The American Interest* 7(2): 66–76.

———. 2010. Food security in an era of economic volatility. *Population and Development Review* 36(4): 693–723.

———. 2008. Our daily bread: A review of the current world food crisis. *Boston Review*, September/October, available at http://bostonreview.net/BR33.5/naylorfalcon.php.

Nielsen, R., C. Tanner, and A. Knox. 2011. *Focus on Land in Africa Brief, Mozambique Lesson 1*. World Resources Institute and Landesa, retrieved from http://focus.wpengine.netdna-cdn.com/wp-content/uploads/2012/04/Brief_Lesson1_Landesa.pdf.

Norfolk, S. and J. Hanlon. 2012. Confrontation between Peasant Producers and Investors in Northern Zambezia, Mozambique, in the Context of Profit Pressures on European Investors. Paper presented at the Annual World Bank Conference on Land and Poverty, the World Bank, Washington D.C. April 23–26, 2012.

North, D. C. 1990. *Institutions, institutional change and economic performance*. Cambridge: Cambridge University Press.

Oakland Institute. 2011a. *Understanding Land Investment Deals in Africa, Country Report: Mozambique*. Oakland Institute, retrieved from http://www.oaklandinstitute.org/sites/oaklandinstitute.org/files/OI_country_report_mozambique_0.pdf.

———. 2011b. *Understanding Land Investment Deals in Africa, Country Report: Zambia*. Oakland Institute, retrieved from http://www.oaklandinstitute.org/sites/oaklandinstitute.org/files/OI_country_report_zambia.pdf.

Oxfam 2012. "Our Land, Our Lives:" Time out on the global land rush. Oxfam International, retrieved from http://www.oxfam.org/sites/www.oxfam.org/files/bn-land-lives-freeze-041012-en_1.pdf.

Pimentel, D. 2011. Legal pluralism in post-colonial Africa: Linking statutory and customary adjudication in Mozambique. *Yale Human Rights and Development Law Journal* 14(1): 59 et seq.

Platteau, J. P. 1996. The evolutionary theory of land rights as applied to sub-Saharan Africa: A critical assessment. *Development and Change* 27: 29–86.

Schoneveld, G. 2011. *The Anatomy of Large-Scale Farmland Acquisitions in sub-Saharan Africa*. Working Paper 85. Bogor, Indonesia: Center for International Forestry Research, retrieved from http://www.cifor.org/publications/pdf_files/WPapers/WP85Schoneveld.pdf.

Tanner, C. 2010. Land rights and enclosures: Implementing the Mozambican land law in practice. In *The Struggle Over Land in* Africa, W. Anseeuw and C. Alden, 105–130. HSRC Press, available at: http://www.hsrcpress.ac.za/product.php?productid=2275.

———. 2002. Law-making in an African Context: The 1997 Mozambican Land Law. FAO Legal Papers Online #26 Rome: FAO, retrieved from http://www.fao.org/fileadmin/user_upload/legal/docs/lpo26.pdf.

Toulmin, C., and J. Quan. 2000. Introduction. In *Evolving Land Rights, Policy and Tenure in Africa*, ed. C. Toulmin and J. Quan, 1–29. London: DFID, IIED, NRI.

USAID. 2007. *Land Use Rights for Commercial Activities in Mozambique*. United States Agency for International Development: Washington, D.C.

———. 2010. *Property Rights and Resource Governance, Zambia Country Profile*. Retrieved from http://usaidlandtenure.net/sites/default/files/country-profiles/full-reports/USAID_Land_Tenure_Zambia_Profile.pdf.

———. 2011. *Property Rights and Resource Governance, Mozambique Country Profile*. Retrieved from http://usaidlandtenure.net/sites/default/files/country-profiles/full-reports/USAID_Land_Tenure_Mozambique_Profile.pdf.

von Braun, J. and R. Meinzen-Dick. 2009. *"Land Grabbing" by Foreign Investors in Developing Countries: Risks and Opportunities*. International Food Policy Research Institute, retrieved from http://www.ifpri.org/sites/default/files/publications/bp013all.pdf.

World Bank. 2009. *Awakening Africa's sleeping giant: Prospects for commercial agriculture in the Guinea Savannah Zone and beyond*. Washington, DC: World Bank.

PART FOUR

Agriculture's Dependence on Resources and the Environment

9

Food, Energy, and Climate Connections in a Global Economy

David B. Lobell, Rosamond L. Naylor, and Christopher B. Field

In July 2012, India experienced a massive power outage that left half its population without electricity for several consecutive days. The prolonged blackout affected hospitals and water treatment plants, shut down trains and road traffic operations, trapped coal miners underground, and left around 670 million people in the northern states—almost one-tenth of the *global* population—in the dark. Numerous factors led to the power outage. The grid was poorly managed and underpowered: some states were drawing 15–20 percent more power than their allotted amount, and hydropower, which typically supplies up to 20 percent of the electricity in the region, was functioning below normal output levels due to low rainfall. The lack of precipitation, caused by a late monsoon, also raised energy demand as farmers in the main rice and wheat belt of the Gangetic Plain relied on electric pumps to irrigate from groundwater resources.

In India, the government has traditionally supported farmers with energy allotments and subsidies, and the summer of 2012 was no exception. With state elections coming up, efforts to capture farm votes—and votes of citizens throughout the country who depend on affordable grains for food security—overshadowed efforts to conserve energy. The grid was further burdened by power demands by the growing middle class for air conditioning in the insufferable summer heat. The sad irony is that while the 2012 power outage wreaked havoc on hundreds of millions of people in electrified northern India, roughly one-third of the country's households lack access to any electricity at all, even for powering a single light bulb. For them, the state of blackout is a constant reality, not a short-term outrage.

The root cause of India's blackout is debatable: was it an energy problem, a water problem, a food problem, a political/management problem, or a climate problem? The answer is "yes" to all. The summer monsoon provides roughly three-quarters of the country's annual rainfall, and its role in agricultural production and freshwater supply is immense. All energy sources—not just hydropower—depend on large volumes of water. When the monsoon is weak or arrives late, as it did in 2012, multiple large-scale problems arise across the energy, agricultural, and water sectors. When rainfall is heavier than normal and causes floods, a different set of problems emerges. The South Asian monsoon has always been unpredictable in timing and strength, and will be increasingly influenced by global warming in the decades ahead. Global warming will also intensify glacial melt in the Himalayan Range, which supplies large amounts of water to South Asia. India thus faces a daunting set of challenges: its population will exceed that of China's by mid-century, its food and energy demands will continue to grow, and climate and water constraints on agriculture and energy production will become more pressing.

This chapter addresses the interconnections between food, energy, and climate—not just for India, but for the global economy as a whole. The links between energy and food systems have always been present, but they have become much tighter since the turn of the twenty-first century as a result of two key factors: biofuels policies in several parts of the world, and growing impacts of and responses to climate change. Over the next several decades, energy and climate policies are likely to have large impacts on global food security, potentially dominating policies focused specifically on food and agriculture.

The reason for energy's rising role is simple: there are biological limits to how much a person can eat, but no such limits constrain a person's energy demand as he or she grows richer. One can always buy another car, take another vacation, or have another delicacy flown in from halfway around the world. As fast as food demand is rising around the world, energy demand in most countries is rising even faster (Box 9.1). And as energy demand continues to

BOX 9.1

Spending Patterns by Income Class

One way to measure the differing natures of food and energy demand is to study how people spend an additional dollar of income (Figure 9.1). In the poorest countries, the average person spends 60 cents of each additional dollar on food, with the poorest within those countries spending significantly more. In the richest countries, only 10 cents of each marginal dollar goes to food, while roughly 50 cents goes to energy-consuming services such as transportation, recreation, and entertainment. As a result, energy demand rises faster over time than food demand in all but the poorest countries.

The challenge for most developing countries in the coming decades is not only to improve food security, or to meet growing energy demand, but to achieve both at the same time. In some cases these goals will be complementary, but in others there will be trade-offs. In deciding which new policies or technologies to pursue, understanding these synergies and trade-offs could mean the difference between success and failure.

build, the impact of climate change is also expected to grow, with food systems affected not only by direct impacts of climate change, but also by energy and land policies intended to slow climate change down.

This chapter is structured in much the same way that one might approach research on this topic. We start with the easy part, which is to identify potential links between the food and energy systems. We then move to more difficult terrain, trying to figure out which of these links really matter for the outcomes of interest, particularly food security, and what the net balance of any innovation or "shock" to the system might be. To do this we explore policies and technologies that promote the use of biofuels in the United States, Europe, and Brazil, and strategies for biofuel development in sub-Saharan Africa, with a specific focus on Ethiopia. For each example, we discuss the implications for food security, and end with an assessment of how the expansion of biofuel use might affect climate change.

Throughout this chapter we emphasize the importance of having a clear reference or baseline comparison when evaluating a new policy or technology. For example, if a government chooses not to provide incentives for biofuel development, then how else will its country's growing demand for transportation be met—with gasoline, with electric vehicles, or with a different energy

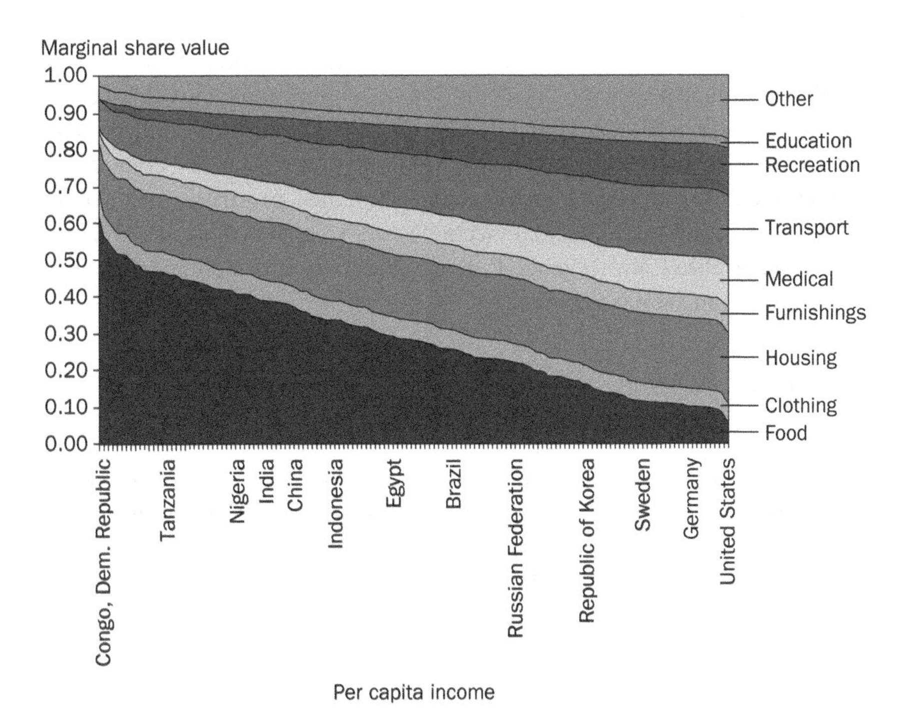

FIGURE 9.1 The fraction of each additional dollar of income spent on different activities by country (and by inference, by level of development)

Source: Figure constructed from data in Muhammad et al. 2011.

technology? And what will be the impacts of these alternatives on climate change and food security? It is difficult enough to measure the impacts of changes that have occurred in the past, let alone changes that might occur in the future. Simply comparing a world before and after a change can be very misleading, because many factors vary over time in ways that could influence the outcomes. To form a meaningful comparison, one also has to develop a clear and viable reference scenario of what would have happened—a "counterfactual" scenario—in the absence of such change.

Overview of Food, Energy, and Climate Connections

Some connections between food and energy are relatively obvious, in that their effects are big enough to be seen within a matter of days to months. For example, a jump in the price of crude oil causes an almost immediate rise in food transport costs, which results in higher retail prices for food. Other effects are more subtle and, in the short term, are typically too small to notice or measure. But if these small effects are persistent over time, they can add up to very large impacts. The gradual rise of atmospheric carbon dioxide (CO_2), for instance, is largely due to energy use, and the CO_2 and climate changes that result can have profound effects on agricultural production.

In sorting out the different connections between food and energy, we find it useful to distinguish between "fast" and "slow" connections. Fast connections are links for which the effect of a change in one system on the other occurs entirely, or at least mostly, within a year or less. In general, the fast connections between the food and energy sectors are asymmetric—mostly in the direction of energy affecting food. Slow connections refer to effects that take at least a year to materialize fully. In general, slow connections include important links both to and from each system.

FROM ENERGY TO FOOD: FAST CONNECTIONS

A household's food security depends largely on its income relative to the cost of acquiring enough food to lead a healthy life. To understand how the energy markets can influence food security, it helps to start by looking at how an increase in energy prices affects both incomes and cost of living. As outlined in Figure 9.2, there appear at least five separate ways that energy markets affect food security:

Global Food Prices

The term "global food prices" (a.k.a., world prices, international prices) refers to the prices of commonly traded commodities such as corn (maize) or wheat on open exchanges. World prices, adjusted for exchange rates and trade policies

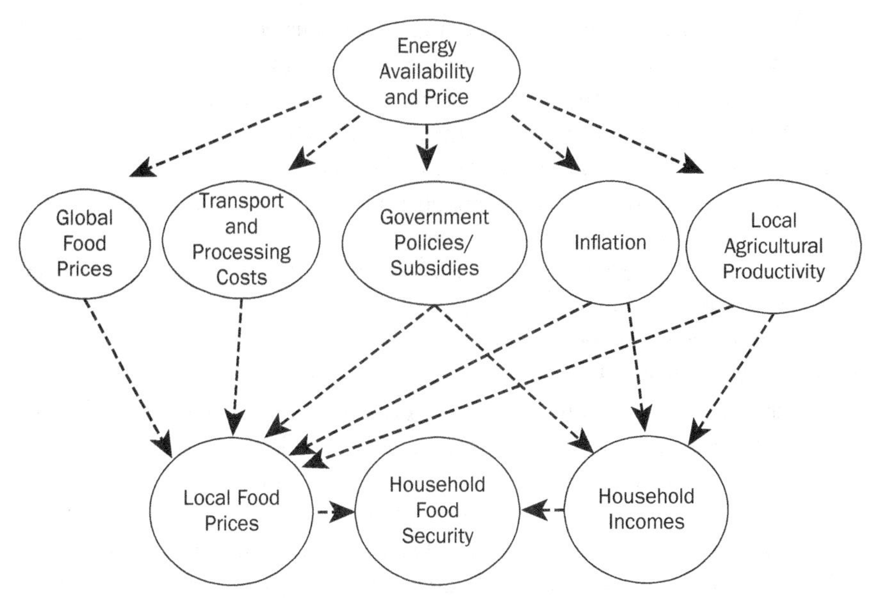

FIGURE 9.2 Short-term impacts of energy prices on food security

Note: For households consuming some portion of their own agricultural production, an increase in productivity is often represented as an effective increase in household income.

such as tariffs, influence domestic retail prices for food, as well as producer prices and farm incentives at the national scale, especially when countries have good transportation infrastructure and highly connected markets (Naylor and Falcon 2010).[1] On the other hand, some of the world's poorest countries are not closely tied to global markets; civil conflict, poor governance, or undeveloped transportation and marketing infrastructure limits the extent to which food producers and consumers in these countries are affected by changes in world prices. The previous section of this volume on sectoral constraints in sub-Saharan Africa underscores this point. Farm households in many parts of the African continent lack broad access to regional or international markets and therefore see little opportunity or risk associated with global price movements.

Energy prices have always played a role in the price of food commodities, in part because higher energy prices raise the cost of crop production. However, the link between energy and food prices has been made considerably stronger in recent years with the Renewable Fuels Standard in the United States

[1] Global prices essentially act as a reference price or opportunity cost (next best alternative) for the domestic agricultural sector. Countries can be fully self-sufficient in agriculture, but it is often less expensive to import some food and more profitable to export some other agricultural commodities (Timmer, Falcon, and Pearson 1983).

and the rapid rise of corn-based ethanol (as discussed also by Cuéllar and colleagues in Chapter 4). Ethanol production consumed less than 10 percent of the U.S. domestic corn supply in 2002, and over 40 percent a decade later (USDA 2012). To put this point in context relative to other changes in demand, roughly half of the *total global* increase in demand for corn during 2005–2010 was accounted for by increased demand for a single use (corn ethanol) in a single country (United States) (Abbott et al. 2011).

As a result, the expansion of the corn ethanol industry has undoubtedly affected corn prices, with some leading analysts estimating a roughly 30 percent increase in overall food prices due to increased ethanol demand (Roberts and Schlenker 2010; Babcock and Fabiosa 2011). Corn has historically been used mainly as animal feed, and thus a price increase in corn induces substitution into other feed ingredients such as wheat, causing wheat prices to rise as well. In the field, U.S. farmers in the Midwest often choose to grow corn or soybeans, and when plantings of corn increase, the supply of soybeans falls and its price rises, holding all else equal. Other crops such as rice and palm oil are in turn similarly linked via substitution in food markets, so major food commodities tend to vary together.

Transport and Processing Costs

In wealthier parts of the world, the price of raw commodities comprises only a small fraction of the price actually paid by consumers. For example, only 10 cents of each dollar spent on food in the United States is typically related to the raw commodity, such as corn or wheat flour (Canning 2011). Most of retail food costs are associated with labor and materials for processing, packaging, and food service, but a considerable amount is also associated with energy used during transport and processing. Increased consumer prices for food in wealthier countries can therefore be tied as directly to energy prices as to the commodity prices themselves.

In poorer, food insecure regions where consumers cannot afford and thus do not demand the same degree of processing, packaging, and service, food is mostly purchased as relatively unprocessed commodities. Prices in very poor areas are therefore less directly affected by higher energy costs than are prices in richer countries. However, consumers must still pay for transport costs, which can be considerable in areas with poor infrastructure, as described by Burney in Chapter 6.

Transport costs are also important for key agricultural inputs in poor regions. Fertilizer in particular tends to be imported in many poor countries, given the large capital costs and energy required for fertilizer (especially nitrogen) production plants. For landlocked regions in Africa, transport costs from the coast to the local market can be as high as $100 per ton; the result being that farmers often have to pay double the cost paid at the dock, and end up using very little fertilizer (Morris 2007). Vitousek and Matson illustrate the comparatively low levels of fertilizer use in sub-Saharan Africa in Chapter 10 of this volume. High transport costs also make it more difficult for farmers to sell products on international markets, and thus create a disincentive to invest in agriculture (see section on local productivity).

Government Policies and Subsidies

Governments widely recognize that food and energy security underlie people's overall sense of security and their perception of government legitimacy—and therefore their propensity to participate in social and political unrest. As a result, many governments subsidize both food and energy consumption to some degree, and commonly increase support during elections or periods of price increases (IMF 2008b). The links between food security and national security are complicated (as discussed by Stedman in Chapter 13), but for many parts of the developing world, food security remains an important concern for national security and political stability.

The overall effect of high energy prices on government finances obviously depends on whether the country is a net exporter or importer of energy. Some food insecure countries rely heavily on energy imports, whereas others are large net exporters. Figure 9.3 shows the net energy imports (as percentage of total energy use) in 2009–2010 for the top 60 food-insecure countries, as measured by the percentage of people judged to be undernourished (values shown in parentheses).

The capacity of any government to subsidize energy and food is limited, however, and poorer countries in particular can face serious fiscal shortfalls when energy prices rise rapidly. During 2006–2008, government actions to cope with rising fuel costs had large fiscal implications throughout the developing world: the International Monetary Fund (IMF) estimated that fuel price increases had roughly four times the impact of rising food prices on government budgets (IMF 2008b). If governments are forced to cut subsidies due to fiscal constraints, then consumers face a situation in which domestic policy changes amplify global food price increases. Such cases have been reported since 2010, with some governments such as those in Bolivia and Malaysia forced to cut food subsidies (Bjerga 2011).

Inflation

Households on the brink of food insecurity spend a large fraction of their income on food. Recall that in poor countries the *average* household spends as much as 60 percent of marginal income on food (Box 9.1). Inflation in other parts of the economy can have important implications for food security because higher prices reduce real incomes, especially if wages are not adjusted accordingly. Higher energy prices can drive overall inflation of prices, not just for transportation or cooking fuel but also for many products that require

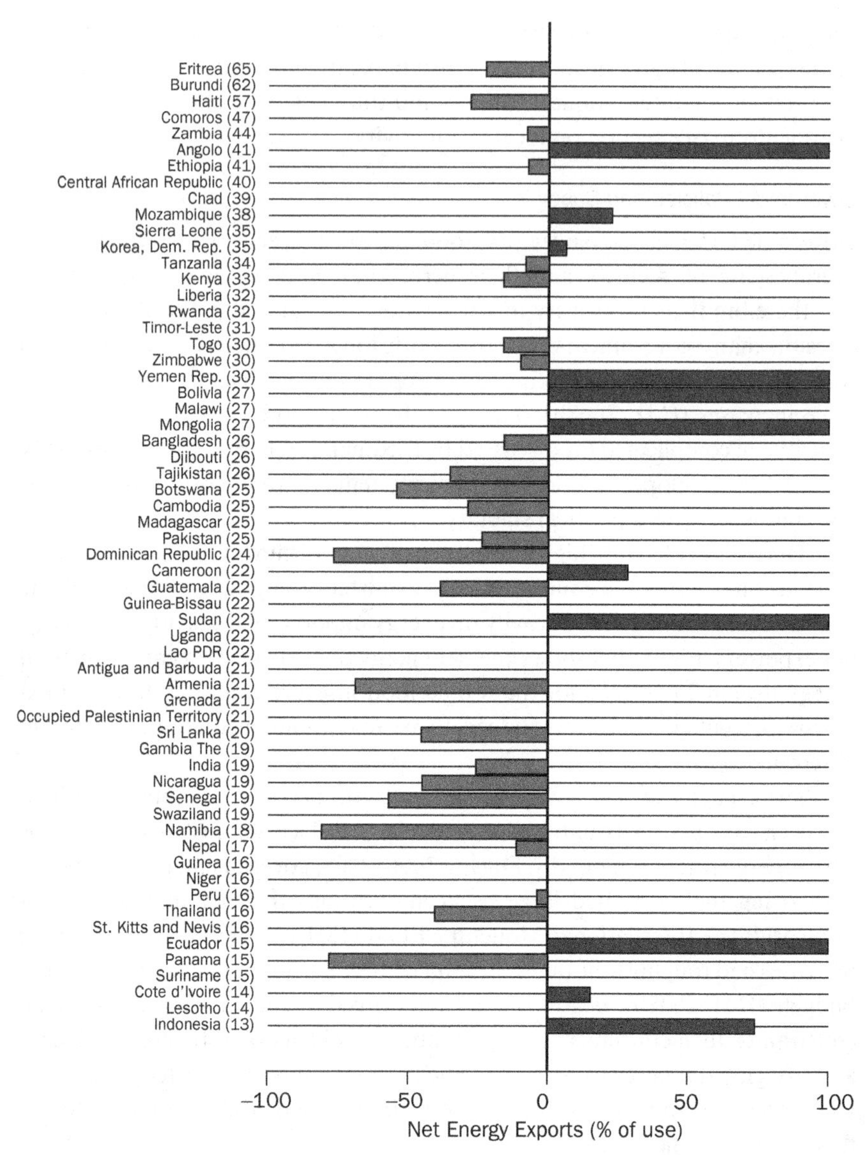

FIGURE 9.3 Net energy exports for the 60 most food insecure countries. Countries are ordered from highest rates of undernourishment at the top (percentage malnourished is shown in parentheses)
Source: World Bank Indicators.

energy or petroleum inputs. For example, an IMF study of Senegal found that the indirect effect of high fuel prices on household budgets was 3.5 times larger than the direct effect on fuel costs (IMF 2008a). For households that spend the bulk of their budgets on food and energy, persistent inflation can have a devastating impact on their consumption, even with food and energy price subsidies.

Local Agricultural Productivity

A majority of the world's poorest people live in rural areas, work in agriculture, and are both producers and consumers of food. Agricultural productivity, measured in terms of net revenue produced per worker or per hectare, thus plays an important role in food security because it affects local food prices, rural household incomes, and levels of home consumption (which is often counted as income). Energy is an important factor in crop production throughout the world, both in terms of direct fuel inputs as well as the embedded energy in machinery, irrigation, and chemical inputs such as fertilizers. The forms of energy, and therefore the relevant energy prices, differ significantly between types of crop inputs. For example, nitrogen fertilizer costs are related to the price of natural gas, which is the main feedstock for synthetic fertilizer production, whereas irrigation typically relies on electricity or diesel fuel.

Even relatively poor areas can rely on external energy inputs for production, especially in countries such as India where large amounts of electricity and diesel fuel are used to pump water for irrigation. However ubiquitous the need for energy in crop production, the overall implications of energy price increases for production are less clear. In the case of India's irrigated agricultural sector, the simulated impact of doubling diesel fuel prices was less than a 1 percent drop in water use, whereas doubling electricity costs led to an 8 percent drop in water use. Both scenarios resulted in less than a 1 percent change in national crop production (Nelson et al. 2009a).

FROM FOOD TO ENERGY: FAST CONNECTIONS

Compared to the ability of energy prices to affect food systems, the impact of food systems on energy consumption and prices in the short term is quite small. Important local influences of food on energy can certainly be found. Most notably, agricultural water consumption can sometimes limit the amount of water available to the energy sector. For example, reservoir managers often balance the need to provide reliable irrigation water with the need to maintain flows for hydropower generation. Water is also used intensively in coal mining and electricity production, and during droughts agriculture often gets political preference over coal for scarce water. Many coal power plants were reportedly shut down briefly in China during a 2011 drought, and similar incidents have occurred in Europe and the United States (van Vliet et al. 2012).

The global influence of food on energy, however, is small. This asymmetry between food and energy prices reflects the basic fact that globally, the amount of energy consumed in power plants and cars is much larger than the energy content of food consumption. Diverting food crops into energy systems therefore has relatively small effects on overall energy supply. For example, average per capita food consumption worldwide was 2,829 kcal per day in 2008 (FAO 2012), which if burned for energy would provide nearly 12 MJ

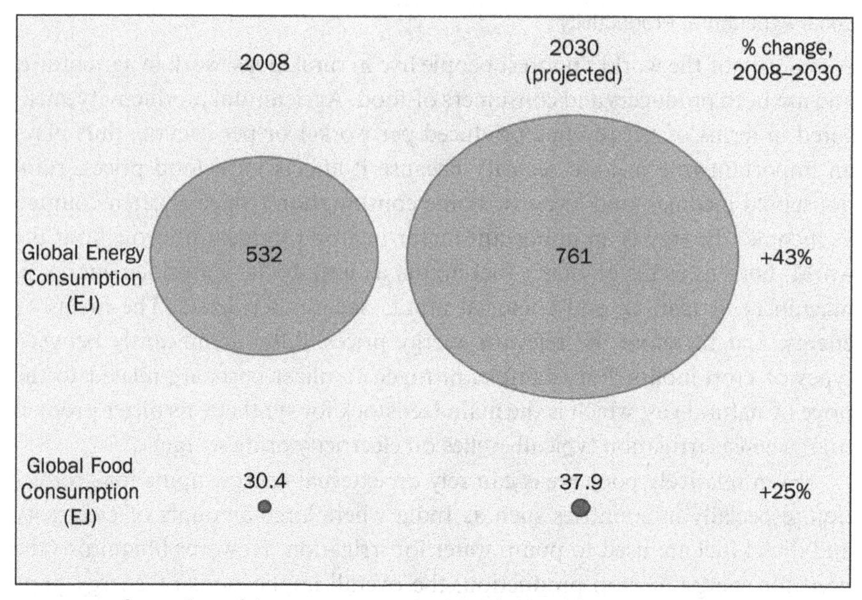

FIGURE 9.4 Comparison of the energy content in global food and energy consumption. Values are shown in EJ, both for current levels (as of 2008) and anticipated levels in 2030

Data source: Author calculations. Current and future energy use from EIA; food consumption in joules computed from FAOSTAT calorie food balance sheets, and projected increases in food consumption from (Bruinsma 2009).

per day.[2] Multiplying by the world's population gives a total energy content of global food consumption in 2008 of roughly 30.3 exajoules (1 EJ = 10^{18} joules) per year. Figure 9.4 compares this amount to the total energy consumption in 2008 of 532 EJ, which is roughly 18 times larger.

This comparison glosses over some details, such as the fact that about 500 kcal per day of an average person's diet comes from animal products, which require more than one kcal of grain for each kcal of meat, milk, or eggs used for human consumption. Even with such corrections, however, per capita energy consumption is at least an order of magnitude larger than the energy content of food consumption. Given that global energy consumption is rising faster than global food consumption, the ratio will have grown to more than 20 to 1 by 2030 (Figure 9.4). Because the energy content of the food system is significantly smaller than that of the global energy system, it has little direct influence on energy prices. The influence of food on energy can be much stronger, however, in the long term via effects on the climate system that shape energy policy.

[2] One joule in everyday life is approximately the energy released as heat by a person at rest, every 1/60th of a second. A megajoule (MJ) is equal to one million (10^6) joules, or approximately the kinetic energy of a one-ton vehicle moving at 160 km/h (100 mph) and an exajoule (EJ) is equal to 10^{18} joules. The 2011 Tōhoku earthquake and tsunami in Japan released 1.41 EJ of energy according to its 9.0 rating on the moment magnitude scale. The annual total energy use in the United States is roughly 94 EJ.

FROM ENERGY TO FOOD: SLOW CONNECTIONS

Some of the economic links discussed here also play out on longer time scales. Probably the most critical link is the effect of persistent high energy costs on agricultural productivity. For example, rural areas that are energy insecure typically do not invest in irrigation, because they do not have a reliable or affordable electricity or diesel source to operate pumps. Conversely, agriculture could be transformed by a new policy or technology that drastically and permanently reduces energy costs, as illustrated by Burney in Chapter 6 in her discussion of small-scale solar-powered irrigation systems.

Emissions related to energy use are arguably the most important factor that links food and energy systems in the long term. When fossil fuels are burned, pollutants including carbon dioxide (CO_2), carbon monoxide, and volatile organic compounds are released into the atmosphere. The latter contribute to the characteristic blanket of smog that surrounds many large cities; high levels of ozone found in smog often extend into neighboring rural areas. Especially in Asia, these near-surface ozone levels are high enough to cause crop damage and lower agricultural productivity, with current global losses estimated at roughly $20 billion USD per year (Van Dingenen et al. 2009).

Ozone effects are short lived; if fossil fuel burning was stopped immediately, then the smog would quickly disappear. But CO_2 from fossil fuels remains in the atmosphere for decades to centuries. As a result, CO_2 accumulates in the atmosphere, much like a bathtub with a slow drain and a fast faucet filling up with water. Because CO_2 is a greenhouse gas (GHG—a gas that traps outgoing heat), its buildup in the atmosphere causes a gradual warming of the Earth and a host of other related changes in climate. These include changes in rainfall patterns, storms, winds, and humidity. Both the CO_2 increase itself and the resulting temperature and other climate changes affect the productivity of agriculture. Meanwhile, the use of nitrogen (N) fertilizers to enhance agricultural productivity also results in N loss to the environment in the forms of near-surface ozone as well as nitrous oxide (N_2O), a potent and long-lived greenhouse gas, as described by Vitousek and Matson in Chapter 10.

Plants use CO_2 from the atmosphere, along with energy from sunlight and hydrogen from water, to produce both their fuel and their chemical building blocks. Increased CO_2 concentrations can have a beneficial effect on crop growth because most crop plants perform this photosynthesis faster at higher levels of CO_2; as a result, crops can grow faster and bigger as long as other factors, such as nutrients and water, are also plentiful. CO_2 has the added benefit of slowing down water loss from the plants themselves. The latter process is because plants can reduce the number and size of leaf openings (called stomata) under high CO_2 conditions.

At the same time, however, climate changes caused by higher levels of CO_2 and other greenhouse gases tend to reduce agricultural productivity. The effects are not as universal as in the case of direct CO_2 stimulation, since cold areas like Siberia or North China actually benefit from warming initially. But where most crops are grown, warming—the hallmark of climate change—tends to reduce yields. Other climate changes associated with high CO_2 are increased frequency of heavy downpours and associated flooding, and less overall rainfall in sub-tropical areas like the Mediterranean or Central America. Extreme rainfall and droughts typically add to the negative effects of warming, further lowering crop productivity.

If one compares the positive effects of CO_2 with the harms of climate change, they tend to balance out in terms of impacts on global scale crop production until an increase in global average temperature of about 2°C above pre-industrial climate. That threshold is projected for sometime between 2030 and 2050. After that point, there is little doubt that further GHG emissions will reduce global food production (Easterling et al. 2007). Before 2030, certain crops and regions will start to feel the effects of warming, and indeed some already likely have. Wheat and corn, for example, are particularly sensitive to warmer temperatures and appear already to be affected negatively by climate change in many locations (Lobell et al. 2011).

The impacts of climate change on food security are likely to occur long before global food production is significantly altered, however. The reason is that the majority of the world's food-insecure people live in tropical and sub-tropical countries where initial crop damages from climate change are expected to be most severe (Battisti and Naylor 2009). One of the most striking inequities of human-induced climate change is that the poorest countries have contributed very little to the buildup of global GHG emissions, yet they will bear much of the human toll associated with the resulting climate change. Although it is difficult to put precise numbers on potential impacts, some studies indicate that climate change could lead to a 20 percent or more increase in the number of food insecure people by 2050 (Nelson et al. 2009b).

FROM FOOD TO ENERGY: SLOW CONNECTIONS

As awareness and concern about climate change have increased, so too has support for some type of international agreement to set targets or limits on greenhouse gas emissions.[3] Herein lies the main long-term impact of food systems on energy. A total emissions cap would include not only emissions from energy use, but from *all* sources. Two of the main non-energy sources of greenhouse gases are deforestation and agriculture, each of which contributes between 10

[3] These targets, if agreed upon, will be based on a combination of scientific and value judgments about how much warming is too much. For example, if the target for the maximum global temperature change (above the long-run, pre-industrial average) is set at 2°C, then the best available science suggests a target for cumulative CO_2 emissions of ~4 trillion tons since pre-industrial times, or ~2 trillion tons after 2009 (National Research Council 2011).

and 15 percent of global emissions annually.[4] The more those sources of emissions can be slowed, the more room there is for the energy sector to emit CO_2 without pushing overall emissions past the target. Conversely, if greenhouse gas emissions from agriculture and deforestation are not curbed, the energy system will be more constrained, and will have to rely increasingly on low or zero-carbon technologies under any global climate agreement.

Without going into detail on all of the ways that food systems contribute to greenhouse gas emissions, it is worth highlighting the critical link between food security and deforestation. In a world in which food prices are high relative to incomes for many people, the marginal benefits of clearing more land for agriculture are large. In a more food-secure world, the benefits of clearing land may still prevail over the costs of emitting carbon (as discussed by Rueda and Lambin in Chapter 12), but not necessarily so. Policymakers may choose to exert stringent controls on deforestation in the latter case, but even the most ardent supporters of climate change mitigation may be reluctant to put climate goals above people's basic need for food.

Improving food security is thus a necessary but not sufficient criterion for slowing deforestation. Slowing deforestation is, in turn, an important factor in determining the amount of greenhouse gases that energy systems can emit without exceeding climate targets. It should be emphasized, however, that even if rapid progress is made in improving food security and slowing deforestation, such progress is unlikely to avert substantial change in the Earth's climate if non-agricultural sources of emissions, such as fossil fuel burning for transport and heating, are not reduced as well.

Food-Energy-Climate Connections as Illustrated by Biofuel Development

Specific examples help to illuminate the true nature of fast and slow connections between food, energy, and climate systems. Although there are countless instances of these connections throughout the world, we have chosen two that relate to the expansion of the global biofuel industry—a recent, high-profile occurrence with relatively well understood causes. The first case focuses on biofuel policies in countries where the technology for biofuel production is well developed (United States, EU, Brazil). The other focuses on biofuel development in a low-income country, Ethiopia, where the industry is in its infancy. Both examples reflect an active debate among experts about what mix of policies

[4] The deforestation contribution is mainly from the release of CO_2 when land is cleared and burned, often for agricultural purposes. Agriculture's contribution comes mainly from non-CO_2 gases: methane related to ruminant belching, manure decomposition, and rice production, as well as N_2O from fertilization and agricultural soils.

is most likely to produce positive outcomes for food security, and which technologies will be good or bad for climate change. With many trade-offs involved, experts often disagree not only on the magnitude of impact on food security, energy security, and climate change, but also on the whether the net effects will be positive or negative. We begin this section with a description of the two cases and their relationship to food security. We then turn to the topic of how biofuel production more generally intersects with climate change, indirectly affecting food security over the longer term.

BIOFUEL POLICIES IN THE UNITED STATES, EU, AND BRAZIL

Although small relative to total energy consumption, biofuels have long been a commercially viable source of transportation fuel in Brazil. Since 2005, they have also become a significant component of the transportation fuel mix in the United States and European Union (Figure 9.5). The rise in the U.S. corn-based ethanol sector has been so rapid that by 2010, it accounted for almost 60 percent of global ethanol production, far surpassing Brazil's sugarcane ethanol output, with one-third of the global total. Nonetheless, sugarcane ethanol comprises up to half of the volume of light-duty transportation fuel used in Brazil, while ethanol makes up only 10 percent of the U.S. total. The European Union (EU), meanwhile, is the largest global producer of biodiesel,

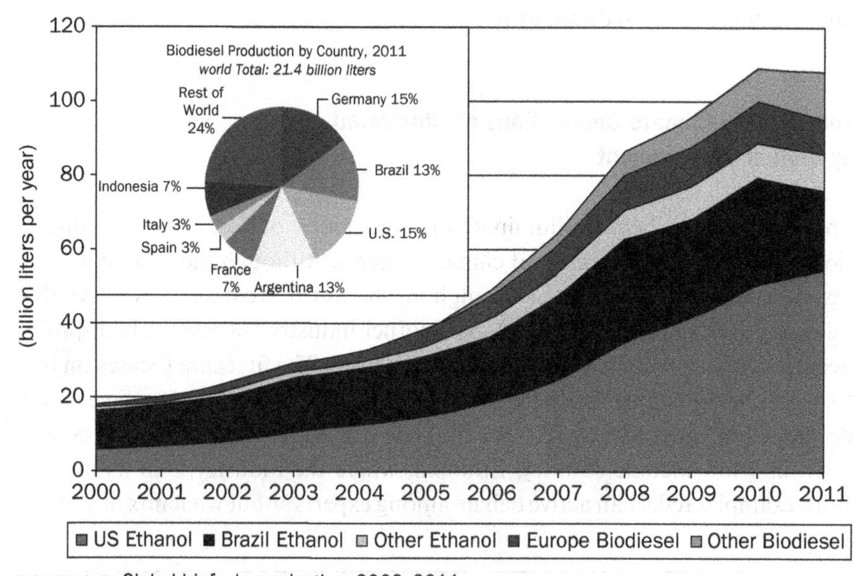

FIGURE 9.5 Global biofuels production, 2000–2011

Sources: Ethanol Production, Europe and Other Biodiesel Production (large chart)—US Energy Information Administration International Energy Statistics (available online from www.eia.gov, accessed August 4, 2012) and REN21 Renewables 2012 Global Status Report (available online from www.ren21.net); Biodiesel production by country (small chart)—REN21 Renewables 2012 Global Status Report (REN21 2012).

made largely from rapeseed (canola).[5] In all three cases, the expansion of the biofuel sector has stimulated the regions' agricultural sectors by creating new and sustained levels of demand. Other large, food-producing countries, such as Argentina, China, Indonesia, and India, have also invested in biofuel production, but to a much lesser extent.

Growth in biofuel production in Brazil, the United States, and EU did not occur through unfettered market forces. Instead, Adam Smith's "invisible hand" had significant help from a wide range of government policies, used to create incentives for biofuel production, fuel blending, and ethanol and biodiesel consumption (Naylor 2014). Brazil developed its sugarcane-ethanol sector in the 1970s and 1980s with government support through direct budgetary spending, subsidized credit, tax relief, and provision of public land and water for sugarcane plantations at below-market value (Valdes 2011a,b). It also uses an aggressive set of fuel blending mandates, set at E20-25 and B5, to ensure a market for its expanding ethanol and biodiesel output.[6] To meet mandates in the future, Brazil is expanding its sugarcane production into the *cerrado* (grassland) region of the country—a move that is effectively pushing soybean production up into the Amazon and creating trade-offs with environmental objectives related to biodiversity protection and greenhouse gas emissions (Loarie et al. 2011).

Like in Brazil, ethanol policies in the United States have taken three main forms: tax exemptions and credits for production, tariff protection against cheaper imports, and mandates for fuel blending (Naylor and Falcon 2011). The first two have their origins in legislation dating back to the 1970s and 1980s, but it is the third element—consumption mandates—that have been critical to the recent biofuel boom.[7] The Renewable Fuel Standard, coupled with the required use of ethanol as an oxygenate additive in gasoline (e.g., E10, a 90 percent gasoline/10 percent ethanol blend), have caused the demand for ethanol to be tightly linked to growth in transportation fuel demand overall.[8] Ethanol not only provides a nationally mandated 10 percent additive to

---

[5] Other vegetable oils, such as soy and palm oil, are also used as biodiesel feedstocks, although their relatively high prices have curbed their use in the EU market. Sustainability issues related to palm oil use in biodiesel have also become important in the EU since the mid-2000s, as discussed subsequently in this chapter. Falcon (Chapter 2) provides the policy context for palm oil development in Indonesia, and Rueda and Lambin (Chapter 12) discuss its implications more thoroughly for land use change.

[6] A blending mandate of E20 implies a fuel mix at the pump of 20 percent ethanol and 80 percent gasoline. For B5, the target is 5 percent biodiesel blended with 95 percent fossil fuel diesel.

[7] U.S. mandates for ethanol and biodiesel fall under the Renewable Fuels Standard (RFS), which requires that the amount of conventional ethanol (e.g., corn ethanol) used in gasoline blends reach a minimum target of 15 billion gallons by 2015, and that advanced biofuels (made from agricultural, cellulosic, and algal materials) reach a minimum of 21 billion gallons by 2022. These biofuels must also be 20–60 percent lower in greenhouse gas emissions than petroleum-based transportation fuels.

[8] Another key policy measure encouraging the use of ethanol in the United States was the phase-out of MTBE (methyl tertiary butyl ether) as a gasoline oxygenate additive in 2005 due to environmental and health risks. While blended ethanol can replace MTBE as an oxygenate, it is important to note that ethanol contains only about two-thirds the energy (BTUs) of gasoline. To be competitive as

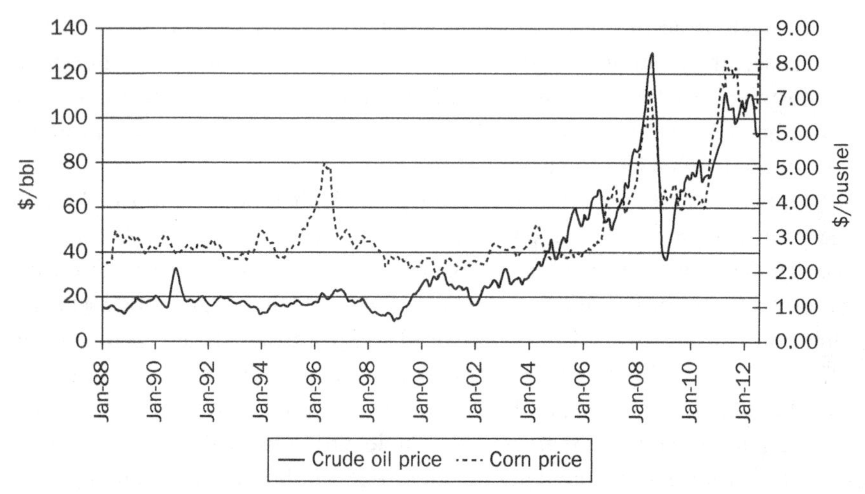

FIGURE 9.6 Monthly prices of crude oil and corn grain (1988–2012)

Note: The correlation was –0.16 for 1988–2005, and 0.81 for 2006–2012.

Source: USDA (corn price); EIA (oil price).

gasoline, but blenders sometimes increase the fraction of ethanol depending on prices (Babcock 2008). Blending decisions are made on a daily basis, with ethanol plants shutting on or off depending on the price of corn feedstocks and ethanol (Abbott et al. 2008). The result is that gasoline prices effectively set a floor to corn prices, with a high correlation between crude oil, ethanol, and corn prices ever since the corn ethanol industry expanded at commercial scale in the middle of the last decade (Figure 9.6).

The ultimate demand for United States ethanol will depend on legislated blending targets for cars and trucks (E10, E15, E85), the competitiveness of ethanol exports in other countries, maintaining mandates, and the future path of the U.S. and global auto fleet. For example, more flex fuel vehicles, which can run on ethanol or gasoline, would likely increase demand for ethanol. A major uptake of electric or natural gas-powered vehicles, on the other hand, would limit the demand for ethanol. Currently, the U.S. fleet is designed mostly with gasoline engines using E10 blends, which means that up to 10 percent of total transportation fuel demand (roughly 140 billion gallons/year), or about 14 billion gallons of ethanol, are demanded each year.[9]

The EU has similarly transformed its policy approach, from an earlier emphasis on tax and trade incentives and non-enforced consumption targets to a more recent focus on blending mandates (Steenbik et al. 2007; Swinbik 2009; Blandford

a direct energy source, ethanol's per-gallon price therefore cannot be more than two-thirds the cost of gasoline. See Naylor and Falcon (2011) and Naylor (2014) for more details.

[9] In 2012, corn-based ethanol and (mostly) soy-based biodiesel supplied around 6 percent of U.S. transportation fuel overall (Beckman et al. 2013).

et al. 2011).[10] After initial mandates led to heavy imports of palm oil from Southeast Asia EU officials added sustainability criteria to their feedstock sourcing policy. These criteria include specific greenhouse gas emissions targets that limit sourcing of crops from deforested and high carbon stock areas. Like the U.S. case, future demand for biofuels in the EU will depend importantly on the development of the transportation fleet, which appears to be trending towards electric vehicles.

EFFECTS ON FOOD SECURITY

The United States, EU, and Brazil are clearly the largest players in the global biofuel sector, and in all three cases, crop-based biofuels (corn, soy, canola, sugar) are currently the mainstay for biofuels production and use. As of 2013, cellulosic fuels have not yet become commercially viable. A key question for this volume, therefore, is what are the consequences of crop-based biofuel development on global food security? The answer remains contentious. On the positive side, growth in biofuel production in the United States and EU has reduced large farm surpluses, raised rural incomes, and alleviated the need for protectionist farm policies that work against low-income farmers in the developing world (see the discussion on U.S. farm policy by Cuéllar and colleagues in Chapter 4). In addition, higher real prices for staple commodities like corn and soy in national and international markets have led to associated price increases for crops that act as substitutes in production and consumption (such as wheat and rice), thus raising farm incomes and land values across the board (Naylor 2014; Abbott et al. 2011). For decades prior to 2005, a declining trend in real agricultural commodity prices created disincentives for investment in agriculture, particularly in capital-scarce countries, where it was often less costly to import cheap food than to invest in rural infrastructure and agricultural productivity growth. The waning of agricultural investments left many rural communities, where most of the world's lowest income households are located, impoverished and chronically malnourished (World Bank 2007).

On the negative side, the biofuel boom has reduced the availability of food (and feed) by diverting more staple crop production—and the land and water used in crop production—to fuel. For example, the United States has typically accounted for 55 to 60 percent of global corn exports, and relatively low prices for corn and corn-based animal feeds have allowed the livestock industry to flourish. By 2011, however, roughly 40 percent of corn utilization in the United States was going to ethanol, exceeding the amount used by the

---

[10] In 2009, the EU passed legislation through its Renewable Energy Directive (RED) that required 10 percent of all transportation fuel to come from renewable resources by 2020 (EU 2009; Flach et al. 2011). Implementation of the EU mandate is in the hands of individual member states, most of which now have legislation in place to meet the targets. The EU biofuels policy is discussed further by Swinnen in Chapter 5.

livestock industry for the first time on record. This sort of demand pressure has contributed to a rise in real food prices in global markets, which has made food less affordable for poor net consumers around the world and reduced their real incomes in the short run. Moreover, the use of mandates as a policy tool to boost the biofuel industry has reduced the price responsiveness of demand for crops used as feedstocks, at least up to the mandated point (that is, the demand curve has become less elastic within the mandated range). As a result, any supply shock caused by extreme weather, natural disasters, or sudden trade restrictions by major food importers or exporters can cause crop prices to be highly volatile (Abbott et al. 2011; Anderson and Martin 2012).

High food prices are particularly detrimental to the world's poor, who spend on average of 60 percent of their marginal incomes on food (Box 9.1). Moreover, households in the lowest income bracket in virtually all countries are net food consumers, and they tend to be the most affected by food price shocks (Naylor and Falcon 2010). Thus a high and unstable food price environment has serious negative consequences for global food security in the short run.

Over the longer run, the balance between biofuels and food security depends on wages adjusted and how individual governments react to the new energy-driven price and policy environment. If developing countries, along with the international community, perceive a world of tight supplies and high food prices, they are likely to invest once again in agriculture and rural development, and the outcome for food security could be positive. The type of investments that governments choose to promote is important, however, and if such investments do not benefit the lowest income brackets of the population, then they will not likely improve food security. Alternatively, if private and public investors see the biofuel boom as temporary, then a renewed focus on agriculture and rural development may not be forthcoming. In either case, the projected outcome could be moot. The energy market itself is changing rapidly despite the growth in biofuels, and relative prices of competing energy sources are fluctuating as more natural gas comes into the market through shale fracking. Over time, biofuels could remain a small segment of the transportation fuels market, eclipsed by electric or natural gas vehicles.

Biofuel Policies in Ethiopia

Ethiopia serves as another interesting example of the complex connections between food, energy, and climate. It is one of the world's poorest countries, with a human development index that ranks 174th out of 187 countries measured globally, and a poverty rate of almost 40 percent (UNDP 2012). Extreme poverty is defined as an individual living on less than $1.25 per day. For these people, food insecurity is a constant threat—a threat that currently affects almost 34 million Ethiopians. At this low level of economic development, it comes as no surprise that almost three-quarters of the country's economically active population engages in agriculture, and that the average household spends over half of its income on food (UNDP 2012). Like many of the poorest countries in sub-Saharan Africa, there is very low diet diversity in Ethiopia, and over 40 percent of children under five years old are stunted. Moreover, Ethiopia is ranked as the most energy poor nation among African countries (as measured by total energy available for transportation, cooking, heating, and lighting) (Nussbaumer et al. 2012).

Still, Ethiopia is growing, in population, urbanization and per capita GDP. The demand for liquid fuels therefore is expected to rise in Ethiopia in the coming decades. The same is expected for all developing economies, but models developed by the World Bank indicate that the rate of increase in demand for transportation fuel will be higher in Ethiopia than in any other sub-Saharan country over the period 2005–2020 (although the average per capita level of consumption will not be high by global standards). The projected per capita fuel consumption growth for Ethiopia is almost 10 percent per annum for gasoline and 12 percent per annum for diesel (Mitchell 2011). As a result of its agrarian economy and its excess fuel demand, Ethiopia has turned toward investments in liquid biofuel production as an energy strategy. In 2011, the government ratcheted up its target for biofuel blends in gasoline from 5 percent to 10 percent by 2015.

Ethiopia's plans to expand biofuels as a key energy source involve three main groups. These are large state-owned molasses ethanol and jatropha biodiesel enterprises; private estates that grow jatropha, castor beans, and other oilseeds for biodiesel and that employ large numbers of poor workers; and private companies that contract with farmers in various grow-out arrangements (Tekle 2008; Negash and Swinnen 2012). Public enterprises remain the largest category to date, although not necessarily the most efficient. In the case of both public and private companies, there are large opportunity costs with respect to biofuel investments, in terms of both natural and financial resources. For example, sugar cane, the most productive biofuel feedstock, requires scarce water and land to grow. Because financial capital is limited in Ethiopia, biofuel development also requires direct foreign investment. However, in order to attract such investment into its domestic biofuels industry, the government must frequently provide land, infrastructure, and supportive policies for growth in the industry, often at the cost of overriding other needed public investments—including health and education, two key pillars of human capital development.

EFFECTS ON FOOD SECURITY

Ethiopia's case is similar to other resource-constrained countries in sub-Saharan Africa that are looking to invest in or seek investors for private sector biofuel enterprises; the question is whether such investments can be productive enough to solve the energy problems while also enhancing rural income growth and food security (Ewing and Msangi 2009). In many cases, inequities mount as communal village land is given to private companies, keeping smallholders on shrinking per capita farm plots, as discussed by Smith and Naylor in Chapter 8. Inequities also develop as water is allocated to fuel versus food crops. Ethiopia is in a particularly vulnerable situation with respect to water, as evidenced by recurrent famines due to drought (Devereux 2009)—a situation that could worsen with climate change. Widespread irrigation activities are in underway in Ethiopia, but there remain serious questions about the sustainability and equity outlook for these new projects (Burney et al 2013). Already Ethiopia receives over one-fifth of all developmental food aid and food security assistance provided to sub-Saharan Africa—the highest share, by far, of any country on the continent (UNDP 2012).

The best way to understand how Ethiopia's energy strategy is likely to affect the country's food security is to trace through a few alternative biofuel scenarios and see how they differ from each other—and even more importantly, how the expected outcomes of the scenarios differ from doing nothing at all in terms of liquid biofuel expansion. The conventional path has been to use molasses from large state-owned sugar operations for ethanol production. Processing plants and supply chains already exist for sugar in the country, and molasses is high in sugar content, which boosts the energy side of the equation. However, as the government aims to expand ethanol production from about 7 million liters in 2010 to over 180 million liters in 2015 (Guta 2012), substantially more land and water will be required for sugar plantations. Some observers, including those in government, view land in Ethiopia as an under-utilized resource.[11]

A contentious issue is whether the distribution of public land to private investors for biofuel development, or even for large-scale food production, improves Ethiopia's food security situation or makes it worse (Smith and Naylor provide a broader discussion on land acquisitions in sub-Saharan Africa in Chapter 8). On the positive side, bringing foreign investment into the agricultural sector has the potential to improve supply chains, enable technology spillovers, and boost rural employment. If successful, such investments could lead to greater feedstock production and ethanol supplies in the country, thus reducing dependence on foreign oil for transportation fuel. The reality is less promising to date, however. Although the state owns all land according to the 1995 Constitution, farming communities and pastoralists who use the land are not always willing to give up their customary land rights, and many have been forced off their traditional lands or relocated to less favorable plots. In these cases, declining assets (livestock and land), incomes, and subsistence food production jeopardize food security. Although employment on plantations has expanded, salary levels and terms vary, and workers can lose their jobs if the

---

[11] As a result, the Agricultural Investment Support Directorate was created in 2009 within the Ministry of Agriculture to identify and transfer land to investors. As of 2011, 6 million hectares were identified for potential large-scale land transfers (i.e., 200 hectares or larger) (Land Matrix Database 2013).

company stops producing feedstocks. In several cases, land identified for private investment in Ethiopia has been unsuitable for production, particularly when it lacks water. As a result, a few large foreign investors, such as Emami Biotech from India, have withdrawn their biofuel projects after prospecting the area. Other companies have similarly left the country after clearing substantial forested areas when the land was deemed unsuitable for long-term productivity.

A more favorable approach for pro-poor growth and food security involves private investments that rely on various contracting arrangements and out-grower schemes with smallholder farmers. Ethiopia has several such distributed private-sector projects, mainly focused on castor beans or jatropha[12] for biodiesel (Negash and Swinnen 2013). In these cases, farmers typically grow the feedstock on a portion of their land—sometimes in hedges around their food plots—while continuing to produce their main food crops. The contracting company often provides the inputs, including seeds, fertilizer, and training, as well as a reliable market for the output. When such arrangements work well, farmers are able to smooth their seasonal incomes through the year-round sale of the feedstock, and they gain from technology transfers, training, and improved supply chains. When such arrangements do not work well, farmers are wed to unfavorable contracts but must still assume all the risks associated with production shocks, uncertain demand, and volatile prices. Because castor bean and jatropha are not food crops and require minimal water for growth, they do not compete directly with staple crop production. However, both crops grow significantly better when water and fertilizers are applied, and if the company ends its contract, the farmer is left with a crop that has no household food value, and potentially no external market (Ewing and Msangi 2009).

In comparing the two approaches, the state-owned sugar enterprise model shows greater promise for improving energy security for the country, while the private sector out-grower scheme is more likely to improve rural incomes, food access, and food security. Contracting arrangements with smallholders in Ethiopia typically involve large transaction costs, and it can be difficult to achieve the desired scale of output, particularly when the population is poorly educated, when water and fertilizer are scarce, and when road networks and other transportation infrastructure are underdeveloped. Despite the drawbacks of both models, turning away from biofuel expansion altogether also has its costs. Without growing supplies of ethanol and biodiesel feedstocks, the Ethiopian economy will remain dependent on imported oil for transportation fuel, and as a result, it will be extremely vulnerable to international oil price volatility. As Africa's most energy-poor nation, biofuels are the only domestic, renewable energy source currently available for transportation. Moving forward, the government faces very high stakes in the energy-food game. If it

---

[12] Jatropha is a small, drought tolerant shrub that produces oilseeds which are poisonous for direct human consumption.

plays the game astutely, Ethiopia could benefit from improved energy security, agricultural productivity, rural incomes, and food security. If it promotes inequities and ill-founded land investments, rural poverty and food insecurity are likely to worsen.

Biofuel Development and Climate Change

As with effects on food security, biofuel development—whether in the United States or EU or Brazil or Ethiopia—can have both positive and negative influence on climate change. On the positive side, biofuels displace fossil fuels, potentially reducing the net quantity of greenhouse gases released to the atmosphere (Box 9.2). In addition, biofuel crops may help cool the regional climate through mechanisms not related to greenhouse gases, as will be explained. On the negative side, development of biofuels can lead to higher food prices and subsequent land use change, which could result in higher total emissions of greenhouse gases through forest clearing and release of carbon from soils. An expanding reliance on biofuels, especially where the vision for future production outstrips the realistic prospects, could also distract attention from efforts to improve fuel efficiency or to develop alternative transportation fuels or fleets.

While a carbon-neutral biofuels sector is theoretically possible, a number of factors along the way can add net emissions, pulling biofuels away from that ideal condition. These net emissions can come from four main sources: (1) preparing the land where the biomass feedstocks are grown; (2) growing the biomass feedstocks; (3) converting the feedstocks from raw biomass into biofuels; and (4) changing the Earth's land cover to grow biofuel feedstocks. Depending on specific (and not always easy to understand) details, particular biofuel projects can be large net sources of greenhouse gases, or they can be quite effective

BOX 9.2
Biofuels, Fossil Fuels, and Carbon

Burning biofuels produces CO_2 emissions, just as fossil fuel combustion does. But there's an important difference in the source of that CO_2 when it comes to climate change. Fossil fuels are formed through a process of carbon burial that takes millions of years. This compression and burial of plant matter—eventually turning it into oil—is also partially responsible for the abundance of atmospheric oxygen, which plays a critical role in the health of the biosphere. Biofuels, in contrast, are derived directly from harvested crops: the cycle from photosynthesis to fuel is less than one year. Conceptually, a vehicle powered by biofuels is running on solar energy, with the biofuels providing a medium for storing the solar energy in a concentrated form that is stable over months or even longer. It is thus possible to have a net zero annual emissions profile from production and use of biofuels, whereas use of fossil fuels results in a net increase in atmospheric CO_2 due to the extremely long time frame for fossil fuel formation.

in offsetting the emissions from the fossil fuels they replace. Here we consider all four potential sources of greenhouse gas emissions.

LAND PREPARATION

Sometimes, a landscape destined for a biofuels project is already in crop production or abandoned. If that is the case, then a transition into agriculture for biomass might not release stored carbon from vegetation and the soil at all. In other cases, land preparation might even lead to an increase in carbon storage, and a net reduction in atmospheric CO_2, particularly if an annual crop is replaced with a high-yielding perennial. But if the biofuel crop is established on a plot with large carbon stocks (e.g., on peat soils in Indonesia), then the clearing of the existing vegetation can lead to large releases of soil and biomass carbon—and a net increase in atmospheric CO_2. The ratio of the carbon release to the amount of carbon that *would* have been released by burning an equivalent amount of fossil fuels can be considered a "carbon debt", or the number of years until the biofuel production breaks even, in terms of emissions, by saving as much carbon as was lost in the land conversion.[13]

The conversion to biofuel crops of land already used for agriculture can potentially produce carbon debts indirectly. The reason is that conversion of food-producing land to biofuels will decrease the supply of food, leading to higher prices and increased pressure for expanding agricultural areas elsewhere (Searchinger et al. 2008). For example, rising demand for Brazilian sugar for ethanol can create a cascade: sugar cane expands into land currently used for soybeans, soybeans move into areas currently used for pasture, pasture takes over areas in subsistence agriculture, and subsistence agriculture expands into forest. The mechanism of indirect carbon debts is conceptually clear, but it is difficult to document empirically.

Conversion of land to biofuel feedstock production does not, however, need to be a source of greenhouse gases. Three kinds of approaches can be practical. One involves starting with lands that have low biomass and low soil carbon (e.g., abandoned land).[14] Abandoned lands that have not reverted to forest could support biofuel crops with little or no carbon debt.

---

[13] Fargione et al. (2008) calculated the "carbon debt" for a number of transitions. Their estimates ranged from 0 years for mixed prairie biofuels on land continuously in prairie, to a whopping 423 years for palm biodiesel on land previously in peatland rainforest in Indonesia or Malaysia. Biofuels projects in areas previously in rainforest created carbon debts of at least 89 years. Biofuels in the Brazilian savannah created debts of 17 to 37 years. Projects to develop corn ethanol in the United States created debts of 48 and 93 years, respectively, on deployments on abandoned cropland or grassland (much of which is currently in the conservation reserve program), highlighting the consequences of large losses of soil and biomass carbon.

[14] Quantifying the amount and productivity of abandoned land is challenging. Field and colleagues (2008) estimated that, worldwide, about 400 million hectares used for crops or pasture in the last 300 years are not used currently.

A second approach to avoiding carbon debts comes from selecting a combination of a biomass crop and agricultural management that results in an increase in soil carbon, aboveground carbon or both. High yields, perennial plants, and careful management can all lead to increased soil carbon, as has been observed in experiments with *Miscanthus* (Zimmerman et al. 2011). Woody biofuel crops like hybrid poplars offer the prospect of combining biofuel production with some level of on-site carbon storage in woody biomass.

A third approach to avoiding carbon debts could involve extracting energy from the biomass removed when a site is prepared for biofuel crops. Currently, many sites are first logged to extract valuable timber, and then burned in order to clear the land with a minimal amount of time and labor. If instead the standing biomass were used to generate electricity or biofuels, the overall demand for fossil fuels (and related carbon debts) would fall.

GROWING THE BIOFUEL FEEDSTOCKS

Cultivating crops for biofuels can potentially generate two sources of greenhouse gas emissions. The first is from the fossil energy that powers the process of farming (machinery, farm operations, irrigation, transportation, and inputs such as fertilizers). In most biofuel operations, the demand for fossil energy is an important but not dominant segment of the emissions budget (Hill et al. 2006). The other potentially important source of emissions from growing biofuel crops is emissions of N_2O, a decomposition product of nitrogen fertilizer. Because N_2O is a greenhouse gas 298 times as powerful as CO_2, even modest releases of N_2O can have a major effect on climate forcing from agriculture. In general, N_2O emissions can be greatly reduced through sophisticated farming practices, such as precision placement of slow-release fertilizers, but the knowledge and incentives to do this successfully are often not in place, even in the most advanced and intensively managed systems (Grassini and Cassman 2012). Vitousek and Matson discuss this topic more extensively in Chapter 10.

CONVERTING BIOMASS TO BIOFUEL

One of the key challenges of large-scale biofuel production is that biomass density is low, resulting in large volumes of feedstock per unit of energy, and hence to high energy costs for transportation of that feedstock. Once at a processing facility, the energy sources and requirements for converting biomass to biofuels vary over a wide range. At the low energy end, simply pressing jatropha seeds can yield an oil suitable for direct use in stationary motors on pumps and generators. At the other extreme, processing corn to ethanol can, with inefficient processes, require more natural gas energy than the final energy content of the ethanol (Pimentel 2005). Such extreme inefficiencies, however, are hard to find in current commercial systems (Liska et al. 2009).

For some biofuel production schemes, the energy for processing comes entirely from fossil sources. For others, part of the crop is burned for the energy to process another part. With Brazilian sugarcane ethanol, this approach is so well developed that heat from burning the bagasse (the crushed stem and leaf material left after the sugar has been removed) is sufficient not only to process the sugar into ethanol, but to also generate electricity for export to the grid. In other cases, such as corn ethanol, parts of the crop not used for ethanol can be recycled into animal feeds (wet and dry distiller grains), which provide additional economic benefits (Taheripour et al., 2010). Across alternative sources and processing technologies, the ratio of biofuel energy output to fossil energy input, or the net energy balance ratio, ranges from less than 1 (a net loss of energy) for some corn ethanol technologies to 10 or more for biodiesel from palm oil. An important trade-off with respect to palm oil use is between the positive energy balance and the potential negative effect on deforestation and carbon release.

LAND COVER CHANGE

Although most studies tend to focus on the role of biofuels in offsetting greenhouse gas emissions, the crops themselves can have important direct effects on local climate through land cover change. Alterations in the land surface can affect the albedo (how much light is reflected), with darker surfaces leading to more energy absorption and higher local temperatures. Albedo changes are the reason, for example, that adding trees in snow-covered areas can cause significant local warming, despite the fact that trees take up atmospheric CO_2. In addition, land cover can influence how much water is transferred from land to the atmosphere (called evapotranspiration). Because the evaporation of water consumes substantial amounts of energy, higher rates of evapotranspiration reduce the amount of incoming solar energy converted to heat, and therefore cool the local climate, in much the same way that sweating helps keep us cool on hot days.

The development of perennial grasses, such as switchgrass or *Miscanthus*, as biofuel feedstocks provides a good example of this local cooling process. Perennials typically have deeper roots than annual crops like corn or soybean, which allows them to take up and transpire greater amounts of water each year. Perennials also grow for a longer portion of the year. Because plant surfaces are generally more reflective than soil, converting land from annual to perennial crops reduces the amount of energy absorbed by the land surface. Combined, these factors cause local to regional cooling and might even be more positive in terms of climate change than the greenhouse saving effect at regional scales. For example, switching from corn to *Miscanthus* in the U.S. Corn Belt could potentially lower surface temperatures by about 0.5°C, far more than any effect of the carbon savings such a switch would deliver (Georgescu et al. 2011).

Overall, there are both benefits and potential risks associated with pursuing biofuels from a climate perspective. The easiest (but still difficult) aspect to consider is whether a biofuel supply chain releases more greenhouse gas emissions each year than a comparable supply chain for gasoline. The net results depend on the specifics of the cropping system and processing technology, and even for a single product such as corn ethanol, it is hard to make general statements about the climate effects. A more difficult aspect not typically incorporated into life-cycle analyses is the opportunity cost of pursuing biofuels. If policies promoting biofuels divert private and public investments away from other renewable energy technologies or energy efficiency efforts, then the climate change effects could be negative. Such an outcome is particularly a concern if ethanol development leads to an entrenched political interest that makes future energy transitions (e.g., to an electric vehicle fleet) more difficult. (See Chapter 4 by Cuéllar and colleagues for a discussion on entrenched political interests in the United States.) In other words, if the counterfactual to biofuel expansion is continued reliance on gasoline, then the promotion of biofuels is probably a modest positive for the climate system. But if the counterfactual is an aggressive pursuit of higher vehicle efficiencies and electrification of transportation, then the net effect of biofuels is likely negative.

Conclusions

With energy demand rising worldwide, and with food prices increasingly tied to the energy system, understanding the links between energy and food is imperative from a policy perspective. As the examples in the previous sections illustrate, the introduction of certain energy and climate policies can have multifaceted impacts on food security, and similarly, changes in the food system can have many spillovers into the energy and climate domains. Because of these complex interactions, the net effects of individual policies are often difficult to anticipate. This conclusion does not mean that an analyst's job is simply to point out both sides and shy away from any definitive statement. It does mean, however, that analyzing the effects of energy or food shocks requires consideration of multiple pathways, and a clear definition of reference or counterfactual scenarios for comparison.

Looking ahead, two very large risks to food security loom at the intersection of energy and food systems. Over the next decade, the continued growth of biofuel demand in both developed and middle-income nations risks putting food prices out of reach for many individuals in least-developed countries. And over the next few decades, the cumulative effects of climate change in tropical and sub-tropical regions could greatly increase the difficulty of raising productivity and incomes among the world's poorest farmers. Trends in energy and climate therefore both have the potential to inflict serious damage to food security in ways that were not possible in previous generations. These risks underscore the importance of ensuring that food security goals are prominent among policy discussions related to both energy and climate, and that today's food policy makers are closely connected with their counterparts in energy and climate policy.

On a more positive note, two important opportunities also lie at the intersection of energy and food. First, advances in energy technologies such as solar panels offer new chances to intensify agricultural systems in remote villages (as discussed by Burney in Chapter 6). Historically, increased energy use has been pivotal to raising agricultural productivity and associated economic development. Second, although the direct effects of climate change will be mostly negative for least-developed countries, concerns about climate change have brought about renewed interest among policymakers and philanthropists in issues of food security. With that interest comes the potential for much higher levels of attention and sustained investment in food security over the next decades.

Turning back to the situation in India where this chapter started, investments in food security will need to be paired with investments in energy and water resources. Renewable energy technology, distributed irrigation infrastructure, and diversified cropping systems that utilize scarce water and energy effectively and profitably are ongoing areas of investment interest by India's burgeoning entrepreneurial sector. New types of management and technology innovations—many of which have yet to be designed—will be required to meet India's many demands in the decades to come. Climate change—and its implications for food, water, and energy security—introduces a new level of urgency and complexity to the equation. A single-pronged approach for addressing food insecurity is doomed to fail, in India and throughout the world, but an integrated approach has the potential to succeed in all crucial domains.

References

Abbott, P. C., Hurt, C. and Tyner, W. E. 2011. What's driving food prices?" Farm Foundation

———. What's driving food prices in 2011? Farm Foundation Issue Report. Accessed June 22, 2012, http://www.farmfoundation.org/webcontent/Farm-Foundation-Issue-Report-Whats-Driving-Food-Prices-404.aspx.

Anderson, K., and W. Martin. 2012. Export restrictions and price insulation during commodity price booms. *American Journal of Agricultural Economics* 94(2): 422–427.

Babcock, B. 2008. How low will corn prices go? *Iowa Ag Review* 14(4): 1–3, 11.

Babcock, B. A., and Fabiosa, J. F. 2011. The impact of ethanol and ethanol subsidies on corn prices: revisiting history. CARD Policy Brief 11-PB 5. Iowa State University.

Battisti, D., and R. Naylor. 2009. Historical warning of future food insecurity with unprecedented seasonal heat. *Science* 323(5911): 240–244.

Beckman, J., A. Borchers, and C. A. Jones. 2013. Agriculture's supply and demand for energy and energy products. United States Department of Agriculture Economic Research Service.

Bjerga, A. 2011. "Risk of Riots Fishing as Governments Cut Food Subsidies, UN's Sheeran Says." *Bloomberg News* (January 25), http://www.bloomberg.com/news/2011-01-25/risk-of-riots-rising-as-governments-cut-food-subsidies-un-s-sheeran-says.html.

Blandford, D., T. Josling, and J. Bureau. 2011. Farm policy in the U.S. and EU: The status of reform and the choices ahead. Washington, D.C.: International Food and Agricultural Trade Policy Council.

Bruinsma, J. 2009. The resource outlook to 2050: By how much do land, water use and crop yields need to increase by 2050? Rome: United Nations Food and Agriculture Organization (FAO), http://www.fao.org/wsfs/forum2050/wsfs-background-documents/wsfs-expert-papers/en/.

Burney, J. A., R. L. Naylor, and S. L. Postel. 2013. The case for distributed irrigation as a development priority in sub-Saharan Africa. *Proceedings of the National Academy of Sciences of the United States of America.*

Burney, J. A., S. J. Davis, and D. B. Lobell. 2010. Greenhouse gas mitigation by agricultural intensification. *Proceedings of the National Academy of Sciences* 107(26): 12052.

Canning, P. 2011. "A Revised and Expanded Food Dollar Series: a Better Understanding of Our Food Costs." Washington, DC: USDA ERS. http://www.ers.usda.gov/publications/err-economic-research-report/err114.aspx.

Devereux, S. 2009. Why does famine persist in Africa? *Food Security* 1(1): 25–35.

Easterling, W., et al. 2007. Food, fibre, and forest products. In: *Climate Change 2007: Impacts, Adaptation and Vulnerability. Contribution of Working Group II to the Fourth Assessment Report of the Intergovernmental Panel on Climate Change* ed. M. L. Parry, O. F. Canziani, J. P. Palutikof, P. J. v. d. Linden and C. E. Hanson, 273–313. Cambridge, UK: Cambridge University Press.

European Union (EU). 2009. Directive 2009/28/EC of the European Parliament and of the Council of 23 April 2009 on the promotion of the use of energy from renewable sources. *Official Journal of the European Union.*

Ewing, M., and S. Msangi. 2009. Biofuels production in developing countries: Assessing tradeoffs in welfare and food security. *Environmental Science and Policy* 12: 520–528.

Fargione, J., Hill, J., Tilman, D., Polasky, S. and Hawthorne, P., 2008. Land clearing and the biofuel carbon debt. *Science* 319(5867): 1235–1238.

Field, C.B., Campbell, J.E. and Lobell, D.B., 2008. Biomass energy: The scale of the potential resource. *Trends in Ecology & Evolution* 23(2): 65–72.

Flach, B., S. Lieberz, K. Bendz, and B. Dahlbacka. 2011. EU-27 Annual Biofuels Report (GAIN Report No. NL1013). Global Agricultural Information Network. The Hague: United States Department of Agriculture Foreign Agricultural Service.

Food and Agriculture Organization of the United Nations (FAO). 2012. "FAO Statistical Databases." http://faostat.fao.org.

Georgescu, M., D. B. Lobell, C. B. Field. 2011. Direct climate effects of perennial bioenergy crops in the United States. *Proceedings of the National Academy of Sciences* 108(11): 4307–4312.

Grassini, P. and K. G. Cassman. 2012. High-yield maize with large net energy yield and small global warming intensity. *Proceedings of the National Academy of Sciences* 109(4): 1074–1079.

Guta, D. D.. 2012. Assessment of biomass fuel resource potential and utilization in Ethiopia: Sourcing strategies for renewable energies. *International Journal of Renewable Energy Research* 2(1): 132–139.

Hill, J., E. Nelson, D. Tilman, S. Polasky, and D. Tiffany. 2006. Environmental, economic, and energetic costs and benefits of biodiesel and ethanol biofuels. *Proceedings of the National Academy of Sciences of the United States of America* 103(30) (July): 11206–11210.

IMF. 2008a. "Food and Fuel Prices—Recent Developments, Macroeconomic Impact, and Policy Responses." Washington, DC., http://www.imf.org/external/np/pp/eng/2008/063008.pdf.

IMF. 2008b. "Food and Fuel Prices—Recent Developments, Macroeconomic Impact, and Policy Responses. An Update." Washington, DC., http://www.imf.org/external/np/pp/eng/2008/091908.pdf.

Land Matrix Database. 2013. Online public database, available at: http://landportal.info/landmatrix.

Liska, A. J., H. S. Yang, V. R. Bremer, T. J. Klopfenstein, D. T. Walters, G. E. Erickson, and K. G. Cassman 2009. Improvements in life cycle energy efficiency and greenhouse gas emissions of corn ethanol. *Journal of Industrial Ecology* 13(1): 58–74.

Loarie, S. R., D. B. Lobell, G. P. Asner, Q. u, and C. B. Field. 2011. Direct impacts on local climate of sugar-cane expansion in Brazil. *Nature Climate Change* 1: 105–109.

Lobell, D. B., W. Schlenker, and J. Costa-Roberts. 2011. Climate trends and global crop production since 1980. *Science* 333(6042): 616–620.

Mitchell, D. 2011. Biofuels in Africa: Opportunities, prospects, and challenges. Washington, DC: The World Bank.

Morris, M. L. 2007. Fertilizer use in African agriculture: Lessons learned and good practice guidelines. The World Bank.

Muhammad, A., J. Seale, B. Meade, and A. Regmi. 2011. "International Evidence on Food Consumption Patterns: An Update Using 2005 International Comparison Program Data." USDA ERS Technical Bulletin No. (TB-1929) USDA ERS. http://www.ers.usda.gov/publications/tb-technical-bulletin/tb1929.aspx.

National Research Council. 2011. *Climate stabilization targets: Emissions, concentrations, and impacts over decades to millennia.* National Academy Press.

Naylor, R. L. 2014. Biofuels, Rural Development and the Changing Nature of Agricultural Demand. In Falcon, W.P. and R.L. Naylor (eds.), *Frontiers in Food Policy: Perspectives in sub-Saharan Africa*. Stanford. CA: Printed by CreateSpace.

Naylor, R. L., and W. P. Falcon. 2010. Food security in an era of economic volatility. *Population and Development Review* 36(4): 693–723.

———. 2011. The global costs of American ethanol. *The American Interest* 7(2): 66–76.

Negash, M. and J. Swinnen. 2012. Biofuels, poverty and food security: Micro-evidence from Ethiopia. Working paper. Belgium: LICOS Center for Institutions and Economic Performance, University of Leuven.

Nelson, G., R. Robertson, S. Msangi, and T. Zhu. 2009a. India greenhouse gas mitigation: issues for Indian agriculture. IFPRI Discussion Paper 00900. Washington, DC: International Food Policy Research Institute.

Nelson, G. C., et al. 2009b. Climate change: Impact on agriculture and costs of adaptation. Washington, DC: International Food Policy Research Institute.

Nussbaumer, P., M. Bazilian, and V. Modi. 2012. Measuring energy poverty: Focusing on what matters. *Renewable and Sustainable Energy Reviews* 16: 231–243.

Pimentel, D. and Patzek, T.W., 2005. Ethanol production using corn, switchgrass, and wood; biodiesel production using soybean and sunflower. *Natural Resources Research* 14(1): 65–76.

REN21. 2012. Renewables 2012 Global Status Report. Paris: REN21 Secretariat.

Roberts, M., and W. Schlenker. 2010. Identifying supply and demand elasticities of agricultural commodities: Implications for the US ethanol mandate. NBER Working Paper No. 15921. http://www.nber.org/papers/w15921.

Searchinger, T., R. Heimlich, R. A. Houghton, F. Dong, A. Elobeid, J. Fabiosa, S. Tokgoz, D. Hayes, and T. H. Yu. 2008. Use of US croplands for biofuels increases greenhouse gases through emissions from land-use change. *Science* 319(5867): 1238.

Steenblik, R., G. Kutas, and C. Lindberg. 2007. Biofuels: At what cost? Government support for ethanol and biodiesel in the European Union. Switzerland: International Institute for Sustainable Development.

Swinbank, A. 2009. EU policies on bioenergy and their potential clash with the WTO. *Journal of Agricultural Economics* 60(3): 485–503.

Taheripour, F., T. W. Hertel, W.E. Tyner, J. F. Beckman, and D. K. Birur. 2010. Biofuels and their by-products: Global economic and environmental impacts. *Biomass and Bioenergy* 34: 278–289.

Tekle, G. 2008. "Local Production and Use of bio-ethanol for Transport in Ethiopia: Status, challenges and lessons." Thesis for the fulfillment of the Master of Science in Environmental Management and Policy IIIEE, Lund University, Sweden.

Timmer, P. C., W. P. Falcon, and S. R. Pearson. 1983. *Food policy analysis*. Baltimore: Johns Hopkins University Press and Washington, D.C.: The World Bank.

United Nations Development Programme (UNDP). 2012. African Human Development Report 2012: Towards a food secure future. New York: UNDP Regional Bureau for Africa.

U.S. Energy Information Administration (EIA). International Energy Statistics. Available online from http://www.eia.gov, accessed August 4, 2012.

USDA. 2012. "Feed grains database." Washington, DC: USDA Economic Research Service. http://www.ers.usda.gov/data-products/feed-grains-database.aspx.

Valdes, C. 2011a. Brazil's ethanol industry: Looking forward. Washington, DC: U.S. Department of Agriculture, Economic Research Service. BIO-02.

Valdes, C. 2011b. Can Brazil meet the world's growing need for ethanol? *Amber Waves* 9(4): 36–45.

Van Dingenen, R., F. J. Dentener, F. Raes, M. C. Krol, L. Emberson, and J. Cofala. 2009. The global impact of ozone on agricultural crop yields under current and future air quality legislation. *Atmospheric Environment* 43(3): 604–618.

van Vliet, M. T., Yearsley, J. R., Ludwig, F., Vögele, S., Lettenmaier, D. P. and Kabat, P. 2012. Vulnerability of US and European electricity supply to climate change. *Nature Climate Change* 2(9): 676–681.

World Bank. 2007. World Development Report 2008: Agriculture for Development. Available online at http://go.worldbank.org/C3TCZPDAJ0.

Zimmerman, J., J. Dauber, and M. B. Jones. 2011. Soil carbon sequestration during the establishment phase of *Miscanthus* × *giganteus*: a regional-scale study on commercial farms using ^{13}C natural abundance. *GCB Bioenergy Special Issues: Land Use Change* 4(4) (July): 453–461.

10

Agricultural Nutrient Use and Its Environmental Consequences

Peter M. Vitousek and Pamela A. Matson

The substantial gains of the Green Revolution had multiple sources—including the widespread availability and use of fertilizers, especially nitrogen (N) and phosphorus (P). By freeing crops from nutrient limitations and by replacing the increased quantities of nutrients removed when high-yielding crops are harvested, enhanced yields could be achieved and sustained across a wide range of soil types. Over the next 30 years there will be 2 billion more people to feed, many people will consume more animal products, and urbanization, land degradation, and climate change will likely diminish the amount of arable land for food production. The use of industrial fertilizers is expected to increase further as a result, as many countries intensify production on existing agricultural land for domestic consumption and trade.

Although essential for food security, increased fertilizer use comes at a substantial cost to human health and the environment at local to global scales. Nonetheless, there are several ways to reduce the conflict between food production, environmental degradation, and human well-being—to continue to increase yields of agricultural products while at the same time reducing the environmental and health damages associated with fertilizer use. In this chapter, we discuss the scale of nutrient use and loss associated with fertilizers; the tensions that exist between food security, human health, and the environment; and the opportunities for resolving such tensions based on our collaborations with national scientists in two of the world's most important grain-producing regions—the Yaqui Valley of Mexico, and the North China Plain.

Understanding the challenges and opportunities surrounding agricultural nutrient use hinges on three main premises:

(1) Added nutrients are essential to high-yielding agriculture, which in turn is essential to global food security. Industrial fertilizer will supply a large portion of these nutrients for the foreseeable future.

(2) The global cycles of N and P have been transformed by human activities—most importantly, by their use as agricultural fertilizer. Transformation of those cycles demonstrably affects human health and well-being, the global climate system, and biological diversity and ecosystem functions on land and in the sea.

(3) The use of nutrients in agriculture, and the human/environmental consequences of that use, differ significantly at different stages of economic and agricultural development. Solutions to human and environmental challenges must take the level of economic and agricultural development into consideration.

The challenge then is to find ways to increase crop yields while simultaneously improving environmental quality. This challenge is starkest in rapidly developing economies like those of Mexico and China. Here the production/environment challenge is closely linked; potential yields of agriculture must be raised close to their biophysical potential to feed large and growing populations on a limited land base, and already-intolerable levels of environmental degradation must be reduced. As these rapidly developing economies become wealthier, trade can make an increasing contribution to their own national food security—but not necessarily to the global production/environment challenge. Internally, China has just 124 million ha of farmland with which to feed over 1.37 billion people (circa 2010), versus 165 million ha and 311 million people in the US (FAO 2010)—and the air and water there are so polluted by industry and agriculture that by one calculation China now experiences 750,000 premature pollution-caused deaths per year (World Bank 2007). Increasing crop yields towards limits of feasibility while at the same time extending many lives represents an immediate and compelling scientific challenge as well as an extraordinary contribution to human well-being.

The Need for External Nutrients

All plants require nutrient elements to survive and grow, and N and P typically are the nutrients in shortest supply. Moreover, some of these nutrients are removed from fields in crops or other harvested products, or lost via other pathways including runoff, leaching, and erosion. If the productivity of an agricultural system is to be sustained, these lost nutrients must be replaced. Some recycling of harvested nutrients is possible, such as via human or animal waste and other organic matter. However, these pathways typically are insufficient, even leaving considerations of cost (which are substantial) aside. New nutrients must be added to sustain yields.

Clearly, it is possible to add nutrients to fields without utilizing industrial fertilizer; humanity added nutrients to cropping systems for millennia before the use of industrial fertilizer became widespread in the mid-twentieth century. Concentrating essential nutrients in space was accomplished through small-scale integrated crop/animal systems that in essence used animals to harvest nutrients from a wide area, and then collected nutrient-rich manure and applied it to intensively cropped areas. Traditional farmers also concentrated essential nutrients in time through shifting cultivation systems, where long fallow periods are used to store dilute environmental inputs of nutrients in biomass. These stored nutrients are then released in a brief time following felling and burning. Either approach (as well as others) can be sustainable, but more land per person must be used than in more nutrient-intensive systems.

Modern organic systems are far more efficient in their use of land; they import nutrients in organic or low-concentration (e.g., rock phosphate) forms, and they manage short fallows and cover crops to supply N in particular through biological fixation. There is an ongoing debate about whether it would be possible to feed the growing, increasingly urbanized and affluent global population using modern organic methods. We do not intend to enter this debate here. However, we believe it is clear that industrial fertilizer is a relatively inexpensive and highly concentrated (hence readily transportable) source of nutrients for intensive agriculture, and that it will continue to provide a large fraction of the nutrients that sustain agricultural yields globally.

Transformation of the Global Cycles

The massive and surprisingly recent alteration of the global cycles of N and P by human activity—predominantly by industrial fertilizers—is one of the clearest and best-documented examples of the influence of human activity on the Earth system, and one that causes substantial human and environmental damage. We can characterize the nature and extent of this human influence by tracking the movement of these elements from their long-term geological reservoirs into circulation in the biosphere.

For N, the main reservoir is dinitrogen gas (N_2) in the atmosphere—a form that most organisms cannot utilize. Prior to human alteration of the global N cycle, all life was supported by the transformation of N_2 in the atmosphere into biologically available forms. Most of this N was "fixed" biologically (combined with carbon, hydrogen, or oxygen into forms most organisms can use) by just a few groups of organisms with a specialized, energy-demanding biochemical pathway. Smaller contributions came from the weathering of biologically available nitrogen from sedimentary rocks, and from fixation of atmospheric N_2 by lightning.

A recent calculation (Vitousek et al. 2013) suggested that background (pre-industrial) biological N fixation totaled about 45 Tg (1 Tg = 1 million metric tons) on land; another 10 Tg was released from organic forms in rock, and about 4 Tg were fixed by lightning (Table 10.1). That ~60 Tg of N supported all life on land; it was balanced by an equivalent loss of N to the atmosphere and

TABLE 10.1
A Global Budget for Nitrogen Fixation on Land

Background Inputs (pre-industrial)	
Biological N Fixation	44
Lightning Fixation	4
Weathering Organic Rock N	10
Reactive N from Oceans	3
Anthropogenic Inputs	
Industrial	122 (100 as fertilizer N)
Crop N Fixation	48
Fossil Fuel	25

Note: All values in Tg/yr (millions of metric tons per year).

Source: Galloway et al. (2008), Herridge et al. (2008), and Vitousek et al. (2013).

oceans. With the invention of the Haber-Bosch process in the early twentieth century (Smil 2004), humanity developed an efficient way to fix N_2 industrially. Industrial N fixation now amounts to ~122 Tg/yr, about 100 Tg of which are used as industrial N fertilizer (Galloway et al. 2008). Another 40–55 Tg of N are fixed biologically in managed cropping and intensive grazing systems (Herridge et al. 2008), and so represent additional human influence. A further 20–25 Tg are fixed inadvertently as fossil fuels are burned for energy. In all, nearly 200 Tg of N are now transferred each year from N_2 to biologically available forms on land by human action—about three times the total background supply.

The cycle of P is altered to a similar extent. The main reservoir is mineral P in rocks, the background flux is the weathering of these rocks, and human enhancement of P supply is caused primarily by the mining and processing of P-rich rocks into industrial fertilizer and other products (Bennett et al. 2001). Indeed, there is concern that accessible reserves of P-rich rocks suitable for fertilizer manufacture are being depleted rapidly. Whether or not an absolute shortage of P is forthcoming, as discussed by Naylor in the introductory chapter, the uneven geographic distribution of P-rich rocks makes economic dislocations in the supply of this non-substitutable resource relatively likely in the next few decades (Elser and Bennett 2011).

The effects of this massive, recent, and ongoing alteration of the N and P cycles are experienced in global climate, in air and water quality, in biological diversity and ecosystem functioning, and in human health and economic systems. A variable (and often large) portion of the agricultural nutrients applied to fields is not taken up by crops, and instead makes its way into the broader environment. Nutrients can also move from the food system (or other pathways) into broader circulation. The greenhouse gas nitrous oxide (N_2O) is increasing in concentration as a result of human activity, most importantly fertilizer application. N_2O makes a several-percent contribution to ongoing human forcing of the climate system. Nitrous oxide can also catalyze the breakdown of Earth's protective stratospheric ozone layer, an impact that could become significant as the influence of anthropogenic chlorofluorocarbons fades.

Reactive gaseous forms of N (ammonia [NH_3] and oxidized N [NO_x] and their derivatives) contribute to photochemical smog, including ground-level (tropospheric) ozone, to the formation of airborne particulate pollution, and to acid rain. Dissolved and suspended forms of N and P drive cultural eutrophication (over-enrichment) in many regions, leading to overgrowth of algae and aquatic plants, and widely distributed hypoxic zones ("dead zones") in lakes, estuaries, and the coastal ocean. In terrestrial as well as aquatic systems, eutrophication greatly reduces the biological diversity of natural and semi-natural ecosystems (Vitousek et al. 1997). Finally, N-related air pollution has substantial effects on human health; the recent European Nitrogen Assessment (Sutton et al. 2011) estimated the costs of N enrichment in Europe to be 70–320 billion Euros/yr, mostly from health costs associated with air pollution. Moreover, as discussed below N-driven air pollution now is substantially worse in China than in Europe.

Many of the environmental costs of altered nutrient cycles also influence agricultural systems, in some cases posing a substantial threat to regional and global food security. Lobell, Naylor, and Field discuss the agricultural consequences of climate change in Chapter 9. In addition to climate-induced impacts, tropospheric ozone caused by increased nitric oxide emissions drives substantial reductions in yields of major crops (Avnery et al. 2011); over-fertilization with N drives soil acidification in Chinese croplands (Guo et al. 2010); and coastal eutrophication can damage fisheries and aquaculture (Diaz et al. 2012).

Regional Differences in the Use and Consequences of Nutrients in Agriculture

Global summaries such as that presented in Table 10.1 gloss over a great deal of variation between agricultural systems in different regions, and indeed within individual nations. Much of this variation is associated with stages of economic and agricultural development. For example, many continuously cropped agricultural systems in sub-Saharan Africa use relatively little fertilizer (or other inputs of nutrients) as discussed in by Burney in Chapter 6. Nutrient removals in harvested crops can be larger than nutrient additions, and in such cases yields are based on drawing down the nutrient capital of what once may have been fertile soils (Table 10.2). These systems have a problem of too few nutrients, not too many, making them relatively unproductive and unsustainable. Land degradation and the conversion of marginal land to agricultural purposes are all but inevitable under these conditions.

In contrast, agricultural policy in many rapidly developing economies, including China, Indonesia, India, and Mexico, has emphasized increasing yields for several decades (as described in Chapters 2 and 3). Many intensive systems in these countries now employ truly massive amounts of fertilizer. On a national scale, China now uses more fertilizer N than the United States and the European Union combined (Figure 10.1). Where many sub-Saharan systems are dependent on "mining" or drawing down the natural store of soil nutrients, many Chinese systems apply far more nutrients than crops can utilize. The unused nutrients are wasted from an agronomic perspective, and can do significant damage from health and environmental perspectives. Fertilizer application levels are high enough to drive land degradation through soil acidification (Guo et al. 2010).

The situation is different in the United States, Canada, the European Union, and other developed economies. Excessive fertilizer application was common in those areas some decades ago, but in most cases the amount of fertilizer applied over and above crop requirements has decreased in recent decades. In some cases, fertilizer application has decreased absolutely; in more cases, farmers are achieving greater yields from similar levels of fertilizer application, largely as a consequence of technological improvements in crop varieties and cropping systems. Either way, the quantity of nutrients removed from fields in harvested products is much closer to the quantity of nutrients added to

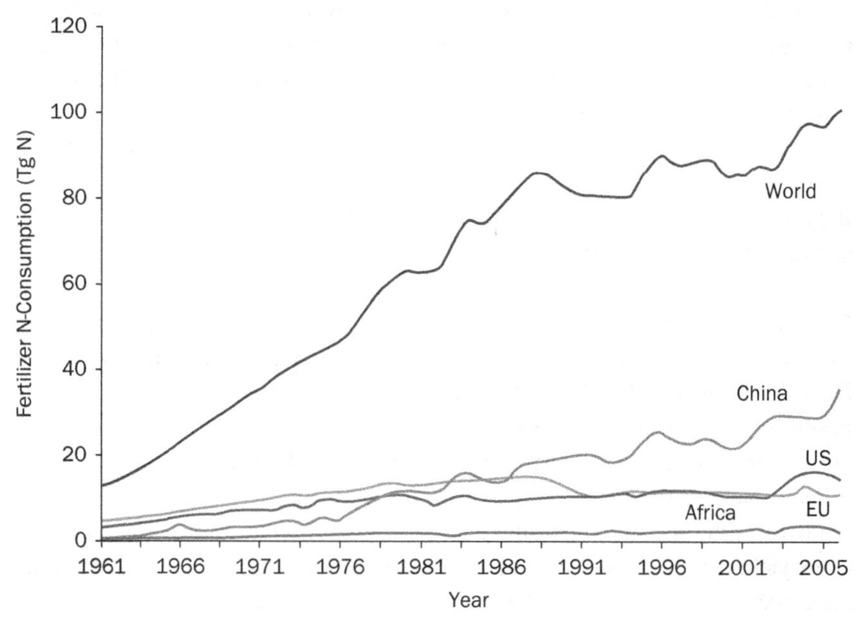

FIGURE 10.1 The Use of Nitrogen Fertilizer, Globally and by Regions, 1961–2006 EU, European Union; Tg, 1 million metric tons (1,000 kg); US. United States.

fields than was the case several decades ago. Air and water pollution from agriculture remain ongoing and expensive problems in Europe and North America, and the current situation is hardly one of harmony between agriculture and environment. However, intensive systems are closer to a balance between nutrient inputs and removal than they are in China, or than they were in Europe and North America 30 years ago (Vitousek et al. 2009). While it is possible to find particular farms, even particular regions, that are similar to any of the columns in Table 10.2 in sub-Saharan Africa, China, or the United States, the table does reflect a strong average difference in nutrient use and deficit or excess that is associated with economic and agricultural development.

The foregoing discussion suggests that industrial fertilizers will continue to be used extensively for the foreseeable future. Indeed, their use will likely increase as many countries develop economically and agriculturally, and seek to increase yields on their more productive agricultural lands. At the same time, industrial fertilizer is the largest source of transformation in the global cycles of N and P, and transforming these cycles has substantial human and environmental costs. At some level this nutrient-based agriculture/environment interaction embodies an irreducible conflict; there is no way to produce sufficient food for over 7 billion people without altering, and in some meaningful ways degrading, the environment locally, regionally, and even globally. However, we believe that it is possible to reduce this conflict substantially—to continue to increase yields of agricultural products while at the same time reducing the environmental damage associated with fertilizer use. The opportunity is clearest in the rapidly developing economies—China, Mexico, and others—where a large fraction of the nutrients added to agriculture never reach their intended

TABLE 10.2

Fertilizer Use in Three Contrasting Regions: A Comparison of the agricultural nutrient economy in three regions at different stages of agricultural/economic development–western Kenya, the North China Plain, and the Corn Belt of the United States.

	Western Kenya		North China		Midwest U.S.	
	N	P	N	P	N	P
Fertilizer	7	8	588	92	93	14
Biological N fixation					62	
Total agronomic inputs	7	8	588	92	155	14
Removal in grain/beans	23	4	361	39	145	23
Removal in other harvested products	36	3				
Total agronomic outputs	59	7	361	39	145	23
Agronomic inputs minus harvest removals	−52	+1	+227	+53	+10	−0

Note: All values in kg ha^{-1} yr^{-1}.

Source: Vitousek et al. (2009).

target. "Win-win," efficiency-based solutions that benefit both farmers and the environment should be possible under these circumstances.

Fertilizer Use in Rapidly Developing Economies—Case Studies

In the following sections, we discuss two regional cases in rapidly developing economies where it should be possible to maintain or enhance yields while reducing the environmental footprint of intensive agriculture: the Yaqui Valley of Sonora, Mexico, and the North China Plain. Our focus here is on fertilizer use and its consequences, but a similar analytical framework can be applied to other inputs in intensive agriculture such as pesticides. We discuss both the development of cropping systems that can maintain or enhance yields while reducing economic and environmental costs, and the opportunities and barriers to implementing these cropping systems in practice, over wide areas. Any strategy for implementation must include both biophysical and social components. A cropping system that uses nutrient inputs highly efficiently will have no impact unless farmers adopt it; conversely, well-designed systems of education and outreach must be based on interventions that make biophysical sense.

YAQUI VALLEY, MEXICO

The Yaqui Valley is the birthplace of the Green Revolution for wheat. In 1943, Norman Borlaug, the American agronomist most closely associated with the Green Revolution, launched a wheat research program in this area. Located in Sonora, northwest Mexico, the Yaqui Valley consists of 235,000 ha of irrigated wheat-based agriculture, and is one of the country's most productive breadbaskets (Naylor et al. 2001). Working with the Mexican government and the Rockefeller Foundation, Borlaug initiated an effort that was the forerunner of the International Maize and Wheat Improvement Center (CIMMYT), which remains active in the region today. Later, a national agricultural experiment center (CIANO) was also established in the Valley. Today, farmers there produce some of the highest wheat yields in the world using a combination of irrigation, high fertilizer application rates, and modern cultivars (Matson et al. 1998). Borlaug selected the Yaqui Valley as an ideal place for the early wheat improvement program because the region is agro-climatically representative of 40 percent of the developing world's wheat growing areas. For the same reason, and due to the long history of cultivation, research and implementation, it makes a useful case study for examining the agronomic and social challenges related to reducing over-fertilization in intensive agricultural systems.

The use of fertilizer N has increased markedly in the Valley in the past four decades. This increasing rate of fertilizer application provided a leading indicator of what is now occurring in other high-productivity, irrigated cereal systems in the developing world (Vitousek et al. 2009; Matson 2012). While the region has a number of important resource and sustainability challenges that interact with one another (Matson 2012), we focus here on issues related to the use of nitrogen fertilizer. In particular, we discuss an integrated analysis of the agronomic, economic, and environmental dimensions of fertilizer application in the Yaqui Valley, and then describe ongoing work to identify and implement fertilizer management practices that could make sense for farmers and the environment.

Beginning in the mid-1990s, a team of researchers led by two of the authors of this volume, Pamela Matson (a biogeochemist then at UC Berkeley) and Rosamond Naylor (an economist at Stanford University), and Ivan Ortiz Monasterio (an agronomist with CIMMYT), began to evaluate crop growth and yields, economic costs and benefits to farmers, and nitrogen inputs and outputs in the Yaqui Valley system.[1] Our surveys indicated that by 1995, the typical farmer applied on average 250 kg/ha of fertilizer per six-month wheat crop, with around 180 kg/ha of the N applied as urea approximately a month before planting, and the rest typically applied as anhydrous ammonia in flood irrigation water later in the growing cycle. Application rates had been steadily increasing since the 1950s, but while wheat yields had increased only marginally between 1980 and 2000, average fertilizer rates continued to increase by over a third—encouraged, up until the early 1990s, by heavy subsidies.

Our surveys suggested that farmers had a number of rational reasons for their fertilizer management approach. Many farmers cited the importance of getting fertilizer on early in the crop cycle for two reasons. There was a perceived need for nitrogen to support plant growth early in the crop cycle, and to reduce the risk that rains would make the fields inaccessible to machinery later in the crop cycle and thus delay fertilization until too late to make a difference in yield. Farmers also cited the need to spread labor through the season by completing fertilizer application before the workers and machinery were needed for planting. Although some farmers had developed machinery for simultaneous seed and fertilizer application, it was not a common practice. Subsidies had maintained low fertilizer prices from the early Green Revolution years through the 1980s. But by 1995 fertilization had become the highest direct production cost in Yaqui Valley farm budgets, exceeding even the costs of tillage and other land preparation (Matson et al. 1998).

Over several years, we evaluated yields, grain quality, soil nutrients, and the flux of important N-containing gases to the atmosphere as well losses of N in solution to ground and surface waters. We carried out our studies in farmers' fields as well as in experimental fields at CIMMYT's station in the Valley. We included several fertilizer management approaches in our experiment: the

[1] For a history and summary of the Yaqui Valley project over the course of two decades of research, see Matson (2012).

conventional farmers' practice for the valley, as determined by farm surveys; three alternative practices that were based on agronomist recommendations and that added less fertilizer N and/or shifted some of the fertilizer application later in the crop cycle; and a non-fertilized control. In our treatment that simulated the farmers' practice, 187 kg N/ha of urea were applied to dry soils one month prior to planting, followed by pre-planting irrigation; an additional 63 kg N/ha of anhydrous ammonia were applied approximately six weeks following planting.

The results of our analyses have been reported and summarized in a number of publications (e.g., Matson et al. 1998; Ahrens et al. 2009; Matson 2012). In short, the typical farmer's practice of fertilizing dry soil and then irrigating prior to planting resulted in extremely high levels of inorganic ammonium and nitrate in the soil, and very large emissions of nitrogen trace gases (nitrous oxide and nitric oxide) to the atmosphere and losses of ammonium and nitrate in water leaching from or running off the site. By the time farmers planted their wheat seeds, 40 percent of the fertilizer had been lost from the top 15 cm of soil and 26 percent lost from the top meter. Our further studies evaluated some of the additional loss pathways of nitrogen and their environmental consequences (Figure 10.2) and provided a picture of fertilizer N losses that cascaded from

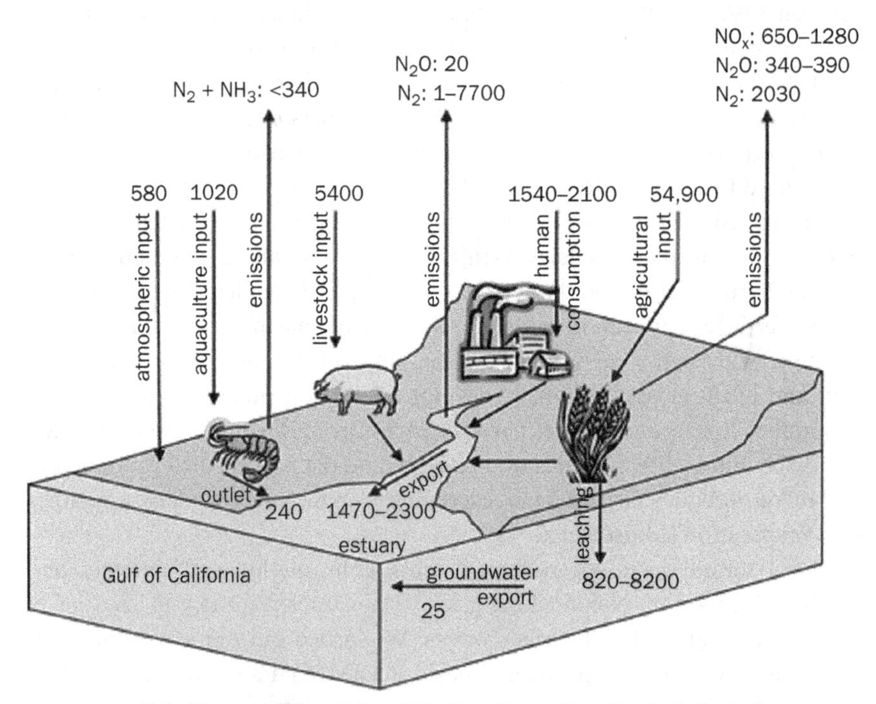

FIGURE 10.2 Major Fluxes of N to and from Agriculture, Aquaculture, Livestock Production, and Human Consumption in the Yaqui Valley Units for all numbers are reported in Mg N/yr.

field to streams and canals and ultimately to the Gulf of California, where it led to vast phytoplankton blooms.

The results of the alternative fertilization treatments indicated that both the amount and timing of fertilizer application matter. Reducing the amount of fertilizer that is applied before planting resulted in significant reduction of losses, even when the same amount of N was applied later in the season to achieve the same total rate of application. The "best" alternative (with respect to reduction in losses) eliminated pre-planting fertilizer application, and instead applied a total of 180 kg N/ha, with 33 percent of the total applied at planting and 67 percent six weeks post-planting. In this treatment, total losses of nitrogen were an order of magnitude less than the farmers' practice. The elimination of pre-planting fertilization and application of most fertilizer later in the crop cycle seemed to make agronomic sense as well.

Fertilizer use and loss are just one component of farm budgets, and farmers typically focus on the balance between costs and expected income from yield of good-quality wheat. Yields reported in our 1994 and 1995 socioeconomic surveys ranged from 3.1 to 7.3 t/ha, with average values of 4.9 and 5.3 t/ha for the two seasons. Mean yields in our simulated farmer practice were 6.08, and the best alternative, which added 180 kg N/ha in contrast to 250 kg N/ha in the farmers' practice, resulted in very similar yields (6.16 t/ha). Likewise, grain quality (estimated as the protein concentration in grain) in the alternative was not significantly different from the farmers' practice (14.87 percent versus 14.83 percent, respectively).

Given fertilizer's important place in Valley farm budgets, we evaluated the extent to which increased fertilizer efficiency would represent cost savings to the farmers (Matson et al. 1998). In contrasting the farmers' practice with our best alternative, we found that the alternative strategy resulted in a savings of N$414–571/ha, or US$55–76/ha at the prevailing prices and exchange rates. The lower application rates and reduced loss of fertilizer could have added 12–17 percent to after-tax profits. Given these good results, this regimen was tested in 28 on-farm trials over the following two crop cycles. The average results supported our conclusion that 180 kg/ha (applied at planting and one time after) was more profitable than 250 kg/ha (applied primarily pre-planting). Altogether, our results suggested that alternative fertilizer practices could reduce trace gas and total losses of fertilizer and maintain yields and grain quality. They also suggested that the best alternative, which matched the timing of fertilizer application to crop demand, ultimately could save money and perhaps allow Yaqui Valley farmers to remain competitive in an era of economic liberalization and expanding free trade.

At the time that we completed and published the first study (and reported our results in farmer workshops in the Yaqui Valley), we wondered if the potential cost savings might actually induce a shift in technology and management toward fertilization later in the wheat cycle. Our next surveys in 1998 and 2001 indicated that, if anything, farmers were applying more N, not less, despite continued increases in fertilizer prices and no overall change in yield. In 2001, the average application rate was 263 kg/ha. Clearly, we were missing something important in terms of farmer decision making!

At this point, we pursued several new research directions in an attempt to understand the barriers to adoption of this new practice despite the significant potential to cut costs, and to adopting more environmentally friendly practices generally. Our analyses indicated that many farmers perceived risks in the potential interaction of rain and fertilization, and thus included pre-planting fertilization. On the other hand, the fact that 10 percent of farmers were no longer doing pre-planting fertilization suggested that all farmers did not view the risks similarly. Team-member research also uncovered the fact that inter-annual variability in growing season temperatures and spatial variability in residual soil nitrogen amounts and availability, among other factors, cause significant uncertainty for farmers in the Valley (Lobell et al. 2004).

Finally, through a number of casual conversations and then a more formal survey, we discovered that the farmer credit unions to which most farmers belonged were encouraging the use of increasing amounts of fertilizer. (Credit unions in this region are farmers' associations that provide members with credit, seeds, fertilizer, other inputs, insurance, postharvest storage, marketing, and technical assistance.) In fact, because credit unions regulate the farm management activities of their members by providing crop insurance that is contingent on compliance with official production recommendations, many farmers had to follow these recommendations. These recommendations no doubt reflected concerns about risks and variability not just in year-to-year climate and field-to-field soil conditions but also in farmer skill levels. By requiring all farmers to over-fertilize, in essence, the credit unions were ensuring that no farmers would under-fertilize and achieve sub-optimal yields. There were also financial incentives for unions to encourage high levels of fertilizer use.

Given the real concerns of farmers and credit unions about temporal and spatial variability, we began to direct our efforts toward the development of diagnostic tools that would assist farmers in decision making under uncertainty. The development of the "GreenSeeker" sensor for site-specific nitrogen management has been described in detail elsewhere (Ortiz Monasterio and Lobell 2012). This technology allows estimation of how much nitrogen a crop will need to reach its yield potential starting at 40 days after planting. Mounted on a fertilizer sprayer, the sensor measures light reflectance in several wavelengths. Combined with information on crop growth patterns and comparisons with plants growing in narrow, highly fertilized strips within the field, these measurements allow estimation of the nitrogen demand of the crop, so that farmers can adjust the application rate accordingly. Thus, farmers have a tool that allows them to determine when and how much fertilizer is needed in specific fields under specific climate conditions.

Having learned the importance of the credit unions in farmer decision making, we realized we needed to engage them as well as individual farmers in trials of this new technology. We engaged six credit unions in the purchase, use and analysis of the system, with each union paying 50 percent of the costs of an instrument that could be dedicated to use by its members. Six years of evaluation in farmer's fields showed an average reduction of 70 kg N ha^{-1} yr^{-1} in fertilizer application without a yield reduction. During the 2008–09 crop cycle, this technology was used on approximately 7,000 ha; further uptake in use is currently being evaluated. This use of relatively inexpensive technology by groups of farmers to help manage uncertainty represents one promising path towards maintaining or enhancing yields, profits, and food security while reducing the environmental costs of fertilizer use. It remains to be seen, however, whether and how quickly it is widely adopted.

NORTH CHINA PLAIN

Like the Yaqui Valley, China has made substantial investments for decades in high-input, high-yield agricultural systems. These investments have paid off in terms of higher yields; for example, total cereal production increased over 30 percent from 2003 to 2011 on a nearly constant area of cropland (Zhang et al. 2013). However, that increase has come at a substantial human and environmental cost. Levels of nutrient pollution, the majority of which is derived from agriculture, are among the highest in the world (Liu et al. 2013). This pollution affects human health and the land, freshwater, and marine ecosystems in and near China (Guo et al. 2010; Kim et al. 2011; Liu et al. 2013).

China's role in global fertilizer applications is striking (Figure 10.1). About one third of all N fertilizer use worldwide takes place there (Robertson and Vitousek 2009), and about 60 percent of the global increase in N fertilizer use from 2000 to 2010 occurred in China (Chen et al. 2011). Equally impressive is the depth of research and analysis on nutrient use and loss in key staple crop systems by Chinese scientists. We have been fortunate to participate in this growing body of research in recent years, with specific attention to the North China Plain.

The North China Plain is one of the breadbaskets of China; much of the area is double-cropped with winter wheat and summer maize grown on farms little more than 0.1 ha in size (Ju et al. 2009). Normal farmers' practices involve remarkably high levels of nutrient addition that exceed even the applications in the Yaqui Valley. Farmers apply nearly 600 kg/ha/yr of N and 90 kg/ha/yr of P directly as fertilizer. Nutrients are also added as manure, in irrigation water for winter wheat, and in atmospheric deposition; these forms contribute another nearly 200 kg/ha of N (Ju et al. 2009). The return on this massive fertilization is low. The partial factor productivity for added N (PFP_N) was just 37 kg grain produced per kg N added to Chinese maize in 2009, for example, versus 66 kg grain/kg N for maize in the United States. This implies that a very large fraction of applied N in China is lost to the environment without contributing to agricultural production (Ju et al. 2009; Chen et al. 2011; Zhang et al. 2013).

Agricultural scientists in China have recognized and documented this overuse of fertilizer and many of its consequences (Gao et al. 2006; Huang et al. 2008; Ju et al. 2009). Indeed, agricultural research and monitoring systems in China are among the best in the world. Following in part on our research in Mexico, researchers in China have developed alternative practices that maintain or enhance yields while greatly increasing PFP_N and so reducing N losses to the environment. For example, Ju et al. (2009) demonstrated that they could halve the use of N fertilizer in a double-cropped wheat-maize rotation, from 588 to 286 kg/ha, without reducing grain yields or grain quality. They further documented much lower losses of N from this more-efficient system. This result was similar to the reduction in fertilizer use (and consequent increase in PFP_N and reduced environmental costs) that we saw in the Yaqui Valley, with the addition that it was carried out at an amazingly larger scale across a broad area of North China Plain. It was also applied successfully to a wheat-rice double cropping system in South China.

Simply maintaining current yields, however, is insufficient to meet the agricultural production targets China seeks. The focus of current research and outreach is the development and adoption of "double-high" systems of production, systems that increase both yields and resource efficiency (including the PFP of fertilizer) (Zhang et al. 2013). This work combines cropping systems that are designed in part through the use of crop models that help to optimize crop varieties, planting times, and planting densities with nutrient management systems that match the supply of nutrients to the demands of crops. In both experimental fields and extensive on-farm trials, they have achieved substantially increased yields of high quality grain without using any more fertilizer than do current practices (Chen et al. 2011). The higher yields achieved by these systems may be more appealing to both farmers and policymakers than are lower-input systems that produce yields equivalent to current farmers' practices.

The barriers to adoption of these information-intensive approaches in China differ from those in the Yaqui Valley. In the North China Plain in particular, barriers include literally millions of farmers with very small farm sizes, the part-time nature of farming, and the relatively low education levels of farmers—which together make it extremely difficult to implement model-based precision agricultural systems in practice. These conditions are not permanent, however; among other factors, ongoing increases in migration to cities and in labor costs likely will contribute to increasing consolidation and mechanization of farms, possibly allowing for more professional management. In fact, designing improved nutrient management strategies for the current small-scale structure of staple crop production in China may amount to addressing the problems of the past, not the challenges of the future.

Despite these challenges, we believe that agricultural research and outreach in China can benefit substantially from the approaches taken in the Yaqui Valley, and we are working with scientists from China Agricultural University in Beijing to facilitate this approach. While the biophysical side of the agricultural research in the North China Plain is superb, it is fair to say that neither the social and policy barriers to implementation of "double-high" agricultural practices nor the knowledge system surrounding the potential adoption of new practices have yet received sufficient attention. Our colleagues in China understand that evaluating intensive agriculture as a coupled human-environment system rather than primarily a biophysical system is likely to prove rewarding.

Conclusion

The Yaqui Valley and the North China Plain are specific examples of a broader phenomenon—substantial overuse of fertilizer in many rapidly growing economies. There are special factors at work in these two examples, and indeed in any particular place. These examples nevertheless illustrate a globally significant issue and suggest ways that the issue may be addressed. Most importantly, any strategy that seeks to address the coupled production-environment challenges and opportunities posed by intensive agriculture in rapidly developing economies must include social as well as biophysical components. A technically beautiful cropping system that uses inputs of nutrients highly efficiently will have no impact unless farmers adopt it. At the same time, well-designed systems of education and outreach must be based on interventions that make sense biophysically.

Understanding both the biophysical and social systems requires time and care. Identifying the key participants in a region's knowledge networks—the interconnected and often overlapping networks of actors and organizations that link knowledge and know-how with action—requires research and dedicated time in the field (McCullough and Matson 2011). Similarly, understanding the incentives, financial resources, institutions, and human capital underpinning an agricultural system requires substantial investment of time and effort (Naylor et al. 2001; Naylor and Falcon 2012). Ultimately, to encourage any positive change we need to evaluate who the real decision makers are in any system, and understand their concerns, motivations, and the historical and cultural context in which they operate.

Coupled human-natural systems are dynamic—and particularly in rapidly developing economies, changes are likely to be substantial and continuous. The ongoing increase in rural wages in China is an example of an important change that will transform the organization of cropping systems in the North China Plain and other agricultural regions of China. Solving the problems raised by the current situation will have little influence on tomorrow's systems, unless these changes are anticipated and incorporated in our analyses, proposals and plans for a transition to sustainable agriculture.

References

Ahrens, T. D., J. M. Beman, J. A. Harrison, P. K. Jewett, and P. A. Matson. 2009. A synthesis of nitrogen transformations and transfers from land to the sea in the Yaqui Valley agricultural region of northwest Mexico. *Water Resources Research,* 44, W00A05.

Avnery, A., D. L. Mauzerall, J. Liu, and L. W. Horowitz. 2011. Global crop yield reductions due to surface ozone exposure: 1. Year 2000 crop production losses and economic damage. *Atmospheric Environment* 45: 2284–2296.

Bennett, E. M., S. R. Carpenter, and N. F. Caraco. 2001. Human impact on erodable phosphorus and eutrophication: a global perspective. *BioScience* 51: 227–234.

Chen, X. P., Z. L. Cui, P. M. Vitousek, K. G. Cassman, P. A. Matson, J. S. Bai, Q. F. Meng, P. Hou, S. C. Yue, V. Romheld, and F. S. Zhang. 2011. Integrated soil-crop management system for food security. *Proceedings of the National Academy of Sciences USA* 108: 6399–6404.

Diaz, R., N. N. Rabalais, and D. L. Breitburg. 2012. Agriculture's impact on aquaculture: Hypoxia and eutrophication in marine waters. *Organization for Economic Co-operation and Development.*

Elser, J. J., and E. M. Bennett. 2011. A broken biogeochemical cycle. *Nature* 478: 29–31.

FAO Statistical Databases. 2010. Agriculture Data. http://apps.fao.org/page/collections?subset=agriculture.

Galloway, J. N., A. R. Townsend, J. W. Erisman, M. Bekunda, Z. E. Cai, J. R. Freney, L. A. Martinelli, S. P. Seitzinger, and M. A. Sutton. 2008. Transformation of the nitrogen cycle: Recent trends, questions, and potential solutions. *Science* 320: 889–892.

Gao, C., B. Sun, and T.-L. Zhang. 2006. Sustainable nutrient management in Chinese agriculture: Challenges and perspective. *Pedosphere* 16: 253–262.

Guo, J. H., X. J. Liu, Y. Zhang, J. L. Sheng, W. X. Han, W. F. Zhang. P. Christie, K. Goulding, P. Vitousek, and F. S. Zhang. 2010. Significant acidification in major Chinese croplands. *Science* 327: 1008–1110.

Herridge, D. F., M. B. Peoples, and R. M. Boddey. 2008. Global inputs of biological nitrogen fixation in agricultural systems. *Plant Soil* 311: 1–18.

Huang, J., R. Hu, J. Cao, and S. Rozelle. 2008. Training programs and in-the-field guidance to reduce China's overuse of fertilizer without hurting profitability. *Journal of Soil and Water Conservation* 63: 165A–167A.

Ju X.-T., G.-X. Xing, X.-P. Chen, S.-L. Zhang, L.-J. Zhang, X.-J. Liu, Z.-L. Cui, B. Yin, P. Christie, Z.-L. Zhu, and F.-S. Zhang. 2009. Reducing environmental risk by improving N management in intensive Chinese agricultural systems. *Proceedings of the National Academy of Sciences USA* 106:3041–3046.

Kim, T.-W., K. Lee, R. G. Najjar, H.-D. Jeong, and H. J. Jeong. 2011. Increasing N abundance in the northwestern Pacific Ocean due to atmospheric nitrogen deposition. *Science* 334: 505–509.

Liu, X.-J., Y. Zhang, W. Han, A. Tang, J. Shen, Z. Cui, P. M. Vitousek, J. W. Erisman, K. Goulding, P. Christie, A. Fangmeier, and F. S. Zhang. 2013. Enhanced nitrogen deposition over China. *Nature* 494: 459–462.

Lobell, D. B., J. I. Ortiz-Monasterio, and G. P. Asner. 2004. Relative importance of soil and climate variability for nitrogen management in irrigated wheat. *Field Crops Research* 87: 155–165.

Matson, P. A. (ed.) 2012. *Seeds of sustainability: Lessons from the birthplace of the Green Revolution*. Washington, DC: Island Press.

Matson, P. A., R. L. Naylor, and I. Ortiz-Monasterio. 1998. Integration of Environmental, Agronomic, and Economic Aspects of Fertilizer Management. *Science* 280: 112–15.

McCullough, E., and P. A. Matson. 2011. Evolution of the Knowledge System for Agricultural Development in the Yaqui Valley, Sonora Mexico. *Proceedings of the National Academy of Sciences USA*.

Naylor, R. L., and W. P. Falcon. 2012. The Yaqui Valley's Agricultural Transition to a More Open Economy," in *Seeds of Sustainability: Lessons from the Birthplace of the Green Revolution*, ed. P. A. Matson, 107–138. Washington DC: Island Press.

Naylor R. L., W. P. Falcon, and A. Puente-Gonzalez. 2001. Policy reforms and Mexican agriculture: Views from the Yaqui Valley. CIMMYT Economics Program Paper No 01-01. Mexico, DF: CIMMYT.

Ortiz-Monasterio, I., and D. B. Lobell. 2012. Agricultural research and management at the field scale," in *Seeds of sustainability: Lessons from the birthplace of the Green Revolution*, ed. P. A. Matson, 139–169. Washington DC: Island Press.

Robertson, G. P., and P. M. Vitousek. 2009. Nitrogen in agriculture: Balancing the cost of an essential resource. *Annual Review of Environment and Resources* 34: 97–125.

Smil, V. 2004. Enriching the earth: Fritz Haber, Carl Bosch, and the transformation of world food production. Cambridge, MA: MIT Press.

Sutton, M. A., C. M. Howard, J. W. Erisman, G. Billen, A. Bleeker, P. Grennfelt, H. van Grinsven, and B. Grizzetti. 2011. *The European Nitrogen Assessment: Sources, effects, and policy perspectives*. New York: Cambridge University Press.

Vitousek, P. M., J. D. Aber, R. W. Howarth, G. E. Likens, P. A. Matson, D. W. Schindler, W. H. Schlesinger, and D. Tilman. 1997. Human alteration of the global nitrogen cycle: sources and consequences. *Ecological Applications* 7: 737–750.

Vitousek, P. M., R. L. Naylor, T. Crews, M. B. David, L. E. Drinkwater, E. Holland, P. Johnes, J. Katzenberger, L. A. Martinelli, P. A. Matson, G. Nziguheba, D. Ojima, C. A. Palm, G. P. Robertson, P. A. Sanchez, A. R. Townsend, and F.-S. Zhang. 2009. Nutrient imbalances in agricultural development. *Science* 324: 1519–1520.

Vitousek, P. M., D. N. L. Menge, S. C. Reed, and C. C. Cleveland. 2013. Biological nitrogen fixation: Rates, patterns and ecological controls in terrestrial ecosystems. *Philosophical Transactions of the Royal Society B* 368: 20130119.

The World Bank and the State Environmental Protection Administration, P. R. China. 2007. *Cost of Pollution in China*. Washington DC: Rural Development, Natural Resources and Environment Management Unit, East Asia and Pacific Region, the World Bank.

Zhang, F. S., X. P. Chen, and P. M. Vitousek. 2013. An experiment for the world. *Nature* 497: 33–35.

11

Water Institutions and Agriculture

Barton H. Thompson, Jr.

Water is critical to food security. In arid regions of the world, irrigation can double or triple yields, making the difference between food shortages and food adequacy. Even in areas of the world with regular rainfall, irrigation can increase overall agricultural productivity. A global expansion of irrigated acreage helped support the Green Revolution of the mid-twentieth century. Over the last decade, however, irrigation expansion has slowed, raising doubts about whether the world can meet the food demands of its growing population. Access to irrigation water does not ensure food security, but in many parts of the world, it is essential to food security.

Effective irrigation requires at least three things: (1) an adequate source of water (physical availability), (2) an economical means of accessing the water and transporting it to the fields (cost-effective technology), and (3) formal or informal institutions that permit farmers to obtain the water that they need on a sustainable and secure basis over time and without adverse side effects (appropriate institutions). To some degree, physical availability and cost-effective technology are substitutes: if local surface water is scarce, for example, technology may still make it possible either to pump groundwater or to import and store surface water from elsewhere. However, appropriate institutions—effective "rules of the road," so to speak—are needed everywhere and at all times. Such institutions consist of all the laws, governmental regulations, distribution organizations, community practices, and other rules and structures that determine what can and cannot be done and influence decision making. Unless institutions provide farmers with reasonable assurance that they can take the water they need, farmers are unlikely to invest in irrigation. If institutions do not protect against excessive groundwater pumping, water tables may ultimately drop to levels that make continued pumping uneconomic. If institutions do not address the need for drainage where water has a high salt content, irrigation can actually undercut agricultural productivity. Plentiful water and technology, in short, are not enough without sound institutions that enable access, provide reasonable security, and address potential problems.

Water institutions also influence the distribution of benefits both within the agricultural sector and between agriculture and other water uses. Water institutions determine which farmers receive how much water and thus how the economic benefits of irrigated agriculture are distributed within a society. Water institutions also determine how water is divided among environmental flows, farmers, cities, and other water users (Figure 11.1). Thus, water institutions are important not only for the sustained effectiveness of irrigation, but also for societal equity.

I first recognized the immense importance of water institutions while studying how farmers in California's Central Valley responded to drought conditions in the late 1980s and early 1990s. Although the farms were all in the same general vicinity and drew water from the same sources, their responses to the drought and the consequences to their farming practices varied widely in response to their institutional settings. Some agricultural districts, for example, permitted their farmers to trade water rights, enabling farmers who could least afford to reduce their water use to lease water from farmers with greater flexibility. Farmers in agricultural districts that had adopted "tiered" pricing structures (where the cost of water went up with use) were more likely to engage in conservation.

This chapter looks specifically at the history of water institutions and their effect over time on food security, equity, and other societal goals. Appropriate water institutions change. Institutions that are adequate for small-scale irrigation often do not work well when irrigated acreage expands and more expensive irrigation techniques are adopted. A society that is focused on expanding food production will favor different water-allocation rules than a wealthier society that is increasingly concerned with environmental protection. The food security

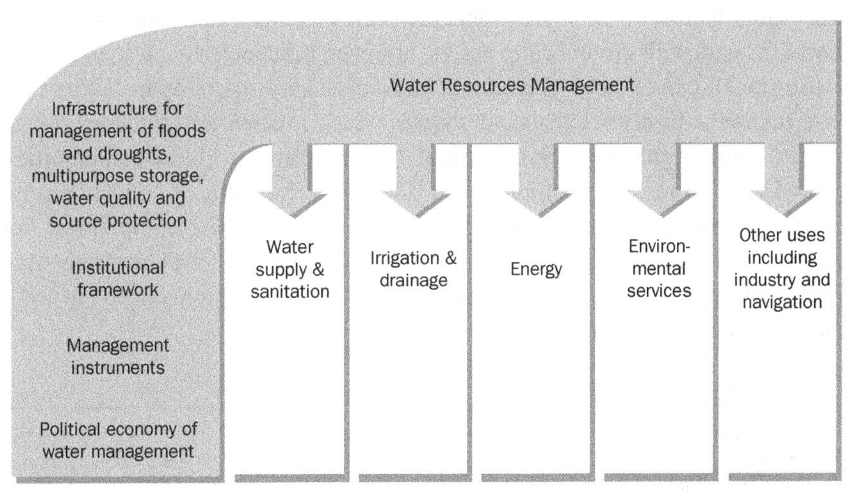

FIGURE 11.1 The Water "Comb"

Source: Briscoe 2013

of a nation thus depends not only on the institutions in place at a particular point in time, but also on how those institutions evolve over time in response to changes in scale and preferences.

The chapter focuses, not surprisingly, on nations that suffer from a "difficult" hydrology, involving combinations of inter-annual and intra-annual variability in surface flows, as well as geographic variability within their borders, since these are the nations where irrigation matters the most. Some of these nations remain in a "low-level equilibrium trap," in which poor water endowments prevent the country from making investments needed to escape water insecurity and poverty (Grey and Sadoff 2007). Other nations, however, adapt at least in part to their poor water conditions through storage and interbasin transfers of surface water, groundwater extraction, or both. Effective water institutions have been crucial to the success of such adaptations.

Because the effectiveness of water institutions is highly context specific, it has become a commonly accepted view that there are few, if any, universal truths to be gleaned (Meinzen-Dick 2007). But this does not mean that there are no broadly experienced lessons to be learned. To draw out those lessons, this chapter abstracts away from many of the specific and unique factors that often determine the adoption and success of particular water institutions in individual nations. However, examining broad trends in water institutions over time, and the relationships of those institutions to agriculture, reveals lessons that are particularly important for developing nations still in the early stages of the evolution of their water institutions.[1]

As described in this chapter, water institutions have tended to evolve through a handful of major phases. Prior to major agricultural development, local customary institutions typically govern the allocation and use of water. To enable larger-scale irrigation projects, governments usually adopt new centralized water institutions that are often at tension with the traditional, local rules. Local farmers with groundwater access, however, have increasingly avoided the institutional, capital, and operational costs of such projects by installing private tubewells to extract the groundwater, which is often not regulated. Over time, dropping groundwater levels lead governments and local communities to find ways to restrict pumping either directly or indirectly. Finally, in recent decades, growing environmental sensibilities and increasing competition for water from cities and industry often have generated new central governmental rules enabling the reallocation of water away from agriculture and to environmental, urban, and industrial use.

[1] This chapter cannot deal with all possible connections between water institutions and agriculture, omitting, for example, issues of water quality. The chapter also touches on issues of transnational water allocation only in passing, even though effective transnational management is critical to water allocation in many part of the world that share watersheds, including poor regions of Africa and Asia (Grey and Sadoff 2007).

Although each nation's water institutions have followed their own unique evolutionary path, a handful of generalizations about the institutions and their impact on agriculture are possible. First, institutional change typically has been politically, socially, and administratively costly and thus difficult to achieve. As a result, change has occurred only fitfully and in response to significant new demands and pressures. Second, institutional rules generally have addressed existing conditions and not anticipated future challenges. This lack of foresight is common to many institutions, but has led to more troublesome consequences in the water field. Third, in an effort to encourage irrigation and thus agricultural production at an early stage, central governments often have overallocated water resources, subsidized water supplies, and eschewed the use of market mechanisms (keeping prices low and forbidding the sale or lease of water). Fourth, water institutions have generally evolved over time toward more centralized (and often less participatory) rules, and these centralized rules have frequently brought with them less flexibility, decreased innovation, greater administrative costs, and sometimes reduced legitimacy among local farmers.

These water institutions in turn have generated a variety of problems. For example, by spurning market mechanisms and overallocating water while undermining customary restrictions on water use, central water institutions have typically undermined incentives to conserve water. Indeed, the evolution from local to central water institutions often has led to decreased agricultural efficiency (in terms of food produced per unit of water used). The high administrative and social costs of institutional rules adopted in connection with large-scale governmental irrigation projects also have contributed, along with other economic and technological factors, to the growth in private borewells and thus to groundwater overdrafting in many parts of the world. Finally, the adoption and evolution of central rules has typically favored large over small farmers; water rights in turn have become more concentrated, paralleling similar trends in land ownership (see Smith and Naylor's discussion on land institutions in Chapter 8).

The Importance of Irrigation and Water Institutions

Irrigation is critical to food security throughout much of the world. In many areas, irrigated acreage is two to three times as productive as non-irrigated land (Namara et al. 2010). As a result, the 20 percent of global agricultural land that is currently irrigated produces approximately 40 percent of the world's food. To meet the needs of the world's growing population over the next three decades, farmers will need to irrigate more land, and irrigation water will need to increase by more than 10 percent globally (Worldwatch Institute 2012).

Irrigation is equally important to many nations' economic development. Low levels of irrigation often relegate a country to low agricultural output and income, although irrigation expansion alone does not guarantee the elimination of widespread poverty as Indian and Pakistan's experiences have demonstrated (Namara et al. 2010). Levels of water development in a country show highly positive correlations with both the Human Development Index and the proportion of people who earn more than US$1 per day (Hanjra and Gichuki 2008).

Water institutions, in turn, affect the feasibility and sustainability of irrigation. At the most fundamental level, water institutions influence how much of a nation's freshwater supply is used by agriculture and how that water is allocated among farmers. (In the case of transboundary waterways, institutions also frequently govern how much water goes to each nation.) Equally importantly, water institutions affect how irrigation is carried out and thus its efficiency (water consumption per hectare or per unit of agricultural production) and sustainability. Increased efficiency can allow agricultural production to grow over time and yet still meet water demands from other sectors. Agriculture is not sustainable if it leads to significant groundwater overdrafting, drainage problems, or other long-term threats. Yet as discussed later, institutions in many regions of the world fail to encourage greater efficiency and pay little attention to sustainability.

A TYPOLOGY OF WATER INSTITUTIONS

Water institutions can be divided into two broad categories: the laws, rules, norms, and policies (generally referred to here as "rules") regarding water allocation and use, and the organizations that manage and distribute water. These two categories of institutions can be further characterized by various defining elements, several of which are relevant to understanding how water institutions have changed over time and their impact on agricultural efficiency, sustainability, and equity.

The first defining characteristic is the level and geography of control: who, in short, makes the water rules or runs the irrigation organization? Either the national government or local entities, for example, might set rules for water allocation: local rule-making bodies, in turn, can be organized around social groups, political boundaries, or geophysical borders such as watersheds. Irrigation districts similarly can be governed by local users or by a national irrigation agency. In some countries, private companies supply irrigation water. In the case of both rules and organizations, the locus of control can influence political power over water allocation and management, as well as the institutions' legitimacy, efficiency, and administrability (Thompson 1993). As discussed later, for example, central rules tend to be less flexible (because of the need for administrative simplicity in a large governance structure) and biased toward larger farmers (who tend to have greater political power).

The choice between certainty and flexibility is another defining characteristic for water rules. For example, allocation rules can range from crystalline (e.g., all farmers receive X cubic-meters per acre) to muddy (e.g., water will be allocated "equitably"). While crystalline rules tend to promote certainty and thus planning and investment by farmers, they often fail to account for nuances and thus can be seen as inequitable in specific settings. Muddy standards, by contrast, allow for subtle gradation in application, but often at the cost of certainty (Rose 1988).

An often-critical distinction among water rules is how they allocate water among classes of users and, within each class, among users themselves. Although an almost infinite variety of allocation rules are imaginable, the six summarized in Table 11.1 are the most common. These allocation approaches are often used in combination (e.g., a country may allocate water among different categories of use based on preferences, but then allocate water among farmers based on either land ownership or price). Choices among the rules can influence not only who can irrigate, but also irrigation investment and security. Some of the allocation rules, such as temporal priority and land ownership, tend toward certainty of water rights and thus encourage investment in irrigation infrastructure. Others, such as reasonableness, do not. Pricing and, to a lesser degree, reasonableness rules promote greater irrigation efficiency by penalizing or prohibiting high water use, while the other allocation rules generally provide no efficiency incentive. Allocation rules also differ in their perceived degree of equity.

TABLE 11.1
Common Allocation Rules

Rule	Description
Use Preferences	Water is allocated by priority among uses. For example, domestic uses might be met before water is allocated for agriculture or industry, or uses that are "riparian" to the waterway might be preferred over non-riparian uses.
Reasonableness	Decision makers determine the "reasonable" amount of water to allocate to each user. Reasonableness determinations often consider the demand from each user, the economic and societal value of alternative uses, and the opportunity that each user has to reduce its water use, among other factors.
Temporal Priority	Water is allocated on a first-in-time, first-in-right basis.
Land Ownership	Water is allocated based on the total amount of land that the user owns or irrigates in the relevant region.
Pro Rata Sharing	In times of shortage, all water entitlements are cut back proportionately.
Pricing	Water is allocated based on the willingness of each user to pay for it. The price might be pre-established by the supplier, or it could float based on overall supply and demand. Auctions are a specific form of a pricing rule.

Water institutions also typically impose a variety of conditions and restrictions on water use. These conditions can promote various goals, including water efficiency (e.g., water "duties" that limit the amount of water that can be used per irrigated hectare of land), environmental protection (e.g., rules that require minimum levels of instream flow), and administrability (e.g., water-metering requirements or administrative fees). Setting limits on groundwater pumping has become particularly important to avoid groundwater overdrafting and its accompanying ills (including increased and sometimes prohibitive pumping costs, salt-water intrusion, and surface subsidence).

In recent years, whether to permit the transfer or sale of individual water entitlements has become one of the most contentious questions in the design of water institutions. Historically, few countries permitted water markets. Water was considered to be too important to human health and development to be traded or even priced at its full cost; political bodies and not markets should determine how water is used. In many cases, moreover, countries did not have the regulatory sophistication or resources to design and implement the type of oversight system needed to ensure effectively operating markets. While most governments still prohibit the buying and selling of water rights, a growing number of countries, most of which are developed, now permit water transactions in at least some contexts (Thompson 2011a).

The marketability of water again affects the ultimate allocation of water in a region (favoring farmers with more valuable crops, as well as cities and industries with higher-value products). Markets also provide another incentive for water efficiency, because farmers who can conserve can sell the water. Markets also give farmers a source of funding with which to pay for conservation (unlike efficiency mandates that require water conservation without helping farmers pay for the needed conservation investments).Where permitted, water markets also can enable environmental interests to increase stream flows by purchasing water and retaining it in the stream. A growing number of "water trusts" are engaged in such environmental purchases of water in the United States (Thompson 2011a).

Turning to irrigation organizations, the two characteristics of greatest importance are generally the geographic locus of control and governance structure. Some irrigation organizations are hierarchical, with an administrative agency or private company determining key policies and water users acting only as customers. Many other irrigation organizations are more participatory or democratic, with water users involved to varying degrees in the formulation of organizational policies. Where water users have a formal vote in the governance of the organization, voting rules vary from one-person, one-vote to approaches that award greater power to users who own more land, provide more financial backing, or receive more water (Thompson 1993). Governance structure can influence the types of rules adopted by an irrigation organization and thus again irrigation practices themselves.

Three Phases of Water Institutions and Agriculture

Historically, water institutions in more arid regions have evolved through three major phases aligning with the principal demands for water—"Early Economic Development," "Expanding Irrigation," and "Multiple Sustainable Uses." Shifts from one phase to another have often required or induced legal changes in water rules and organizations, leading in turn to both political controversy and socioeconomic displacement.

In the Early Economic Development phase, farmers irrigate, if at all, only through small-scale, low-volume systems. Farmers remain highly vulnerable to variability in precipitation, and agricultural productivity and GDP per capita are both low (Hanjra and Gichuki 2008). Informal customary water institutions, which typically rely on local indigenous structures and emphasize flexibility and sharing, govern water issues and generally respond well to local needs and disputes. Formal state water laws either do not exist or remain largely ignored and unenforced. Many regions of sub-Saharan Africa (SSA) currently typify this phase. Despite having significant water resources, for example, Ethiopia has very little reservoir infrastructure (less than 1 percent of the reservoir capacity per person enjoyed by the United States) and consequentially low agricultural productivity and income (Grey and Sadoff 2007).

During the second phase—Expanding Irrigation—nations increase the quantity of land under irrigation, the amount of water applied per unit of land, or both, with a resulting increase in food production and a decline in poverty rates. The percentage of the total water supply used by agriculture increases, while water efficiency often declines as water becomes less expensive compared to other agricultural inputs (Rock 1998 and 2000). Indeed, the most common form of irrigation in the world is flood irrigation, despite its low efficiency rate (Worldwatch Institute 2012). Subsidization of water supplies (or of the energy needed to extract, transport, and deliver the water) can further undermine water efficiency.

Regions tend to take one of two routes (some take both paths) to expand irrigation during this second phase of water-institution evolution: the development of large-scale infrastructures such as dams, reservoirs, and canals; and the local, self-initiated adoption of distributed water technologies such as borehole wells. Under the first approach, nations develop large-scale programs to divert, store, and distribute surface water to farmers, generally leading to more centralized and formal water institutions. Many areas of Asia, the Middle East, and South America undertook this shift in the latter half of the twentieth century, leading in some cases to sizable increases in agricultural productivity (Hanjra and Gichuki 2008). Bangladesh, Iran, Mexico, and Turkey are current examples.

Under the alternative distributed-technology approach, farmers install groundwater wells or adopt other low-cost technologies to tap local sources of irrigation water that do not require new institutions or governmental funding

TABLE 11.2

Phases of Water Institutions and Agriculture

Phase	Early Economic Development	Expanding Irrigation		Multiple Sustainable Uses
		Large-Scale Irrigation	*Distributed Irrigation*	
Agricultural and Socioeconomic Characteristics	✓ Low water use per capita ✓ Irrigation is small-scale and low-volume ✓ Low water efficiency and agricultural productivity ✓ Low per capita income	✓ Increasing irrigation ✓ Increased water use per capita ✓ Water efficiency typically drops		✓ Growing urban regions and economy ✓ Percentage of water used in agriculture decreases ✓ Often balanced by increases in water efficiency
Characteristics of Water Institutions	✓ Local and customary ✓ Emphasis on equitable sharing and flexibility ✓ No marketing	✓ Greater centralization ✓ Emphasis on certainty and security of rights ✓ New hierarchical water organizations ✓ Increased administrative costs ✓ Larger farmers are advantaged ✓ Formal water marketing often remains illegal ✓ Overallocation ✓ Water is subsidized ✓ Increased rent seeking and corruption ✓ Often generates political and legal conflict ✓ Shortsighted focus on immediate issues	✓ Irrigation increases outside formal institutions ✓ Local informal markets can arise to meet needs of farmers lacking their own local supplies ✓ Stress can grow on local groundwater supplies ✓ Local water institutions sometimes must address issues of noncompetitive behavior	✓ Further centralization ✓ Urban uses often receive preference ✓ Water marketing is authorized ✓ Existing laws used to protect riparian ecosystems ✓ New legal protections of the environment sometimes adopted ✓ Direct and indirect efforts made to eliminate groundwater overdrafting ✓ Political and legal conflict again can emerge over conflicts with agriculture

to implement. Rather than face the high administrative, political, and economic costs that accompany large-scale irrigation projects built by the government, farmers adopt less costly means of self-help. Government generally plays no role at all (except indirectly in some cases through subsidized energy). Technological advances, along with poor experience with large-scale irrigation projects, have led to a discernible shift toward such self-initiated irrigation in recent years (Hanjra and Gichuki 2008). Rural areas of China, India, and Pakistan exemplify the trend.

Finally, in the Multiple Sustainable Uses phase, changes in demographics, markets, fiscal policies, and societal norms lead to attempts to shift at least some water from agriculture to other uses. These uses include domestic water for growing urban areas, industrial water, hydropower, and environmental flows. Efforts to restrict groundwater overdrafting and other unsustainable water uses often come into play at the same time. As a result, the percentage of the total water supply used by agriculture drops, while water efficiency necessarily rises. New water institutions are often needed to reallocate water and manage water withdrawals. Centralization again often increases, as does the use of administrative agencies with hierarchical structures. Existing farmers generally oppose such shifts in water institutions, creating a significant barrier to their adoption and enforcement. Developed jurisdictions such as the United States, Australia, and arid regions of Europe best illustrate this final phase, although many developing nations such as China, India, and South Africa also are responding to rapid urban growth, intensifying environmental demands, and sustainability concerns while still trying to meet unmet needs for irrigation water.

The description of these phases, by necessity, is highly stylized. In actuality, phases often blend together with no clear demarcation between them. Not every nation, moreover, passes linearly through each phase. Historically, phases could last decades, if not centuries, providing relatively clear differentiation in the characteristics of each phase and allowing gradual institutional evolution. In recent years, however, rapid societal change has compressed the evolutionary progression, so that countries today sometimes experience multiple shifts simultaneously or in close temporal proximity. This makes the task of developing effective water institutions all the more difficult (Grey and Sadoff 2007). China, India, and South Africa again illustrate nations that are seeking to increase irrigation at the same time that they must deal with growing urban needs, environmental demands, and sustainability concerns. The three-phase construct, nonetheless, is useful in understanding the interrelationship of water institutions and agriculture.[2]

That improved understanding can perhaps best be put to use in SSA, where the Early Agricultural Development stage of water institution evolution still predominates. Better understanding the lessons of the latter two phases

---

[2] For a suggestion that water institutions go through similar phases in tandem with advances in socioeconomic development, see Rock (1998).

is critical to ensure the successful and sustainable expansion of irrigation and agricultural productivity there. Significant improvements in agricultural productivity will be essential in reducing poverty levels in the region; according to some studies, water use in SSA must increase 10 percent by 2015—and 25 percent by 2025—to meet estimated demand for food production and other uses (Hanjra and Gichuki 2008). Such improvements, however, will demand effective water institutions that can avoid problems experienced by other countries and better anticipate future challenges.

The Evolution of Water Institutions

EARLY ECONOMIC DEVELOPMENT

In many countries with low levels of irrigated agriculture and economic development, customary local institutions manage the allocation and use of water for irrigation and other purposes. Local religious or village leaders administer the institutions, based on rules and standards that have developed over time in tandem with water conditions and needs. As a result of this co-evolution of conditions, needs, and institutions, customary water institutions often operate very effectively in managing local water issues. In some cases, local governments incorporate elements of customary water systems into statutory provisions or formalize the authority of local officials or organizations to apply and enforce the customary rules. Governmentally generated authority is not needed, however, so long as the traditional community of water uses remains intact and does not significantly expand. Paralleling many indigenous common-property regimes, customary water institutions often incorporate principles of equitable sharing and cooperation and provide governing officials with significant flexibility in addressing questions that might arise; disputes are often resolved through negotiation among community members rather than through fixed rules (Meinzen-Dick and Nkonya 2007). Water entitlements, to the degree they are held individually, are not marketable.

Customary water systems can harbor and facilitate the development of small-scale irrigation and community-level infrastructure (Baker 2005; Bruns 2007; Namara et al. 2010). Irrigation supported by customary institutions, moreover, can be reasonably widespread. In Tanzania, for example, customary water systems provide water for two-thirds of the irrigated acreage, although only 5 percent of agricultural land in the country is irrigated (Namara et al. 2010). Other examples of customary water systems that have supported significant but localized irrigation include the Subak systems of Bali, the Chhatis Mauja of Nepal, and the zanjeras of the Philippines (Lansing 1987; Meinzen-Dick 2007).

The acequias of Spain and the southwest United States are a particularly stable and long-lasting example of irrigation supported by local, customary institutions (Rivera 1998; Cox and Ross 2011). Acequias are community-based water organizations that distribute water from local rivers or snowpacks to their member farmers. Locally elected commissioners and mayordomos run the acequia's canals and distribution systems and allocate its water by prior agreement based on rules of pro rata sharing. In the Taos Valley of New Mexico, acequias provide irrigation water to approximately 40 km^2 of land. The acequias have helped local water users successfully adapt to inter-seasonal water variability and drought for several centuries. Starting in 1907, the State of New Mexico formally integrated acequias into the state government (Cox and Ross 2011).

Customary water institutions, however, generally cannot support the development of large-scale—or trans-boundary irrigation infrastructure (van Koppen 2007). The effectiveness of customary water institutions declines as the community of water users grows and becomes more heterogeneous. Where customary water institutions are dominant in arid nations, they effectively limit the extent of land that can be irrigated through the construction and operation of large dams, reservoirs, and conveyance facilities.

Despite their value, acequias are also an example of the limits of customary water institutions. In their study of acequias in the Taos Valley, Cox and Ross (2011) found that acequias with larger numbers of users had lower levels of crop production than smaller acequias. Fragmentation of the land area of an acequia by urban development also negatively impacted crop production within the acequia. Regions that wish to expand significantly the amount and area of irrigation therefore must generally turn to new institutions beyond customary systems.

IRRIGATION EXPANSION

Agricultural communities have used two primary methods, as noted earlier, of expanding irrigation beyond what is achievable under customary water institutions: the construction and operation of large-scale irrigation infrastructure, and the local adoption of distributed water technologies. Both approaches have led to significant global increases in irrigation, but have had very different institutional impacts. As discussed later, large-scale irrigation has led to more formal, central, and secure water entitlements—in the process raising administrative costs, often redistributing water holdings, and sometimes posing obstacles to efficient and effective management. Distributed irrigation, by contrast, has often occurred outside formal water institutions.

Large-scale Infrastructure

Many regions have invested in irrigation projects that divert, store, and distribute water to large groups of agricultural users. When successful, large-scale irrigation can help significantly reduce poverty. In India, for example, poverty rates are 70 percent in un-irrigated districts, but only 25 percent in irrigated districts (Grey and Sadoff 2007).

More formal and predictable water-allocation rules often accompany large-scale irrigation projects. International funders such as the World Bank, as well as national governments that provide irrigation funding, often insist on clear and secure rights to source water in order to ensure that the project can deliver the benefits for which it is built. New rules also are needed for the allocation of the project water itself. Customary water rules, where they exist, are generally designed for a smaller and more homogeneous group of users than the project will serve; as noted earlier, the trademark flexibility of customary rules is less effective as numbers and heterogeneity grow (Bruns 2007; Meinzen-Dick and Nkonya 2007). Water users, who are typically expected to pay for the project water, also will often insist on more secure rights, as will the water funders who will want to encourage new agricultural investment (Worster 1985; McCool 1987). The flexibility and negotiation that previously characterized water rules thus switches to an emphasis on certainty and security. This shift often brings with it new administrative requirements, including permitting and monitoring of water use, and thus increased administrative costs (Komakech et al. 2012).

Historically, large-scale irrigation projects also have led to more centralized water rules (van Edig et al. 2003; Meinzen-Dick 2007). Several factors have pushed toward greater centralization. First, when national governments either provide the initial funding or serve as the conduit for international funding, they generally expect a greater say in water rules so to ensure the project's success. Second, national governments often are the only authorities with the administrative expertise needed to develop rules for a large, complex infrastructure project. Finally, the hydraulic boundaries of the project are often larger than or different from the community or customary boundaries within which water was historically governed (Bruns 2007). Increased centralization, in turn, generally brings more uniform and inflexible rules, since they are easier for central authorities to implement and enforce than are flexible and context-specific rules (Boelens et al. 2007). The rules themselves generally focus on achieving national (or international) policies, with little attention to local norms, conditions, or concerns (Cremers et al. 2005; Stein et al. 2011; Cullet 2012).

Just as large-scale irrigation has led to more centralized rules, it also historically has led to the creation of new, state-run entities to operate and maintain the irrigation projects. Existing customary organizations again frequently lack the geographic breadth and expertise to serve as project operators. Even where customary organizations could be effective, the central government frequently has demanded direct oversight of the project (Boelens et al. 2007). Central governments want to control the irrigation project both out of concern that local organizations may prove incompetent to run the project successfully and to distribute the economic benefits flowing from the project.

These shifts in water rules and organizations generally change the status of water users too. While water users are generally active participants in customary water institutions, they are frequently mere customers of state-created irrigation districts (van Koppen 2007). Water users thus switch from being a principal in a self-governing community to a constituent of a hierarchical administrative structure. This shift further reinforces the centralization of water rules; with less input from local users, project authorities become even less likely to incorporate local lessons or knowledge into their rules and processes (Cremers et al. 2005). Even where water users have a say in the operation and management of an irrigation project, control often is allocated among water users according to land holdings or the size of their water deliveries (which is often correlated again with land holdings), rather than by the one-person, one-vote system more common to customary systems (van Koppen 2007).

These various shifts can pose problems for the management of large-scale irrigation projects. Recipients of project water may object to new rules, requirements, and processes that had no counterpart in prior customary institutions. As already noted, the new water institutions may impose added costs or administrative responsibilities on water users and often focus on national or international policy interests to the exclusion of local concerns. To the degree that the new rules vary in fundamental aspects from customary rules, water users also may view them as illegitimate. Similarly, national and project authorities are often viewed as outsiders and may have difficulty enforcing the new water rules and collecting water payments (Stein et al. 2011). In centralizing water institutions, the national government displaces traditional enforcement authorities and social networks that have been key to effective management (Bruns 2007; Meinzen-Dick and Nkonya 2007; van Koppen 2007; Namara 2010). Yet national authorities, at least initially, often lack the local legitimacy needed to replace the customary enforcement system (van Edig et al. 2003). This is particularly true where local users have little say in the setting of water rules or the operation of the irrigation project (Cremers et al. 2005). Centralized water organizations also tend to be relatively inflexible and methodical in their approaches; as a result, innovation and effective adaptation to new conditions are often stifled. Finally, the more attenuated control, along with the large infusion of funding that often accompanies large irrigation projects, can lead to both rent-seeking and corruption (van Edig et al. 2003).

In combination, these problems have sometimes undermined the success of large-scale irrigation projects in developing nations. Studies of irrigation projects over the last 40 years have found that many have failed to perform up to original expectations (Meinzen-Dick 2007; Mehari, van Steenbergen, and Schultz 2007). In response, many nations have moved to devolve authority over projects back to the local level (Easter 2000; van Edig et al. 2003; Cremers et al. 2005; Bruns 2007; Meinzen-Dick 2007).[3]

---

[3] Other factors, including fiscal pressures and more general trends toward decentralization, have also encouraged the effort at devolution.

Such devolution has had mixed success (Garces-Restrepo et al. 2007; Swallow et al. 2007). In many cases, the devolution has been incomplete. Some countries, for example, have tried to gain the benefit of local involvement while still dictating uniform, fixed rules and processes that undermine the flexibility needed to account for local conditions and concerns (Meinzen-Dick 2007). Incomplete devolution also sometimes has generated confusion over the relative authority and responsibilities of the multiple levels of government and the users (Cremers et al. 2005). In other cases, efforts at devolution have failed to ensure that local institutions have the ability to manage the projects. Many national governments, for example, transfer authority to local users without the funding mechanisms or training needed to maintain and run the irrigation works effectively (Cremers et al. 2005; Swallow et al. 2007). Finally, devolution sometimes does not succeed because farmers have become dependent on subsidies previously provided by central authorities (Swallow et al. 2007).

The changes in water institutions that accompany large-scale irrigation projects also present a significant risk to existing water users (Ferguson and Mulwafu 2007). New centralized water institutions often favor large landowners and other politically powerful interests over smaller farmers who often enjoyed significant entitlements under customary water institutions (Cremers et al. 2005; van Koppen 2007; Namara et al. 2010; Komakech et al. 2012). As a result, small farmers are often excluded from the direct benefits of increased irrigation (Boelens et al. 2007).

Poor farmers are at a disadvantage in the creation and awarding of new state-created water rights for several reasons. First, small farmers often enjoy less political power than large landowners in the negotiation and setting of the rules for new water entitlements (Bruns 2007; Meinzen-Dick and Nkonya 2007). Second, nations and project funders may see a policy advantage in awarding water rights to large landowners because they are more likely to be reliable customers who make water payments on time, maximize agricultural production, and thus best ensure the long-term success of the project. For these reasons, new water rules often allocate water by land ownership rather than by pro rata sharing, and only formally titled land is generally considered in the setting of entitlements (Bruns 2007; van Koppen 2007).[4] Even where the new rules recognize and protect existing uses, they often acknowledge only individual rights, while many customary institutions hold water communally (van Edig et al. 2003; Bruns 2007).

Small farmers also can be at a disadvantage in claiming water rights to which they are entitled under the new rules. Small farmers and other existing

[4] Requirements of land ownership also can lead to gender discrimination in the allocation of water rights (van Koppen 2007).

users often do not have the knowledge or resources to pursue formal rights (Bruns 2007; van Koppen 2007). For example, where new rules protect existing water entitlements, existing users generally must prove their historic use—often in a relatively short window of time. However, existing users often do not learn of or understand the requirements or are deterred from proving their historical use by the significant administrative or legal costs involved in the process. For example, when Chile established new private rights to water in the latter half of the twentieth century, along with processes for claiming them based on existing uses, peasants often did not learn of the processes until it was too late (van Koppen 2007). New water entitlements also frequently carry application fees or annual water charges that poorer farmers may not be able to afford (van Koppen 2007; Komakech et al. 2012).

The displacement of existing rights and institutions can generate political or legal opposition. National laws sometimes protect existing rights from confiscation and, in a small but growing set of developing nations, explicitly protect indigenous rights (Cremers et al. 2005). However, small farmers seldom have the resources or expertise to pursue such rights, even where they exist, without external assistance. As a result, discontent is more likely to take the form of political protest or violence. In Tanzania, for example, the installation of gates to control and monitor water withdrawals as part of a new irrigation project in the mid-1990s led to protests by local farmers and the destruction of many of the gates (Komakech et al. 2012).

Two other aspects of the water institutions that have emerged as part of large-scale irrigation projects bear emphasis. First, the institutions often are economically inefficient. For example, central governments often subsidize water use, which discourages conservation by the irrigating farmers. Most governments also ban water marketing, preventing high-value water users from paying other water users to conserve or reduce their diversions in order to transfer the water to the high-value users (Thompson 2011a). As a result, water use is generally higher within the irrigation projects than is economically efficient.

The new water institutions historically have also been shortsighted. Rules have focused on maximizing the short-term success of the irrigation projects, with little anticipation of or provision for longer-term consequences or issues. For example, the water rules historically have sought to maximize water use despite potential environmental consequences. In the view of political leaders, water that is not used is "wasted"; by contrast, the leaders assume that maximization of use will cure famine or food shortages (Gandhi 1946). Indeed, in the rush to promote irrigation and local economic development, many governments have overcommitted available water supplies, assigning or promising more water than is actually available (Molle and Berkoff 2009). Water institutions also have often not anticipated other negative externalities of irrigation, such as drainage problems (Thompson et al. 2013).

Distributed Irrigation

In recent years, farmers in developing nations have looked increasingly to local, "distributed" systems for new irrigation water in place of the large-scale irrigation projects that dominated earlier eras (Hanjra and Gichuki 2008). At least three factors help explain this shift. First, technological developments in the late twentieth century, particularly the development of affordable electric pumps for borewells, have made new distributed options cost effective compared to large-scale irrigation programs. Second, the institutional problems just discussed have effectively raised the price of large-scale irrigation both for local farmers and for the central government. Finally, the distributed options generally do not require any institutional changes by the central government, avoiding the need to mobilize political support for new institutional rules and organizations and minimizing the transition costs and social risks that historically have accompanied changes in water institutions.

The informal borewell markets of eastern and southern Asia are an example of such local self-initiated efforts. Groundwater is an open-source commons under most customary water systems (Meinzen-Dick and Nkonya 2007). Farmers who want to expand their irrigated acreage or apply additional water to their fields need only construct their own wells or purchase groundwater from a neighbor with a well. Dramatic improvements in well technology have made it far more affordable to withdraw groundwater, and a rapid rise in informal markets has increased the ability of farmers to purchase groundwater from others when they are unable to afford their own well (e.g., Wang et al. 2007). Governments also have sometimes promoted groundwater development by subsidizing the energy cost of pumping (Shah et al. 2007; Giordano 2009).

Numerous studies have demonstrated the short-term success of local, self-initiated efforts to expand irrigation through groundwater wells and markets. Through local development of groundwater, both China and India have brought millions of hectares under irrigation (Siebert et al. 2010). In 2002, approximately three-quarters of villages in India irrigated crops, and the vast majority of those villages used groundwater (Shah 2007). Bangladesh has experienced a similar, but smaller growth in irrigation through local groundwater development (Hemson 2008).

Groundwater borewells are not the only example of small-scale, self-initiated approaches to increased irrigation. For example, farmers in a growing number of regions are constructing small storage facilities to provide irrigation water throughout the irrigation season. Farmers in other regions are turning to rainwater harvesting and storage (van Edig et al. 2003; Hanjra and Gichuki 2008; Lein 2008).

As noted, none of the local distributed approaches generally requires the central government to establish new water institutions. Indeed, that is one of the advantages of a distributed approach to local farmers. Administrative, political, and social costs are nil and therefore do not pose an obstacle to adoption. Because distributed approaches involve individual decisions by a multitude of local actors, many central governments would find it difficult with their limited resources to bring distributed approaches into a centralized institutional system even if they wished to do so (Komakech et al. 2012).

Distributed approaches sometimes require changes in informal or local water institutions, but these appear to be far less problematic. For example, borewell markets can provide large landowners with oligopolistic power over poor farmers who cannot afford to drill wells themselves. Studies have found that large landowners have sometimes used this power to gain non-competitive profits in the sale of groundwater (Shah 1993), although this is not always the case (Wang 2007). Where non-competitive behavior has occurred, institutional changes have often successfully addressed it. In some cases, informal norms have emerged within the community in favor of sharing and against water charges greater than the cost of producing the groundwater. In other cases, local water-user groups or villages have adopted rules prohibiting profit-making in the sale of groundwater (Rawai 2002).

That new water institutions are not needed is both a cause and a benefit of distributed irrigation. This apparent advantage can also sow the seeds of long-term failure of distributed irrigation in cases where the irrigation poses any significant negative externality. In the absence of any institutional control over groundwater pumping, for example, local farmers often withdraw increasing amounts of water from local aquifers and ultimately begin to overdraft the aquifers. When farmers become concerned about dropping groundwater tables, there is no institutional mechanism to which to turn for relief. New institutions must be developed, and as discussed later, the process of adopting such institutions at so late a stage can be difficult and lengthy. As a result, the groundwater overdrafting often goes on unabated. While the centralized rules accompanying large-scale irrigation are often short-sighted, at least institutions have been created that can try to address new problems when they arise. That is often not the case with distributed irrigation.

The Indus Basin: A Case Study[5]

The Indus Basin, which has played a critical role in the history of irrigation expansion, provides a valuable case study of many of the lessons outlined above. The Indus Basin is now the largest contiguous irrigated agricultural system in the world—some four times larger than the irrigated area of California. Rising in Tibet, the Indus River flows some 1,800 miles from the Himalayas to the Arabian Sea. The Indus Basin covers some 350,000 square miles of mostly

[5] This case study draws heavily on the experiences of Professor Walter Falcon, who was a member both of the U.S. Presidential Commission on Land and Water Development in the Indus Plain (1964) and the Indus Basin Assessment Research Group (IBARAG 1978). The author is deeply indebted to Professor Falcon, who provided a draft of the case study for the author's use in this chapter. The case study also draws significantly from the excellent analysis by Briscoe and Qamar (2009).

arid land, of which some 75 million acres are potentially suitable for agriculture and forestry (IBARAG 1978). Irrigable land is scarce, in other words. But water for irrigation is the resource that most limits agricultural production there. Pakistan's irrigated area within the Indus Basin currently totals only about 40 million acres.

The Basin includes four rivers in addition to the Indus: the Jhelum, Chenab, Ravi, and Sutlej. Collectively, the irrigation systems that draw on these rivers embrace 40,000 miles of canals and 90,000 watercourses that bring the water to farmers' fields. A substantial amount of groundwater also underlies the Basin and is used conjunctively with the surface water to produce a wide variety of crops.

What makes the Indus of special interest for food security is its location in Pakistan, one of the poorest countries in the world. Despite the Green Revolution in wheat and rice, Pakistan's farmers remain mostly small and poor. Wheat, the most extensively grown crop, exhibits yields that are only about 40 percent of those in comparable ecosystems in northern Mexico. In Pakistan as a whole, nearly 75 percent of the population earns less than $2.50 per day (World Bank 2007), and 35 million Pakistanis—approximately 20 percent of the population—are now "food insecure" (based on the FAO classification). The Indus Basin is a case study of unrealized potential and the role that irrigation institutions play in causing this potential to go untapped.

Irrigation in the Indus Basin traces to 3500 BC. Sizable irrigation, however, did not begin until the British, between 1860 and 1900, engineered massive canals and diversion structures (called barrages) that transformed much of the arid Indus terrain that is now Pakistan into a semi-modern irrigated landscape. The undertaking was unbelievably bold, and whole new communities, in the form of canal colonies, emerged.[6] By the turn of the twentieth century, the new infrastructure had "commanded" five million acres of land (Lieftink 1968).[7]

The success of colonial construction, however, carried with it three seeds of destruction that continue to plague the system today. First, the areas commanded by the canals exceeded the amount of land that could be irrigated with the available water. The British intended the system to prevent local famines and to promote cotton-growing for British mills. For these purposes, the wide and thin spreading of water seemed logical. The system, however, proved unsuitable later from the perspective of intensive agricultural development in the newly formed Pakistan. Particularly in light of increased population densities, surface water was simply insufficient to support efficient modern agriculture.

[6] For a detailed account of the Colonial period, see Ali (1988).

[7] "Commanded" refers to the land potentially coverable by an irrigation system from an elevation perspective. The acreage commanded, however, does not ensure that there is enough water to produce crops on the entire area.

Second, the institutional organizations created to operate and maintain the irrigation system were slow moving and top-down. The organizations focused on allocating scarce water through a methodical rotational approach. There was little room for agricultural innovation in the cropping systems. Farmers at the head end of waterways tended to fare better than those at the tail end. And the continuing water scarcities provided strong incentives for corruption.

Third, both irrigation design and water institutions failed to anticipate future problems, leading to almost-disastrous long-run consequences. Systems and institutions, for example, failed to provide for adequate drainage. Although the surface water was generally of high quality, the water still contained small quantities of dissolved salts (60–300 parts per million, depending on the season). Irrigation organizations, however, chose to spread the limited available water over a wide number of acres. Although this decision might have made good short-run sense from a welfare perspective, it also meant that there was inadequate water to flush the salts through the root zone. Lacking water and adequate drainage, salts built up in soils over time, severely limiting their productivity. Canals also were unlined, leading the water table to rise to near the surface over time as water soaked in (see Figure 11.2). In numerous areas, the root zone for crops became waterlogged, and the capillary action from the very high water tables led to even more salt buildup on the soil surface.

The violent process that created Pakistan in 1947 affected both the irrigation system and its accompanying institutions. Most of West Pakistan's irrigation water flowed first through India, raising questions of what waters each government controlled and fears that India could shut off water to Pakistan in periods of strife. After more than a decade of negotiations, India and Pakistan signed the Indus Water Treaty of 1960. An extraordinary triumph of wisdom and patience, the treaty guaranteed Pakistan 75 percent of the irrigation water while giving India the opportunity to add substantially to its hydroelectric capacity. More importantly, the treaty enabled both countries to obtain huge

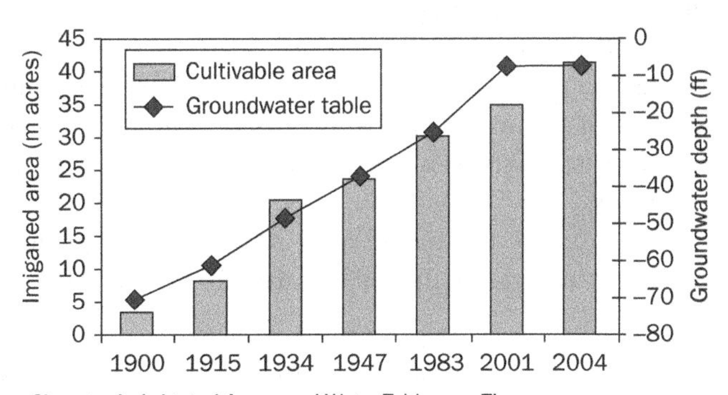

FIGURE 11.2 Changes in Irrigated Acres and Water Table over Time

Source: Briscoe and Qamar (2009).

water loans and investments from the World Bank.[8] Between 1960 and 1980, the World Bank provided nearly $15 billion (in 2004–05 dollars) in support of the Indus Treaty infrastructure (Briscoe and Qamar 2009). While largely outside the scope of this chapter, the Indus Treaty has been highly stable and has positively affected security in South Asia. Although Pakistan and India have continued their war of words (and sometimes of guns), both countries have observed the Treaty for more than 50 years. Even in times of armed hostilities between the two nations, the irrigation system has been off limits to warfare.[9]

By the early 1960s, however, the Indus Plain was white with salt—a clear sign of serious soil degradation. In 1961, the Kennedy Administration (which had developed a close relationship with Pakistan) therefore appointed a commission to examine land and water challenges in the Plain (U.S. Presidential Commission 1964). The commission concluded that there was no way in which the large World Bank loans for irrigation improvements under the Indus Treaty could be repaid without sizable improvements in agricultural productivity. Productivity improvements, however, in turn required major revisions in the institutions of water and agriculture.

The recommended solution involved the installation of large electric public-sector tubewells. These wells were to use the existing canal system for water distribution. By tapping into the large areas of low-saline groundwater, the tubewells could produce extra water to flush salts down through the root zone, lower groundwater tables to reduce capillary action to the surface, and most important of all, provide more water for crop irrigation, particularly during the winter season when surface flows were historically low. The new approach generated much optimism for a win-win-win solution: however, the optimism proved premature.

The new approach foundered again on organizational structure. As Falcon (Chapter 2) has noted, this was an era when regional project authorities and integrated agricultural development were coming into vogue. The proposed creation (and actual implementation) of several reclamation projects (SCARPs—Salinity Control and Reclamation Projects) essentially proved unworkable. The overlay of multiple governance agencies created conflicts. Moreover, having large electric irrigation wells in the public sector was itself a problem. The lack of reliable power flows, resulting in brownouts, burned out irrigation motors; the rotational system for irrigation was still far too rigid for flexible agriculture; and the increasing tightness of public funds caused a major slow down in SCARP investments.

Just as earlier irrigation and organization problems led to public tubewells and SCARPs, the challenges of the public tubewells and SCARPs led to yet

---

[8] For more on the Indus Water Treaty and the World Bank's role in its negotiation and implementation, see Lieftinck et al. (1968) and Briscoe and Qamar (2009).

[9] Legal challenges to the Treaty, however, continue to today (Briscoe 2013).

another set of institutional and technological innovations. Farmers began to realize what supplemental irrigation water could do for their crops and began to develop private groundwater systems. Local production started of shallow centrifugal pumps using rope as strainers. Even more remarkably, in a modern-day sword-into-plowshare story, several northern towns long noted for their ability to produce local firearms began to produce rudimentary diesel engines. These power units were slow, inelegant, and roughly engineered. But they did not require the complicated paperwork that imports did during a period of foreign-exchange control, were locally repairable, and ran on readily available diesel fuel. Rural electrification was no longer a prerequisite for effective irrigation.

From 1960 to 2010, the number of wells increased from 2,000 to 700,000. In 1960, groundwater accounted for only 8 percent of farm-gate water supplies. By 1985, the figure was 40 percent, and currently more than 60 percent of the water delivered, and more than 75 percent of the value of the water delivered, is delivered from tubewells. The great unlined canals of the earlier irrigation expansion have become less a water delivery system and more of a groundwater recharge mechanism, in which water seeping from the canals refills the aquifers (Briscoe and Qamar 2009). In a twist of fate, the canals that began as a problem are now part of a higher-productivity solution.

The era of private tubewells, however, is now generating its own challenges. The question, yet again, is whether a new set of institutional changes can address these emerging challenges. The incentives to drill ever-increasing numbers of wells are already causing water tables to drop significantly, thereby raising the cost of pumping. Wells are also being placed in areas with more saline groundwater, threatening delicate salinity balances. The incentive-driven private tubewells of the Indus Basin, in short, may "over succeed" and necessitate a new set of institutional innovations to limit pumping. Such a change, however, will be exceedingly difficult, especially within the context of a country still suffering considerable food insecurity and internal conflict.

MULTIPLE SUSTAINABLE USES

While water institutions in the middle phases are all about making water available for more extensive irrigation, the issue in the final phase is how to meet other competing societal demands for water and ensure water sustainability without undermining food security. Irrigation needs become a constraint rather than the driving force. Institutions must respond to two separate issues in the final phase. First, demographic, economic, or normative changes generate pressure to reallocate water away from agriculture to other consumptive uses, including for industrial and municipal supplies. Second, environmental or sustainability concerns can lead to demands to restrict the total amount of water being withdrawn from surface or groundwater sources.

Reallocation

Over time, nations that are already consuming a large percentage of their annual available water supply will frequently face constraints in meeting the needs of growing urban populations or economies. If new supplies are not readily available, the least-cost option often will be to reallocate water from other uses, and where farmers engage in significant irrigation, agriculture often will use the highest percentage of water and thus be the most obvious source. A growing number of countries therefore are experiencing reallocations of water away from agriculture to domestic and commercial urban users, and manufacturing. Reallocations to hydroelectric facilities may also occur where power facilities are downstream of agricultural diversions or the optimal timing of agricultural diversions and hydroelectric production conflict with each other (Rock 1998).

Reallocations can take place gradually, without the benefit of conscious planning, or reflect a purposeful change in governmental policy (Molle and Berkoff 2009). Reallocations often begin during extreme droughts, when there is insufficient water to meet all the existing demands for domestic supplies, energy, industry, and agriculture. In Thailand, for example, a drought in 2005 led to the reallocation of water from agriculture to both domestic and industrial use, and Cyprus cut agricultural water use in favor of domestic users in response to a 1998 drought (Molle and Berkoff 2009). While transfers away from agriculture in drought periods are normally temporary, they often require the construction of new infrastructure, such as wells or aqueducts, that unintentionally enables and reduces the cost of longer-term reallocations. With these barriers to reallocation lowered, what begins as an emergency response to drought conditions often leads to the permanent reallocation of water in response to continued population and manufacturing growth.

Cities and other non-agricultural water users have used at least three strategies to obtain water from agriculture. Two of these generally require explicit changes in formal water institutions. In the first and less common of these strategies, non-agricultural users lobby for new water rules that formally reallocate water entitlements away from agriculture—for example, new preference rules that give domestic uses first priority in times of shortage. Various governments ranging from China to Tanzania have announced that, when water is short, agriculture will be the residual claimant: farmers will get what is left after other users satisfy their needs (Molle and Berkoff 2009; Komakech et al. 2012). While such reallocations can be uncompensated, most governments find it politically impossible to either adopt or enforce such reallocations without some form of compensation acceptable to local farmers. Thus, in Chennai, the government has provided compensation to local farmers whose wells are now being used to expand the urban water supply in the Indian state of Tamil Nadu (Molle and Berkoff 2009). Where agricultural water is still managed by customary institutions, efforts to transfer water through new national water rules can create political controversy similar to the problems, discussed earlier, associated with the movement to large-scale irrigation (Komakech et al. 2012).

Second, non-agricultural users can attempt to purchase water entitlements from farmers. Because water institutions historically have not permitted formal water marketing, such trades typically require changes in the water institutions to explicitly authorize water trades in at least limited situations.[10] Even where water institutions do not formally prohibit water trades, prospective non-agricultural purchasers may want new rules to facilitate water transfers—for example, to reduce the time and expense of any needed governmental approvals, eliminate restrictions (such as prohibitions on the transfer of water outside of an irrigation district or a particular geographic region), or to enable the use of existing infrastructure to transport acquired water to the new place of use (Thompson et al. 2013). However, effective water markets require sufficient administrative sophistication to monitor trades and protect against negative externalities to other users and the environment. They also often rely on current infrastructure to move purchased water to the new user. As a result, effective water market institutions are currently limited to a handful of developed nations, including Australia, Spain, and the United States, as well as Chile and South Africa (Meinzen-Dick 2007; Thompson 2011a; Komakech et al. 2012). Proposals to develop transferable market systems in other developing nations, including Indonesia, Sri Lanka, and Thailand, have generated significant opposition (Bruns 2007). Market concepts clash with key norms historically underlying water institutions, including the view that water is not a commodity, but a public-trust resource that should be managed to provide equitable water distribution and maximize broad societal goals (Thompson 2011a).

Finally, because of the time and transaction costs associated with the first two options, non-agricultural water users also sometimes simply take needed water without seeking any formal reallocation through changes in institutional rules or markets. Non-agricultural users may take the water legally or illegally. For example, if groundwater is a legal commons, cities and other non-agricultural users can lawfully pump from an aquifer that is currently being used by agriculture, even though any resulting overdraft of the aquifer effectively is taking water from the existing agricultural users. Cities in countries as diverse as Egypt, Peru, the Philippines, Tanzania, and Thailand have increasingly encroached on agricultural water supplies in this fashion (Molle and Berkoff 2009; Komakech et al. 2012).

As the impact of such unofficial reallocations becomes obvious to agricultural users and begins to grow, political conflicts can emerge and even turn violent (Molle and Berkoff 2009; Komakech et al. 2012). Where there is no

---

[10] In some cases, informal markets may arise that do not require governmental authorization or that effectively ignore existing institutional rules. Informal tanker markets in India, for example, have effective moved groundwater from agriculture to urban domestic use (Molle and Berkoff 2009).

legal authority for the reallocation, non-agricultural users may seek institutional sanction; where the reallocation is legal, as in the case of groundwater withdrawals, farmers may seek an institutional prohibition. Because of their greater political power in national and even local governments, however, cities, energy companies, and industry typically, but not always, prevail in battles over such institutional rules. [11]

Constraints on Water Withdrawals

Over time, nations also typically confront concerns about the sustainability of large irrigation withdrawals. For example, large diversions of surface water can threaten native fish species, shrink lakes, and destroy estuarine ecosystems, with serious repercussions not only for the environment but also for such local ecosystem services as commercial fisheries and weather stabilization (Pearce 2007). As illustrated by the earlier case study of the Indus Region, areas that rely significantly on groundwater to meet irrigation demand also can confront problems from groundwater overdrafting, including the increased cost of energy needed to pump from falling water tables, the depletion of the aquifer's "fossil" water that could take hundreds or thousands of years to replenish naturally, salt-water intrusion into aquifers, subsidence of the overlying surface, and loss of groundwater-dependent ecosystems (Thompson 2011b). Irrigation has contributed to overdrafting in significant areas of both developing nations (for example, China, Egypt, India, Iran, Libya, Saudi Arabia, and Yemen as well as Pakistan) and developed countries, including Australia, Israel, Spain, and the United States. In some regions, groundwater tables are dropping at a rate of one meter or more per year (Giordano 2009).

In some cases, existing laws, either water-related or generic, may provide a mechanism for addressing environmental concerns. For example, in the United States, the national Endangered Species Act has provided environmental interest groups with a tool to reduce both surface-water and groundwater withdrawals that are jeopardizing the continued existence of fish and other water-reliant species (Thompson et al. 2013). More frequently, sustainability concerns require changes in existing water institutions, such as the establishment of environmental flow requirements or the capping of groundwater withdrawals. South Africa, for example, adopted new environmental flow reserves for its river basins that must be met before water can be withdrawn for irrigation (Grey and Sadoff 2007). Tanzania also has provided that, where there are water shortages, the protection of ecosystems has a higher preference than

---

[11] Concerns about urban encroachment on agricultural groundwater supplies led to significant changes in United States water institutions in the early twentieth century that effectively prohibited cities from using groundwater (Thompson, Leshy, and Abrams 2013). Agriculture, however, had greater political power at that point in the United States than it has today.

any water uses other than domestic (Komakech et al. 2012). Australia, Mexico, and portions of the United States have required governmental approval for groundwater extraction and limited the total withdrawals that can take place (DuMars and Minier 2004; Sandoval 2004; Giordano 2009; Thompson, Leshy, and Abrams 2013).

A major obstacle to addressing sustainability concerns is that, by the time the problem is publicly salient, farmers typically have become highly reliant on their water use, raising the political and economic costs of addressing the concerns. For example, groundwater overdrafting often worsens over many years, if not decades, before governments take action to address it. Although groundwater tables drop and energy needs increase in the interim, both farmers and the government often see overdrafting as tolerable so long as there are no serious and immediate side effects; where the government charges flat rates for energy, farmers may not even feel the financial pinch of increasing pumping costs. Farmers also may fear that governmental regulation will undermine their ability to increase groundwater use in periods of drought or higher agricultural demand. Many regions therefore do not consider institutional responses until major crises occur—typically salt-water intrusion or the loss of a significant percentage of local wells from falling groundwater levels.

Efforts to eliminate or restrict groundwater overdrafting at this stage often prove extremely difficult. Local groundwater users have grown reliant on their current withdrawal levels even though they are unsustainable; in many cases, groundwater users will have invested in land and infrastructure based on the expected continuation of those current extraction levels. Political opposition to changes in existing water institutions therefore will be greater than it would have been at the time that overdrafting first began or earlier.

Groundwater restrictions also can pose significant administrative hurdles for nations that do not have systems currently in place for groundwater oversight. Effective groundwater regulation requires records of existing extractions and the ability to police the behavior of typically hundreds if not thousands of individual agricultural users. Neither of these capacities generally exists in nations that historically have treated groundwater as a commons. In such settings, national adoption of groundwater restrictions is unlikely to be successful. For example, China has adopted both national and local laws to regulate groundwater wells, impose fees on groundwater extraction, and prohibit unsustainable groundwater use. Yet these laws are seldom enforced and have had little impact on groundwater extractions (Wang et al. 2007). Developed nations may be somewhat more successful in making the needed transition to the multiple sustainable uses phase, both because they often have significant regulatory experience and capacity and because farms there tend to be larger and already regulated for other purposes (Giordano 2009).

Lacking an effective central institution that can directly regulate groundwater withdrawals, some nations have turned to one of two other approaches.

First, some governments have tried to indirectly influence groundwater withdrawals. Several Indian states, for example, have used energy policy to increase the cost of groundwater extraction and thus reduce pumping. West Bengal has reduced groundwater withdrawals by metering agricultural energy deliveries and thus increasing the marginal cost of groundwater extraction (Giordano 2009). Gujarat has accomplished much the same result by rationing agricultural energy supplies (Shah et al. 2008). Employing a different strategy, parts of California and India have subsidized imports of surface water from other regions in order to reduce the local demand for groundwater (Giordano 2009; Thompson, Leshy, and Abrams 2013). While the latter strategy has often been successful in the short run, it is costly to the government and often stimulates a longer-term increase in irrigation and related investments and development, which can lead to groundwater overdrafting to supplement the imported water (Thompson et al. 2013).

Second, some governments have sought to increase the capacity of local groundwater users to manage groundwater withdrawals themselves, on the theory that local groundwater users want to avoid overdrafting but lack the ability to do so. These efforts can include the authorization of new local institutions, awarding of new local powers, and processes for the effective negotiation of local rules (Blomquist 1992). In some cases, existing local institutions can be enlisted. While local farmers may still resist reducing groundwater withdrawals, local groundwater management districts in areas of the western United States and Australia have demonstrated significant creativity in finding locally acceptable solutions to groundwater sustainability (Nelson 2011).

Agricultural Impacts

Reallocation of irrigation water to other sectors, as well as constraints on diversions to ensure long-term sustainability, have raised concerns over the potential impact on agricultural production. A growing number of experts, for example, worry that poor groundwater management in China and India ultimately will force cutbacks in water withdrawals and thus local food production (e.g., Seckler 1999; Brown 2007; Giordano 2009). To date, however, reductions in irrigation water have typically led to increased irrigation efficiency rather than reduced production. Increasing water demand for the environment and other non-agricultural sectors, in short, has imposed a market discipline on agricultural use that was largely lacking in earlier eras, forcing agriculture to adopt economically more efficient practices.

In response to scarcer water, farmers have switched to less water-intensive crops, installed improved irrigation technology, and adjusted the timing of their irrigation to reduce water use (Molle and Berkoff 2010). Indeed, gross water efficiency has increased over the last several decades, primarily in developed nations, through the adoption of more efficient irrigation techniques such

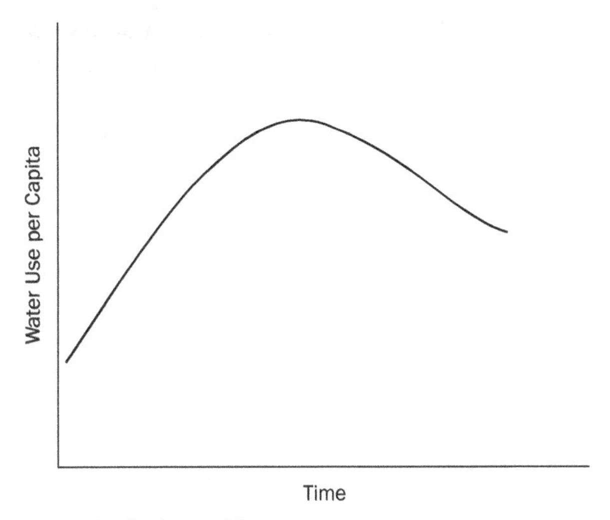

FIGURE 11.3 Water Use Per Capita over Time

as drip or micro-irrigation (Worldwatch Institute 2012).[12] As a result, water use per capita in this final phase often drops—leading to an inverted U-shaped relationship between water withdrawals and economic development where water use per capita begins relatively low in Early Agricultural Development (when irrigation water is scarce), rises during Irrigation Expansion, and then declines again in the final phase of Multiple Sustained Uses (Rock 1998 and 2000) (see Figure 11.3).

Conclusions

A variety of themes emerge from the study of how water institutions and agriculture historically have co-evolved and are important in the development of institutions in developing nations seeking to increase their irrigation capacity. The first, and most obvious, theme is the importance of institutional design to the sustainability of large-scale irrigation and thus to food security. Appropriate institutions, moreover, change over time, so governments need to think about water institutions as an evolutionary and adaptive process.

A second theme is the need to capture the benefits of both central and local control. Local, particularly customary, institutions often are well designed to address local conditions and culture, display significant flexibility and adaptiveness, and enjoy significant support from local users. While large-scale irrigation

[12] Increases in gross water efficiency, however, do not necessarily translate into similar improvements in net efficiency and thus real water savings. Net efficiency will not increase, for example, where the water that crops have not consumed has flowed to underlying aquifers or neighboring rivers and been reused by others (Hanak et al. 2011).

often requires more central authority, new central institutions have often failed to incorporate elements of the historical local institutions, even when they could be incorporated readily, leading to suboptimal results and risking local rejection. As discussed, central institutions also are typically inflexible and mechanistic and find it difficult to adapt to changing conditions. Melding of the two institutional levels could improve irrigation results.

A third theme is the value of greater foresight in addressing the needs of the later phases of development. This gets back to the importance of thinking of water institutions in evolutionary terms. Governments often create new water institutions, or allow new practices such as extensive groundwater pumping to develop, without considering the long-term need for sustainability and greater flexibility in water use. Efforts to address these predictable problems at a later stage can be more difficult and less effective, particularly where farmers have grown dependent on existing conditions. Anticipating problems before they arise can be politically difficult, particularly when they require institutional steps that involve administrative expense or threaten to constrain water users. Experience, however, shows the importance of effective anticipation.

A fourth theme is the importance of recognizing the inequities that can creep into efforts to change water institutions. Changes often present an opportunity for politically powerful interests to reallocate water entitlements in their favor. The rise of large-scale irrigation, in particular, has often led to a reallocation of water from small local farmers to large landowners. While this might benefit overall production levels, it also results in less equitable allocation of water entitlements. In developing new institutions, countries must take care to avoid such reallocations to the degree that it does not undermine the goal of increased food supply.

A final theme is the need to consider market approaches at an earlier stage of institutional development. Many of the problems of the Expanding Irrigation phase stem from the subsidization of water supplies and the prohibition of market transactions. Reducing subsidies and allowing well-regulated markets can encourage greater efficiency in agricultural water use and enable flexible reallocation of water within the agricultural sector. Markets also can help solve the challenges of increasing water scarcity in the Multiple Sustained Uses phase. Market approaches, moreover, can be designed so that they do not undermine equity and can actually improve environmental conditions (Thompson 2011a). Care must be taken because the introduction of markets before a country has the regulatory resources and expertise to design and implement an appropriately managed market system can undercut equity, lead to greater water diversions and withdrawals, and create political turmoil. Markets, however, remain an underutilized institutional tool in the efficient management of scarce water resources.

References

Ali, I. 1988. *The Punjab under Imperialism*. Princeton, NJ: Princeton University Press.

Baker, M. J., 2005. *The Kuhls of Kangra: Community-Managed irrigation in the Western Himalaya*. Seattle, WA: University of Wash. Press.

Blomquist, W. A., 1992. *Dividing the waters: Governing groundwater in Southern California*. San Francisco, CA: ICS Press.

Boelens, R., R. Bustamante, and H. de Vos, 2007. Legal pluralism and the politics of inclusion: Recognition and contestation of local water rights in the Andes. In *Community-based water law and water resource management reform in developing countries*, ed. B. van Koppen, M. Giordano, and J. Butterworth, 97–113. Oxfordshire, UK: CAB International.

Briscoe, J., and U. Qamar. 2009. *Pakistan's water economy: Running dry*. Washington, DC: Oxford University Press and World Bank.

———. 2013. Winning the battle but losing the war. *The Hindu*. Feb. 22.

———. 2014. Water and agriculture in Africa: The politics of the belly or the politics of the mirror? In Falcon, W.P. and R.L. Naylor (eds.), *Frontiers in Food Policy: Perspectives in sub-Saharan Africa*. Stanford, CA: Printed by CreateSpace.

Brown, L. 2007. Water tables falling and rivers running dry: International situation. *International Journal of Environmental Consumption 3*: 1–5.

Bruns, B. 2007. Irrigation water rights: Options for pro-poor reform. *Irrigation and Drainage 56*: 237–246.

Cox, M., and J. M. Ross, 2011. Robustness and vulnerability of community irrigation systems: The case of the Taos Valley acequias. *Journal of Environmental Economics and Management 61*: 254–266.

Cremers, L., M. Ooijevaar, and R. Boelens. 2005. Institutional Reform in the Andean irrigation sector: Enabling policies for strengthening local rights and water management. *Natural Resources Forum 29*: 37–50.

Cullet, P. 2012. Is water policy the new water law? Rethinking the place of law in water sector reforms. *IDS Bulletin 43*(2): 69–78.

DuMars, C. T., and J. D. Minier, 2004. The evolution of groundwater rights and groundwater management in New Mexico and the Western United States. *Hydrogeology Journal 12*: 40–51.

Easter, K. W. 2000. Asia's irrigation management in transition: A paradigm shift faces high transaction costs. *Applied Economic Perspectives and Policy 22*: 370–388.

Ferguson, A., and W. Mulwafu. 2007. If government failed, how are we to succeed? The importance of history and context in present-day irrigation reform in Malawi. In *Community-Based Water Law and Water Resource Management Reform in Developing Countries*, ed. B. van Koppen, M. Giordano, and J. Butterworth, 211–227. Oxfordshire, UK: CAB International.

Gandhi, M. 1946. Speech at Prayer Meeting, Bombay, February 18, 1946. In *Collected Works*, vol. 89.

Garces-Restrepo, C., D. Vermillion, and G. Munoz. 2007. Irrigation management transfer: Worldwide efforts and results. Rome: FAO.

Giordano, M. 2009. Global groundwater? Issues and solutions. *Annual Review Environmental Resources 34*: 153–178.

Grey, D., and C. W. Sadoff, 2007. Sink or swim? Water security for growth and development. *Water Policy 9*: 545–557.

Hanak, E., J. Lund, A. Dinar, B. Gray, R. Howitt, J. Mount, P. Moyle, and B. Thompson. 2011. *Managing California's water: From conflict to reconciliation*. San Francisco, CA: Public Policy Institute of California.

Hanjra, M. A., and F. Gichuki. 2008. Investments in agricultural water management for poverty reductions in Africa: Case studies of Limpopo, Nile, and Volta River Basins. *Natural Resources Forum 32*: 185–202.

Hemson, D. 2008. Water for all: From firm promises to "New Realism"? In *Poverty and water: Explorations of the reciprocal relationship*, ed. D. Hemson, K. Kulindwa, H. Lein, and A. Mascarenhas, 13–46. London: Zed Books Ltd.

Indus Basin Research Assessment Group (IBARAG), 1978. Report of the Indus Basin Research Assessment Group. Islamabad, Pakistan: Planning Commission of the Government of Pakistan.

Komakech, H. C., P. van der Zaag, and B. van Koppen, 2012. The last will be first: Water transfers from agriculture to cities in the Pangani River Basin, Tanzania. *Water Alternatives 5*: 700–720.

Lansing, J. S. 1987. Balinese "water temples" and the management of irrigation. *American Anthropologist 89*: 326–341.

Lieftinck, P. 1968. *Water and power resources of West Pakistan: A study in sector planning*. Baltimore, MD: Johns Hopkins Press.

McCool, D. C. 1987. *Command of the waters: Iron triangles, federal water development, and Indian water*. Tucson, AZ: University of Arizona Press.

Mehari, A., F. van Steenbergen, and B. Schultz, 2007. Water rights and rules, and management in spate irrigation systems in Eritrea, Yemen, and Pakistan. In *Community-based water law and water resource management reform in developing countries*, ed. B. van Koppen, M. Giordano, and J. Butterworth, 114–129. Oxfordshire, UK: CAB International.

Meinzen-Dick, R. 2007. Beyond panaceas in water institutions. *PNAS 104*: 15200–15205.

Meinzen-Dick, R., and L. Nkonya, 2007. Understanding legal pluralism in water and land tights: Lessons from Africa and Asia. In *Community-based water law and water resource management reform in developing countries*, ed. B. van Koppen, M. Giordano, and J. Butterworth, 12–27. Oxfordshire, UK: CAB International.

Molle, F., and J. Berkoff, 2009. Cities vs. Agriculture: A review of intersectoral water re-allocation. *Natural Resources Forum 33*: 6–18.

Namara, R. E., M. A. Hanjra, G. E. Castillo, H. M. Ravnborg, L. Smith, and B. van Koppen, 2010. Agricultural water management and poverty linkages. *Agricultural Water Management 97*: 520–527.

NASA Earth Observatory. 2010. "Flooding in Pakistan." Accessed September 9, 2013, http://earthobservatory.nasa.gov/IOTD/view.php?id=45200.

Nelson, R. 2011. Uncommon Innovation: Developments in Groundwater Management Planning in California. Water in the West Working Paper 1. Stanford, CA: Woods Institute for the Environment.

Pearce, F. 2007. *When the rivers run dry: Water—the defining crises of the twenty-first century*. Boston, MA: Beacon Press.

Rawai, V. 2002. Non-market interventions in water-sharing: Case studies from West Bengal, India. *Journal of Agrarian Change 2*: 545–569.

Rivera, J. A. 1998. Acequia culture: Water, land, and community in the Southwest. Albuquerque, NM: University of New Mexico Press.

Rock, M. T. 1998. Freshwater use, freshwater scarcity, and socioeconomic development. *Journal of Environmental Development 7*: 278–301.

———. 2000. The dewatering of economic growth: What accounts for the declining water-use intensity of income? *Journal of Industrial Ecology 4*: 57–73.

Rose, C. M. 1988. Crystals and mud in property law. *Stanford Law Review 40*: 577–610.

Sandoval, R. 2004. A Participatory Approach to Integrated Aquifer Management: The Case of Guanajuato State, Mexico. *Hydrogeology Journal 12*: 6–13.

Seckler, D., R. Barker, and U. A. Amarasinghe. 1999. Water scarcity in the twenty-first century. *International Journal of Water Resource Development 15*: 29–42.

Shah, T. 1993. Groundwater markets and irrigation development: Political economy and practical policy. New York: Oxford University Press.

———. 2007. Issues in reforming informal water economies of low-income countries: Examples from India and elsewhere. In *Community-based water law and water resource management reform in developing countries*, ed. B. van Koppen, M. Giordano, and J. Butterworth, 65–95. Oxfordshire, UK: CAB International.

Shah T., S. Bhatt, R. K. Shah, and J. Talati. 2008. Groundwater governance through electricity supply management: Assessing an innovative intervention in Gujarat, Western India. *Agricultural Water Management 95*: 1233–1242.

Shah T., et al. 2007. Groundwater: A global assessment of scale and significance. In *Water for food, water for life: A comprehensive assessment of water management in agriculture*, ed. D. Molden, 395–419. Earthscan: London, UK.

Siebert, S., J. Burke, J. M. Faures, K. Frenken, J. Hoogeveen, P. Doll, and F. T. Portmann. 2010. Groundwater use for irrigation—a global inventory. *Hydrolology and Earth Systems Science 14*: 1863–1880.

Stein, C., H. Ernstson, and J. Barron, 2011. A social network approach to analyzing water governance: The case of the Mkindo Catchment, Tanzania. *Physics and Chemistry of the Earth 36*: 1085–1092.

Swallow, B., L. Onyango, and R. Meinzen-Dick. 2007. Coping with history and hydrology: How Kenya's settlement and land tenure patterns shape contemporary water rights and gender relations in water. In *Community-based water law and water resource management reform in developing countries*, ed. B. van Koppen, M. Giordano, and J. Butterworth, 173–195. Oxfordshire, UK: CAB International.

Thompson, B. H., Jr., 1993. Institutional perspectives on water policy and markets. *California Law Review 81*: 673–764.

———. 2011a. Water as a public commodity. *Marquette Law Review 95*: 17–52.

———. 2011b. Beyond connections: Pursuing multidimensional conjunctive management. *Idaho Law Review 47*: 265–323.

Thompson, B. H., Jr., J. D. Leshy, and R. H. Abrams. 2013. *Legal control of water resources: Cases and materials*, 5th ed. St. Paul, MN: Thomsen Reuters.

U.S. Presidential Commission, 1964. Land and Water Development in the Indus Plain. Washington, DC: The White House.

van Edig, A., S. Engel, and W. Laube. 2003. Ghana's water institutions in the process of reform: from the international to the local level. In *Reforming Institutions for Sustainable Water Management*, ed. Neubert et al., 31–51. Bonn, Germany: German Development Institute.

van Koppen, B. 2007. Dispossession at the interface of community-based water law and permit systems. In *Community-based water law and water resource management reform in developing countries*, ed. B. van Koppen, M. Giordano, and J. Butterworth, 46–64. Oxfordshire, UK: CAB International.

Wang, J., J. Huang, S. Rozelle, Q. Huang, and A. Blanke. 2007. Agriculture and groundwater development in Northern China: Trends, institutional responses, and policy options. *Water Policy* 9(Supp.): 6–74.

World Bank. 2007. World development report: Agriculture for development. Washington, DC: World Bank.

Worldwatch Institute. 2012. Global Irrigated Area at Record Levels, but Expansion Slowing. Vital Signs Online. http://vitalsigns.worldwatch.org/vs-trend/area-equipped-irrigation-record-levels-expansion-slows.

Worster, D. 1985. *Rivers of empire: Water, aridity, and the growth of the American West.* Oxford, UK: Oxford University Press.

12

Global Agriculture and Land Use Changes in the Twenty-First Century

ACHIEVING A BALANCE BETWEEN FOOD SECURITY, URBAN DIETS, AND NATURE CONSERVATION

Ximena Rueda and Eric F. Lambin

Introduction

Agricultural output has grown rapidly over the last decade, and continues to expand to meet the needs of a larger and more affluent population. In the last 50 years, mostly through intensification, agriculture has been able to produce the food required to satisfy increasing demands, albeit with significant distributional and environmental challenges. Agricultural intensification—defined as higher levels of inputs and increased output (in quantity or value) of cultivated or reared products per unit area and time—permitted the doubling of the world's food production from 1961 to 1996 with only a 10 percent increase in arable land globally (Tilman 1999). In considerable part, these results were the consequence of the Green Revolution (as discussed in Chapters 2 and 3). Although the contribution of agricultural land expansion to this production increase has been relatively modest, the conversion of natural ecosystems, and in particular the clearing of tropical forests, has nonetheless been rapid and has affected negatively the provision of ecosystem services such as carbon sequestration and watershed protection. This outcome has led some authors to emphasize a trade-off between food security and natural ecosystem conservation: feeding the hungry comes at a high ecological cost (Foley et al. 2011). Close observation shows that the reality is more complex. Actually, the effects of economic globalization on the geography of global agriculture has given rise to another trade-off: between ensuring the food security of the poor and meeting the ever growing food demand in rich countries and emerging economies, with major implications for land use.

Most global-scale studies of the issue have explored the competition for land between cropland, pastures and natural ecosystems at an aggregated level—all crops versus all pastoral uses, versus all forest types. Distinguishing between different crop types and how central they are to meeting the calorie and nutrient needs of a community leads to a more nuanced picture, however. Actually, different crops are grown for different markets—local to international—and meet different needs—subsistence to luxury, and food to bioenergy. As a result, for different crop types, decisions of what to grow where and using which technology are made by different types of actors, with different impacts on both food security and natural ecosystems.

Simply distinguishing between staple and non-staple crops, for example, reveals that truly feeding the hungry has lower negative trade-offs than supplying global markets. Staple crops are those that form the foundation of people's diets in a specific region. They are usually cereals, coarse grains, or starchy root vegetables. The most important staple crops worldwide include rice, maize, wheat, and potatoes. Staple crops are the ones that support the food security of the poor. By contrast, non-staple crops constitute a smaller portion of a standard diet and supply a lower proportion of the energy and nutrient needs of the traditional diet of a given population. They comprise crops such as coffee, cocoa and sugar cane, as well as those from which vegetable oils are extracted. They also include crops that are used as animal feed and to produce biofuels. They are often considered as "cash" crops, but also comprise fruits, vegetables, and legumes greens (i.e., non-staple but nutrient-dense foods grown for home or local consumption). Several crops have multiple uses, such as maize, which, in addition to being a staple food, is used as animal feed and to produce sweeteners, ethanol, starch, corn oil, and chemical products such as plastics.

We argue that, over the last decade, land conversion for agriculture has been associated less with cereals and other staple crops but increasingly more with non-staple crops and pastures. For example, between the years 2000 and 2010, the number of hectares globally devoted to the production of the main, permanent, non-staple crops increased by 7.6 million hectares (Mha). In the process, oil palm, cocoa, coffee, and banana plantations replaced other crops and natural ecosystems (FAO 2012). Between 1980 and 2000, pastures expanded by 42 million hectares just in Latin America, the fastest–growing supplier of meat globally, mainly at the expense of forests (Gibbs et al. 2010).

Access to land with a high agro-ecological suitability is an essential component of food security. The success of the Green Revolution led analysts to take land for granted. The stable land base dedicated to cereals at a global scale, over the last few decades, hid important changes with respect to agricultural expansion for other crops and for pastures, the geography of agricultural production, and the changing nature of the demand for food products. The relatively steady acreage data for cereals were markedly different than acreage data associated with other food, fuel, and beverage categories. In particular, areas allocated to the production of non-staple food crops such as oil palm, soy, and sugar, and for animal feed and bioenergy crops have grown fast recently. Moreover, the structure of the global food system has changed dramatically with a reorganization of supply chains. Since the turn of the century, the growth in agricultural production has been accompanied by a greater concentration of the agribusiness sector: a few countries produce most of the food commodities traded globally; a small number of transnational corporations control many links in the value chain for these goods; and a small number of retailers have a global reach, delivering the same types of products across the globe.

The reorganization came about as global barriers to trade were lowered, and agribusiness corporations sought greater efficiency and higher profits. One could argue, however, that this reorganization of the global food system has done little to improve global food security, while mostly benefitting consumers in wealthy, urban regions and in emerging economies. These consumers have changed their diets, demanding more meat and high-value products such as sugar, vegetable oils, coffee, and cocoa. By better connecting tropical agricultural regions that today produce the bulk of these non-staple crops with urban consumers worldwide, the highly concentrated supply chains for agricultural commodities have accelerated the conversion of natural ecosystems at one end, and delivered food products that are rich in fats and sugars at the other end, without delivering much improvement for the food security of the "bottom billion." It makes possible the ironic result of a world where, in 2010–12, some 1.5 billion overweight and obese adults shared the planet with nearly 870 million people who were chronically undernourished. Did our modern food system really destroy our tropical forests to contribute to the obesity crisis in rich countries without feeding the hungry?

The objectives of this chapter are to describe recent changes in the global agricultural system, explain how they came about, and explore the implications of these trends for both land use change and food security. Geographers and land change scientists traditionally approach food-security issues through a supply-side perspective: where is agricultural production expanding, and how does it compete with nature conservation and other land uses. But that leaves out the influence of economic globalization on the global food system, without which a thorough understanding of the impact of demand-led growth on the geography of food production and land use is impossible. The first section provides an overview of the growth in production, geographic distribution and trade of the main agricultural commodities during the 2000–2010 period, based on FAO's production and the United Nations' trade data (FAO 2012; UN 2012). The second section identifies the main driving forces behind the emerging spatial concentration and specialization of food production. The third section discusses the magnitude of global demand for agricultural land and the competing land uses—for bioenergy crops, timber extraction, conservation, and urban growth—for which demand for land is also increasing. We conclude with a discussion of the food security repercussions of a globalized food production system and associated global land use changes.

The recent wave of economic globalization and the growing demand for non-staple crops are reshaping the geography of food production at a global scale, leading to more demands on land in the handful of countries that produce the bulk of the non-staple crops that dominate today's agricultural trade. While globalization has the potential to increase food security, because food imports can compensate for local agricultural shortfalls, it also creates a greater exposure to supply shocks in grain-growing regions and accelerates the conversion of the increasingly scarce potential available cropland. Data suggest that improving global food security does not cause much environmental harm; rather, meeting the increasingly sophisticated food demands of the wealthy is the main cause of natural ecosystem conversion—such as through palm oil expansion into tropical forests, or soybean expansion into tropical woodlands or former pastures that are themselves pushed into forests. Thus, there is little trade-off between food security and environmental conservation; there is only a trade-off between feeding the poor and meeting the food desires of the consumers in rich countries and emerging economies.

A Changing Geography of Food Production

AGRICULTURAL EXPANSION: FROM STAPLE CEREALS TO NON-STAPLE FOOD

Recent estimates indicate that close to 49 million km^2 are cultivated or under pastures (FAO 2010). This is roughly one third of the planet's land surface. For decades after the Second World War, food crops expanded mainly for the production of staple food. Growth of the area under cereal production, such as wheat and rice fields, has slowed in the first years of the twenty-first century. Instead products rich in proteins; those that are used for vegetable oils, animal feed, or biofuels (soy, maize, sugar cane, and other sugar crops, oil palm, and rapeseed); or that contribute to pleasurable but dispensable consumption (coffee, cocoa) are experiencing more rapid expansions in both area planted and volume harvested (Figures 12.1 and 12.2).

Losing the appetite for cereals. The bulk of agricultural output (in volume) still comes from cereals. Wheat, maize, and rice represent 75 percent of the world's cereal production and a large portion of human caloric intake (Cordain 1999; Cordain et al. 2005); but for the most part, their production and areal extent are stagnant (Table 12.1). Wheat production has increased mostly from productivity increases. Area expansion has been minimal and occurring on previously abandoned cropland, particularly in Eastern Europe after the collapse of the Soviet Union (Nefedova 2011). Rice production is equally stagnant. Per

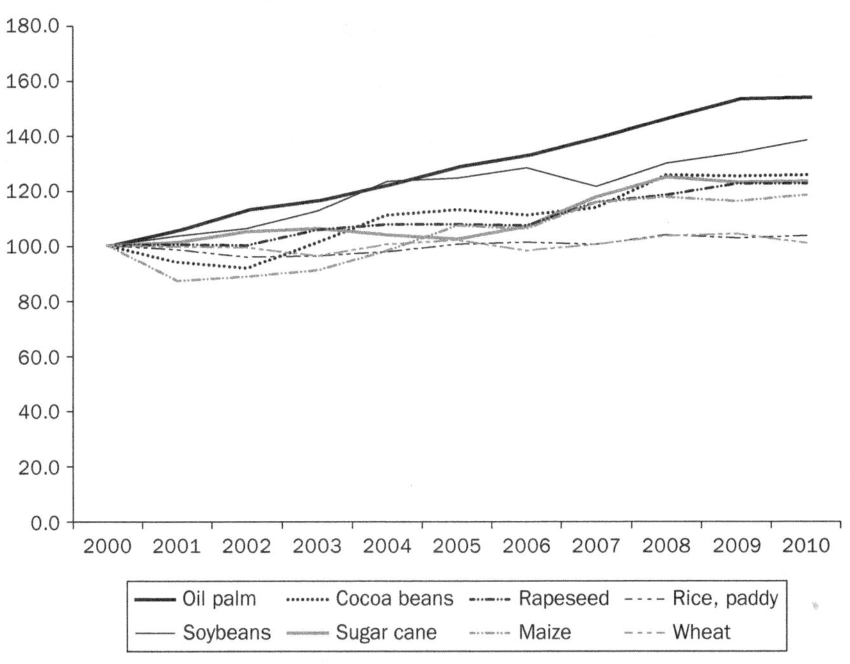

FIGURE 12.1 Growth in Area Harvested for Selected Staple and Non-staple Crops

Source: United Nations Food and Agriculture Organization (2012)

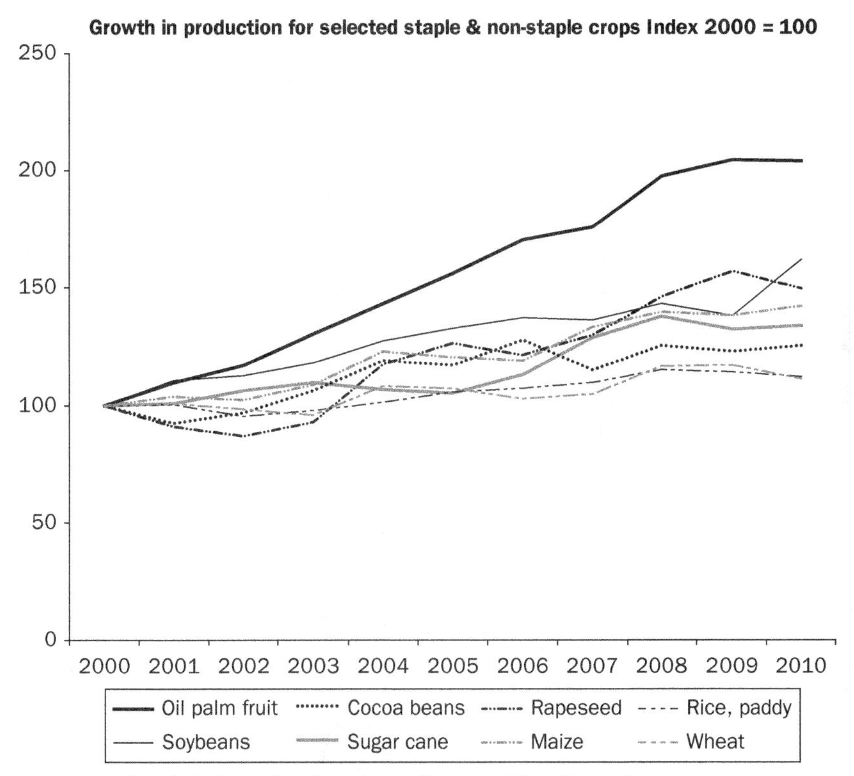

FIGURE 12.2 Growth in Production for Selected Staple and Non-Staple Crops

Source: United Nations Food and Agriculture Organization (2012).

TABLE 12.1

Global area harvested, production and main global producers for the main staple and non-staple agricultural crops 2000–2010

Crop	Area harvested (ha)		Avg. annual growth (%)	Production (tons)		Avg. annual growth (%)	Yields (tons/ha)		Change 2000–2010 (%)	Main producers (countries & %)
	2000	2010		2000	2010		2000	2010		2010
Cereals										
Wheat	215,436,907	217,219,395	0.09	585,690,370	653,654,525	1.23	2.72	3.01	10.69	China (18), India (12), US(9), Russia (6), France(6)
Rice (paddy)	154,059,904	159,416,542	0.38	599,355,455	696,324,394	1.68	3.89	4.37	12.28	China (29), India (18), Indonesia (10)
Maize	137,004,755	161,765,388	1.86	592,479,279	840,308,214	3.96	4.32	5.19	20.12	US (37), China (21)
High-value commodities										
Soybeans	74,363,900	102,556,310	3.64	161,289,911	264,991,580	5.67	2.17	2.58	19.13	US (35), Brazil (26), Argentina (20)
Palm oil	10,029,567	15,410,262	4.89	22,227,777	43,573,470	7.77	2.22	2.83	27.58	Indonesia (48), Malaysia (38)
Rapeseed	25,843,903	31,640,756	2.27	39,526,198	59,007,753	4.55	1.53	1.86	21.94	China (22), Canada (20), India (11)
Sugar cane	19,395,810	23,877,378	2.34	1,257,498,759	1,711,087,173	3.48	64.83	71.66	10.53	Brazil (43), India (16)
Beef	NA	NA	–	59,078,318	67,602,794	1.51	–	–	–	US (18), Brazil (13), China (10), Argentina (4), India (4), Australia (3)

Crop	Area harvested (ha)		Avg. annual growth (%)	Production (tons)		Avg. annual growth (%)	Yields (tons/ha)		Change 2000–2010 (%)	Main producers (countries & %)
	2000	2010		2000	2010		2000	2010		2010
Cereals										
Meat (including beef)	NA	NA	–	234,057,062	295,462,319	2.62	–	–	–	China (27), US (14), Brazil (8)
Milk	NA	NA	–	578,967,321	723,143,305	2.50	–	–	–	India (17), US (12), China (6), Pakistan (5), Russia (4), Brazil (4)
Non-staple food crops with potential for differentiation										
Coffee	10,725,243	10,234,363	–0.52	7,564,401	8,228,018	0.94	0.71	0.80	13.99	Brazil (34), Vietnam (13), Indonesia (10)
Cocoa	7,608,187	9,541,698	2.55	3,373,727	4,187,587	2.43	0.443	0.439	–1.03	Côte d'Ivoire (29), Indonesia (19), Ghana (15)

Source: FAO (2012).

capita rice consumption is falling in Asia, a region that represents 90 percent of global rice production and consumption (Pingali 2007). Production has increased due to better technology while the area harvested has remained virtually unchanged (Table 12.1).

Maize production is the only exception, growing at an average rate of 4 percent annually. The versatility of the crop to produce sugar substitutes for human consumption, animal feed, and ethanol explains the rapid increase in production, dominated by the United States, China, and Brazil. Except for China, where most of the production stays in the country as an input for processed products, the same countries dominate most of the global trade in maize. Although global statistics on the final use of maize are hard to come by, data for the United States show that about 50 percent of production goes to animal feed, 40 percent to ethanol, and only 10 percent to food, seeds, and industrial uses (USDA 2012a). These figures vary slightly from year to year, depending on the relative price of maize and its close substitutes, but they provide a rough estimate of the importance of non-food uses of the cereal. The area harvested for maize has increased by 25 percent between 2000 and 2010 (Table 12.1). Most of the increase is concentrated in China and the United States, mostly on previously cultivated lands.

Higher value commodities in high demand. For reasons we will explore in the next section, the demand for agricultural products has diversified from the staple grains for food into animal feed and oils. Particularly important is the growth in soybean and its by-products (vegetable oil and meal for feed); and in palm oil and rapeseed for the production of biodiesel and other food and industrial uses. Global soybean production increased 64 percent between 2000 and 2010. The main producers are the United States, Brazil, and Argentina. Five South American countries make up 51 percent of total production: Brazil, Argentina, Paraguay, Uruguay, and Bolivia. The area planted in soybeans has also increased rapidly since 2000, although important gains have also been made in yields (Table 12.1). Expansion in South America has come at the expense of large portions of the Brazilian Cerrado, and the Amazon and Chaco forests (Grau et al. 2005; Morton et al. 2006; Jepson et al. 2010). Indirectly, soybean is also contributing to the expansion of cattle ranching into the Amazon frontier, as it has displaced existing pastures. For example, from 2006 to 2010, 91 percent of the expansion of soy production in Mato Grosso state was on previously cleared pasture areas; pasture expansion, meanwhile, accounted for most new deforestation (Macedo et al. 2012). Soybeans and their by-products are exported to China, Mexico, Japan, and Western Europe, mostly for the production of soy-based feed and oils, as well as food products.

Palm oil production doubled between 2000 and 2010, supplying more than 30 percent of the world's edible oil consumption in 2010. The land area of oil palm plantations also increased more than 50 percent between 2000 and 2010 (Table 12.1). After its expansion in Malaysia in the 1990s, oil palm production is now increasing in Indonesia, as well as elsewhere in Asia, Africa, and Latin America. The environmental consequences of this expansion have raised concerns as valuable rainforest and peat lands have been replaced to make room for oil palms, with negative consequences for biodiversity and greenhouse gas emissions (Koh et al. 2011; Carlson et al. 2012). Asia is the principal market for palm oil (58 percent), where it is the main source of cooking oil. An important fraction also goes to Europe (27 percent), as a cheap source for biodiesel.

Rapeseed, which is also used in the production of biodiesel, grew 50 percent in 2000–2010. China, Canada, and India are the main producers (Table 12.1). Canada, France, and Ukraine make up 60 percent of all exports. Although Japan and China are important markets for rapeseed, close to 40 percent of the trade happens within Europe, to satisfy biofuel mandates. Land area in rapeseed expanded by almost 6 Mha globally, mainly in previously abandoned agricultural lands.

Production of sugar cane increased by 34 percent between 2000 and 2010, led by Brazil with 40 percent of global output, then India and China (Table 12.1). The most dynamic growth is occurring in Brazil and Argentina. Land area in sugar cane expanded by 25 percent between 2000 and 2010, mainly at the expense of pastures in Brazil (Martinelli et al. 2011).

The production and consumption of beef and dairy products has also increased rapidly over the last decades, particularly in emerging economies. The world's beef production has grown 14 percent between 2000 and 2010; the United States, Brazil, and China together the supply 40 percent of the global output (Table 12.1). Milk production increased by 22 percent during the same period. Most milk is consumed locally, with only 14 percent of the world's production entering the global market. The majority of that trade occurs within Europe. Consumption of milk products, however, is expanding rapidly in every emerging economy, even in East Asia, where consumption of dairy products had traditionally been very low.

Global-scale data on acreage devoted to pasture are not readily available. FAO estimates, however, show that the area in pastures peaked in the 1990s and declined somewhat during the 2000s. Still, FAO estimated that by 2000, pastures occupied an area 50 percent larger than in 1960 (FAO 2006). This new pasture is largely a result of the conversion of different types of natural ecosystems, from forests to marginal lands. It follows an increased demand for meat, particularly in emerging economies. Deforestation in the Brazilian Amazon is, to a large extent, the result of the increased global demand for meat: directly, as land is cleared to make room for pastures, and indirectly, as this pasture is transformed to produce soybeans that are used in part as animal feed (Steinfeld et al. 2006)

Food commodities with a high potential for differentiation. The change in global consumption patterns that is most clearly associated with increasing global affluence and least associated with increasing food security is the increased demand for non-staple "gratifying" food products. The consumption of these products does not contribute significantly to meeting basic energy or nutritional needs, but is considered a dispensable pleasure. Coffee, tea, and chocolate are examples of the most widely consumed gratifying foods, but global trade is also expanding for olive oil, wine, and some of the less common fruits and fresh vegetables. These products are also subject to an increasing trend towards product differentiation based on quality (Saitone and Sexton 2010). The result is a growing range of niche commodities that are traded over long distances, and often at a significant price premium.

Attributes such as taste, appearance, convenience, brand appeal and healthfulness are among the qualities consumers look for and markets provide (Saitone and Sexton 2010). Consumers are also expressing preferences for production narratives that are resonant with their values. These include products that show care for nature (i.e., ensuring environmentally friendly practices); care for others (i.e., fairness in trade); or care for traditions (i.e., geographical indications) (Le Polain and Lambin 2012a). For example, argan oil, which is produced from a tree that is endemic to southwestern Morocco, is sold abroad as cooking oil and in cosmetic products for prices up to over US$400 per liter. Part of the argan oil production is certified organic; its high price is thought to encourage ecosystem conservation; the oil is associated with gender empowerment and rural development narratives, as it is mostly produced by women organized in cooperatives; it has desirable health properties; and it is associated with Berber identity and tradition, and with a unique "*terroir,*" as recognized by a geographical indication (Le Polain and Lambin 2012a).

The cases of cocoa and coffee are particularly significant. Their production stretches over more than 60 tropical countries, and affects the livelihoods of millions of rural dwellers, many of whom are also food insecure. The production of cocoa, the basic ingredient in chocolate, increased 24 percent between 2000 and 2010. Most of the global supply comes from Côte d'Ivoire, Indonesia, and Ghana. Although production has been stagnant in Côte d'Ivoire, due in no small part to the country's civil unrest, other points of origin such as Indonesia, Cameroon, and Togo have shown dynamic growth. Area planted expanded 25 percent between 2000 and 2010, adding almost 2 Mha during the decade. Growth in cocoa plantations has come at the expense of natural forest in the lowland tropical regions of the world. Limited husbandry, poor soils, fungal and insect plagues, and the use of hybrid varieties that are resistant to damage from direct sunlight and pests have all encouraged or enabled the expansion of cocoa into natural systems (Rudel 2002). As productivity decreases over time, farmers clear additional hectares of forested lands, removing most of the canopy and planting hybrid varieties, which are more productive in the short run but also more homogeneous and susceptible to the attacks of pests and diseases (Rice and Greenberg 2000).

Coffee production shows two parallel trends: significant growth in the top producers (Brazil, Vietnam, and Indonesia) and slower growth in the other points of origin. Brazil, Vietnam and Indonesia account for most of global coffee output (Table 12.1). They are also the global leaders in the production of robusta coffee, used mainly for instant coffee preparation. Instant coffee is the entry product for new coffee consumers, and its consumption is increasingly popular in Eastern Europe and Asia. Latin America and East Africa produce more Arabica coffee, preferred in the matured markets of Europe and North America. Growth rates of coffee consumption are significant in the emerging economies of Asia and the Americas. After a big push in the 1990s to expand the production in Vietnam and Indonesia, the area devoted to coffee has not increased substantially during the first decade of the twenty-first century; about 8 Mha are devoted worldwide to this tropical crop, located mainly in mountainous regions with a long history of human settlement and cultivation in Latin America and East Africa.

GLOBAL PATTERNS

The agricultural landscape has shifted in the first years of the twenty-first century in response to changes in the demand for but also in the supply of many food products. Although cereals continue to be the primary basis of the human diet, they are not expanding substantially in either volume or area. Non-staple crops with multiple uses in the food and energy sectors, and that meet the (mostly) urban demand for a more gratifying and diversified consumption are the main proximate cause of land use change, with stark implications for food security and climate change (see Lobell et al., Chapter 9). Land-rich countries with adequate infrastructure and large domestic markets supply the world with these products and are becoming the breadbaskets of the world. The United States, China, Brazil, India, Indonesia, and Argentina account for most of the global output of the crops that represent the main food products that are consumed and traded globally. These six countries account for 91 percent of global soybean output, 70 percent of both maize and sugar cane, 61 percent of rice, and almost 50 percent of beef, palm oil, and coffee; they represent lower, and yet still significant proportions of the global output for wheat and cocoa (39 percent and 25 percent, respectively). These countries are agricultural powerhouses, also involved in the production and trade of many more food products that are less prominent in global markets. Smaller countries with similar biophysical conditions, and recent injections of capital, are following this trend: Paraguay, Uruguay, and Bolivia in soybeans; Nigeria, Colombia, Papua New Guinea, and Côte d'Ivoire for palm oil; Ukraine for maize and soybean. Peru has become the fifth largest exporter of coffee, which has become its main agricultural export crop (Tulet 2010). Ecuador has specialized in the production of bananas, which represent 65 percent of its agricultural exports, and cacao, with a recent focus on the export of high quality cacao beans.

Hubs of Supply for a Globalized Lifestyle

Before exploring the implications of these recent trends in global agricultural production, we need first to apprehend how they came about. The drivers of the growing prominence of non-staple crops help explain who benefits most from this recent and largely unnoticed change in the global agricultural system. Similarly, the emerging pattern of spatial concentration and specialization in food production has both supply and demand causes. On the supply side, reduced trade and investment barriers, a greater integration of supply chains, and technological advances in production, processing, and transportation facilitated the geographic concentration of agricultural production. On the demand side, urbanization, income growth in emerging economies, and the adoption of a Western lifestyle are responsible for the massive uptake in the consumption of higher value commodities and the relative disregard for staple foods. This section explores each of these drivers in turn. What emerges is that the sources of investment, the centers of decision, and the places where demand for new products resides are neither among the food insecure, nor in the producing countries, but largely in the urban centers of the rich countries and emerging economies.

TRADE LIBERALIZATION AND INCREASING FOREIGN INVESTMENT

Agriculture remains protected and subsidized in most of the industrialized countries. Nevertheless, the decades between the 1980s and the 2000s saw dramatic changes in the trade policies of many countries, as part of the structural adjustment agenda put forward by the multilateral banks (Bleaney 1993; Greenaway and Morrissey 1993). Tariffs and subsidies were reduced, bilateral and multilateral trade agreements signed, and agriculture was included for the first time in World Trade Organization agreements in the Uruguay Round of 1994. Global trade agreements for agriculture have been elusive, but still, production and trade of all global commodities has expanded over the last decades. Agricultural imports, as a share of apparent consumption, have grown from 11 percent globally in 1960–64 to 18 percent in 2000–04 (Anderson 2010).

During the 1990s and 2000s, many land-rich countries reduced restrictions to foreign direct investment (FDI). As a consequence, FDI flows for agricultural production, while still modest, tripled to $3 billion annually between 1990 and 2007 (UNCTAD 2009). There was a significant reduction from 2008 onwards due to the global recession. Today, most developing countries rely on direct foreign investment as their main source of external funding for the overall economy, followed by remittances from citizens working overseas and international aid (UNCTAD 2009). These investments have been driven by different factors in different contexts. For Africa, they have been driven mainly by concerns over food security by non-African countries (Hallam 2009), as land-scarce countries, mostly in Asia, attempt to secure arable lands abroad to meet the future demands of their own population. According to Borras (2012), for Latin America, these investments have been driven by other dimensions of the food-land-agriculture system. These include bioenergy crops, climate mitigation strategies as part of multilateral treaties and international financial mechanisms (i.e., control over land assets that produce ecosystem services, and in particular carbon sequestration). Recently, these drivers have been joined by demands for resources from large companies within the region, for example, sugar cane in Brazil for ethanol production.

Recipient countries have crafted FDI-friendly policies that have allowed investment in specific crops or sectors to spur development. Starting in the 1990s, for example, Brazil reduced restrictions on foreign investment in the soybean market, restructured farm income taxes; lowered import tariffs on machinery, fertilizers and pesticides; and eliminated whole soybean export taxes (Schnepf et al. 2001; Hawkes 2006). Since 2011, however. Brazil has implemented severe restrictions on rural land acquisition by foreign companies and Brazilian companies with a majority of foreign capital; Argentina introduced similar restrictions at the same time. Mexico reduced restrictions to foreign investment and foreign ownership as part of NAFTA (Hawkes 2005). In the mid-1980s, Chile allowed financial groups to invest in agricultural land for the production of fruits and vegetables for the international market (Arce and Marsden 1993). In Latin America, the state has also played a central role in providing a favorable climate for domestic investments by such mechanisms as remapping and reclassifying potentially available cropland (Borras et al. 2012). As a result, foreign and national direct investment in agriculture has soared in the region over the past decade, with all but 4 of the 17 countries studied by FAO and reviewed by Borras et al. (2012) showing high levels of capital flows in agriculture and in land.

Large transnational agribusiness corporations (TNC) have become a major source of FDI in developing countries. Some of them, especially those that are directly related to the production stages, are based in developing countries and are emerging as important players in the global agricultural circuit (UNCTAD 2009). TNCs in the food and beverages and retailing sectors hold sizeable assets overseas: Nestlé and Walmart, the leading actors for each category, hold more than $60 billion in foreign assets each, representing 49 percent and 39 percent of their total assets, respectively (UNCTAD 2009).

CONTROL AND COORDINATION: THE ROLE OF TRADERS, PROCESSORS, AND RETAILERS

Global agribusinesses and transnational food companies have shaped many of their value chain arrangements to respond to the new economic environment. Their role has changed from direct ownership of plantations (typical of the

BOX 12.1:

The Colombian Coffee Chain

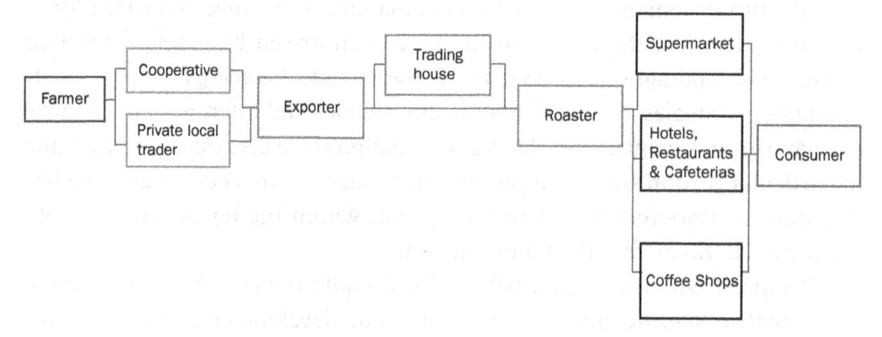

The Colombian Coffee Value Chain: An Illustration

Global value chains have been used to analyze the actors and steps involved in bringing a product from the place of production to the consumer. The Colombian coffee value chain serves to illustrate the concept as well as the relative concentration of companies in trade and processing (Reproduced with permission of *World Development* For details and citations see Rueda and Lambin (2013a).).

1. The farm: The coffee value chain starts with the coffee growers who harvest the cherries; de-pulp and dry them; and, once dried (i.e., parchment), take them to purchasing point. The coffee market is highly asymmetrical, as more than 500,000 farmers sell their coffee to co-operatives and local traders.
2. The purchasing point: Farmers in Colombia have the option of selling their coffee to local traders or to FNC, the Colombian Coffee Growers Federation, through local cooperatives. FNC has over 500 purchasing points distributed across the country, covering 95 percent of the coffee-growing municipalities where farmers can sell their coffee. Other buyers include formal and informal traders who can have permanent or itinerant points of purchase in any given municipality at any given moment in time. Buyers evaluate the quality of the coffee and pay farmers accordingly. They then sell the coffee to an exporter and deliver it to milling facilities.
3. Milling, transporting, and exporting: Exporters mill the parchment, transport, and export it through ports in the Atlantic and Pacific coasts. For 2010 there were 80 active exporters of Colombian coffee (FNC 2011). FNC was the largest, with 26 percent of the total share, while five national companies represent 24 percent of the market and four multinational trading firms make up 22 percent. The rest are smaller local and international companies.
4. Importing and roasting: Traders or roasting houses import the coffee. Most roasters in North America and Japan employ traders who deal with importing and provide hedging from the risks involved in coffee trading. In most cases, European roasters import coffee themselves. Three companies dominate international trade (Kaplinky 2004), and five roasters control 69 percent of the market (Daviron and Ponte 2005). Data for FNC confirms this structure: 10 clients make up 73 percent of the sales volume (FNC 2010a). Roasting companies and large coffee traders make up most of FNC's portfolio of clients. Both roasters and traders participate in the specialty coffee segements, and they vary from large corporations to small coffee shops that roast coffee in small batches. The level of coordination between roasters/importers and FNC varies from very loose, in which agreement is reached based solely on the differential above the C contract at which Colombian coffee will be traded for specific month of delivery; to very tight, in which clients and local experts identify specific cup profiles, linked to particular communities in the field.
5. Distribution: Roasters distribute coffee via three channels: coffee shops, supermarkets, and hotels, restaurants, and cafeterias.

pre-WWII period) to one of control of the production-processing-distribution circuit. Current roles of these large food companies vary from leasing land, to contract agriculture, to determining and/or enforcing "export quality" standards for products, depending on the opportunities and challenges presented by each commodity in each particular market.

For export-oriented commodities with little direct processing and many providers, the preferred type of involvement has been a very loose association in production, with much stronger control over quality standards. This is the case with cocoa and coffee, for example. Both crops are very labor-intensive and produced by smallholders in most of the countries where they are grown. Domestic processing is not indispensable, and a large portion of the value created for consumers consists of brand, packaging, and processing technology, typically added close to the centers of consumption. Five companies—Nestlé S.A., Kraft Foods Inc., Sara Lee Corporation, J.M. Smucker Co. and Tchibo—control 69 percent of the coffee market (Daviron and Ponte 2005) and own the leading brands offered to consumers.

For crops with substantial economies of scale but little domestic processing, such as cut flowers and bananas, the preferred form of engagement of TNCs appears to be through direct contracts with farmers, or with suppliers who group farmers together. As part of the supply contracts, TNCs may provide seeds, fertilizers, technical assistance, and other inputs and services, such as credit, but have no ownership over land and no direct contractual responsibility over labor. Contractors might source part of their supply from their own facilities, as in the case of bananas, for which about 50 percent of the sales of Chiquita, Dole and Del Monte (three of the four dominant global suppliers) come from their own plantations (UNCTAD 2009).

For commodities for which transformation at origin is key, the involvement of TNCs seems to be more direct. For sugar production in Brazil and Western Africa, TNCs and new investors are involved in the production and processing of sugar and ethanol (UNCTAD 2009). For soybeans, TNCs dominate all stages of production, except for farming. In Brazil, four TNCs—ADM, Bunge, Cargill, and Louis Dreyfus—control over 40 percent of the country's crushing capacity (UNCTAD 2009).

TECHNOLOGY AND INFRASTRUCTURE DEVELOPMENT

Technological changes in crop science, which were largely the product of research and development by TNCs and publicly-funded agronomic research centers, have reduced costs and increased efficiency along the value chain. Besides productivity gains in the field, new processing technologies have allowed the development of multiple uses for single crops, as in the case of maize and soy. Researchers have labeled these poly-functional crops "flexible crops," observing that they have in effect produced a diversification in the portfolio of outputs based on a single commodity and a single land use, thus spreading the risk related to investments in a single crop (Borras et al. 2012). TNCs, which in many cases supply producers with the required technology and know-how, can transfer those assets from one country to neighboring ones, thus accelerating agriculture intensification. Changes in transportation and storage technologies, such as controlled atmosphere and refrigerated containers, vacuum packing, and flexible intermediate bulk containers, were already apparent at the beginning of this century (Reardon and Barrett 2000). Together with larger and faster ships, they have reduced transportation costs and duration, enabling greater international trade in agricultural products, particularly perishable ones.

CHANGING DIET

As societies become wealthier and populations become more urbanized, they go through a "nutrition transition" (Popkin 2003). Diets that are high in complex carbohydrates and fiber (such as whole grains and tubers) give way to diets with a higher proportion of fats, animal proteins, and sugars. They also become more varied, as food availability does not depend merely on locally produced calories and nutrients but can increasingly be obtained through trade. For the recent past, Drewnowski and Popkin (1997) pointed out that this transition was partially decoupled from income growth as a high-fat diet became available for poor and rich countries alike. This was due to the availability of relatively inexpensive oilseed-based fats, the consumption of which has increased globally (Drewnowski and Popkin 1997). This decoupling is the result of dramatic changes in the production and processing of oilseed-based fats in their different forms, which have grown from 86.1 million tons in 1999–2000 to 147.9 million tons in 2010–11 (USDA 2012b). Soybean, sunflower, rapeseed, palm, and peanut oils are the main sources for vegetable oils, with palm and soy representing 89 percent of global output (USDA 2012b). The growth in demand for these vegetable oils has been strongest in the newly industrialized nations of East and Southeast Asia.

Rapid urbanization has contributed to the acceleration of this transition. Urban dietary changes include a move towards cereals such as rice and wheat away from maize and millet, and a higher and more diverse consumption of animal products (including meat, fish, milk, eggs, and cheese), sugar and processed foods, but also fruits and vegetables. This has given rise to what Pingali (2007) called the globalization of the human diet, as populations around the globe, and particularly in Asia, adopt the diet of the Western world—a diet in which fat provides about 30 percent of total calories. At the same time, wealthier consumers around the globe have access to high-value commodities such as baked goods, chocolate, and wine. These highly differentiated commodities are among the most dynamic trading sectors (Pingali 2007). Gratifying food products such as coffee and cocoa have experienced an increase in their demand, fuelled by consumption in emergent markets and product differentiation (Rueda and Lambin 2013a). The adoption of the Western diet has thus created what Hawkes has called a "convergence-divergence duality in which high income groups accrue the benefits of a more dynamic market place and lower income groups experience convergence towards a poor quality [. . .]diet" (Hawkes 2006).

Linking this change of diets directly to major socioeconomic forces is not an easy task. However, Drewnowski and Popkin (1997) have found compelling evidence, based on regression analyses of more than 90 countries for 1962 and 1990, that urbanization can be credited for most of the increase in the consumption of sweeteners and vegetable oils, while higher incomes might be responsible for the transition towards animal fats and proteins.

SYNERGISTIC INTERACTIONS

Growing economic liberalization of trade and foreign investment, increasing concentration of commodity value chains, technological and infrastructure improvements, and rising incomes and diet diversification have acted synergistically to reshape the geography of food production at a global scale. Economic liberalization has allowed increased trade in products, which are shipped in ever-greater quantities over longer distances. From 1961 to 2001, the cross-border trade in food commodities increased more than fivefold (Lambin and Meyfroidt 2011). Liberalization has also allowed high-yielding technologies to spread globally. A small number of global agribusiness corporations have increased their control over most food production, processing, and distribution. This concentration has improved coordination and control over value chains that now encompass several countries, ensuring the production of standardized, homogenous products, available for global consumers. Higher incomes and changes in the availability of food have fostered a diversification of the diets of hundreds of millions of people around the world, mostly in urban areas. This trend is reducing the share of consumption devoted to staples and increasing the intake of animal products, oils, processed foods, and fruits and vegetables.

Impacts on Global Land Use

The growing concentration of the global food system, as driven by the recent wave of globalization and urban demands for food products, has led to a rapid expansion of non-staple crops at the expense of natural ecosystems, with little or no benefit for the food security of the poor. These recent trends do not bode well for the future of global land use: as more people are lifted out of poverty by economic growth and adopt a more Western diet, the demand for non-staple crops will continue to increase, thus accelerating the conversion of natural ecosystems. The supply of suitable land for agriculture is becoming tight, and increasing competition between non-staple and staple crops for the remaining underutilized, productive land has the potential to have adverse effects on food security in the long run. In this section, we discuss the land use dimensions of the trends highlighted earlier. We show that multiple pressures on land lead to a growing competition for the best lands and a looming threat of land scarcity for agriculture.

CROPLAND EXPANSION

The expansion of cropland and the changes in geographic patterns of production that took place over the last decades have converted natural ecosystems at a large scale, causing significant losses of ecosystem services. Declines in croplands in industrialized and transition countries were more than outweighed by increases in developing countries. Between 2000 and 2005, about 27 Mha of forest were cleared in tropical regions (Hansen et al. 2008). This deforestation was caused by a combination of timber extraction and forest conversion to agricultural crops or plantations. In sub-Saharan Africa, between 1975 and 2000, agriculture largely expanded at the expense of forests, which decreased by 16 percent over the same period (Brink and Eva 2009). Between 1980 and 2000, 83 percent of the agricultural expansion in the tropics came at the expense of intact or disturbed forests (Gibbs et al. 2010), mostly in the Amazon Basin, Southeast Asia, and to a lesser extent West and Central Africa. The recent wave of large-scale, cross-border land transactions highly affects forests: about 24 percent of the land deals are located in forested areas, representing 31 percent of the total surface of land acquisitions (Anseeuw et al. 2012). The recent trends continuing, the deforestation from 2000 to 2030 might represent an additional 152–303 Mha (Lambin and Meyfroidt 2011). In theory, trade-offs between conserving forests and feeding the world's population could be minimized given the small contribution of deforested areas to the recent increases in food production (Angelsen 2010).

FUTURE LAND AVAILABILITY

Multiple demands for land combine to drive rapid conversion: not just demand for more cropland to increase food and biofuel production, but also for industrial forestry to produce timber; for fast-growing tree species for carbon sequestration; and for urban and recreational spaces to accommodate a growing urban population. Moreover, demand for protected areas for nature and biodiversity conservation, and for natural or managed ecosystems to provide a range of ecosystem services further contribute to a potential conflict between various land uses. The additional land demand for all agricultural, bioenergy, tree plantation, urban, and nature conservation uses was estimated to range from 285 to 792 Mha by 2030, compared to the 2000 baseline of 4,200 to 5,560 Mha allocated to all these uses combined (Lambin and Meyfroidt 2011), under a scenario that includes both significant land productivity increases and abandonment of degraded lands.

Urbanization as land cover, in the form of built-up or paved-over areas, occupies less than 0.5–2 percent of the earth's land surface. But this area is increasing fast: urban land cover grows on average at more than double the rate of growth of the urban population (Angel et al. 2005). This rapid raise is likely to continue in parallel with urban population and income growth. If current trends continue, urban land cover might increase by up to 120 Mha by 2030 (Seto et al. 2012). This is equivalent to nearly tripling the global urban land area compared to the 2000 baseline. Other studies estimate an urban expansion ranging between 66 and 153 Mha for the 2000–2030 period (Angel et al. 2005; Seto et al. 2011). In the developed world, large-scale urban agglomerations and extended peri-urban settlements fragment the agricultural landscapes (Seto et al. 2011), thus decreasing the possibility or profitability of large-scale mechanized farming. In some regions of the less-developed world, urbanization outbids all other uses for land adjacent to cities, including prime croplands under intensive agriculture. In other regions, land-use planning prevents the loss of large agricultural estates, which benefit from their proximity to urban centers. Preliminary results by Angel et al. (2011) suggest that, on average, half of the projected urban expansion will take place on cultivated land. This expansion is likely to take place mostly on prime agricultural land under intensive use, located in coastal plains and in river valleys.

These additional land demands often compete for the most productive and accessible lands, that are least vulnerable to natural hazards and climatic variability, and where a skilled labor force is available and political stability favors productive uses and long-term investments in infrastructure. Additional land demand also targets suitable lands that are currently underutilized. At the same time, various forms of land degradation are causing continuous losses of productive land, and climatic change is likely to alter the comparative advantage of land in many parts of the world (see Lobell et al., Chapter 9).

Much uncertainty remains on the area of potential available cropland (PAC). Only about a quarter of the total ice-free land globally is suitable for rain-fed agriculture, given temperature, rainfall, and soil constraints. Early estimates of the amount of land available for future expansion of cultivation ranged from 1,670 to 1,900 Mha, depending on the method used (Young 1999). More recent, global-scale estimates of PAC have revised this figure downward significantly. A new global assessment of agro-ecological zones identified 3,330 Mha of land that are moderately to very suitable for rain-fed crop production (IIASA/FAO 2012). Of the uncultivated fraction of these lands (1,800 Mha), only 445 Mha globally were non-forested, non-protected, and populated with less than 25 persons/km^2, and therefore assumed to be available for potential cropland expansion, if one attempts to minimize ecological costs of land conversion (World Bank 2010).

A more recent study suggests that there is much less PAC than is generally assumed, and that converting land is always associated with social and ecological trade-offs (Lambin et al. 2013). In this more detailed study, we adopt a spatially explicit, country-by-country approach, based on a limited number of key regional or country case studies where significant detail is available including the Chaco, Cerrado, and Amazon arc of deforestation in South America, as well as the Democratic Republic of Congo, Indonesia, and Russia. There are few remaining places with "free and easy" lands once social, institutional, economic, and physical constraints are taken into account. The ecological costs of converting this remaining suitable land would be significant, both in terms of carbon emissions and biodiversity and ecosystem services losses. A more realistic estimate of the availability and geographic distribution of potentially available cropland—or land reserve—is a priority for land use planning, greater policy foresight, and more adequate information for potential investors.

Based on global trends, productive land could become a more limited resource in most developing countries by mid-century. Under that scenario, lands with a lower productivity would be brought into use, and forests would continue to be converted for agriculture. Indeed, market and policy responses associated with land scarcity are likely to stimulate the future adoption of more efficient land management practices. Innovations that could prevent a global land shortage include technological breakthroughs on genetically modified crops, investments for restoration of degraded lands, adoption of more vegetarian diets in rich countries, a mix of land taxes, subsidies and land use zoning to preserve prime agricultural land, or new industrial processes to produce synthetic food, feed and fibbers. Moreover, improving the adjustment of actual land use to land potential offers a way to spare land. For example, with rising food prices, some lands with a high potential for food production may be taken away by land managers from growing fuels and animal feed, and may become better protected from urban sprawl and land degradation by local land use planning agencies.

FOREST TRANSITIONS AND GLOBALIZATION

Although global rates of tropical deforestation remain alarmingly high, they have decreased over the past decade, and a handful of developing, tropical countries have recently been through a forest transition, thus shifting from shrinking to expanding forests (including tree plantations) at a national scale (e.g., Vietnam, Costa Rica, China, India) (Meyfroidt and Lambin 2011). Forests and plantations of mostly exotic species have expanded primarily on abandoned agricultural lands. The ecological value of this reforestation depends on the residual deforestation of old-growth forests, the proportions of natural regeneration of forests and tree plantations, and the location and spatial patterns of the different types of forests. Forest transitions occur through different pathways that are contingent upon the local socioeconomic and ecological contexts. A few generic processes of forest transition have been identified (Rudel et al. 2005). These include agricultural intensification and concentration in the most suitable lands, driven by industrialization and the ensuing labor scarcity in agriculture; tree plantation, forestry intensification, and forest protection by private and public actors, driven by scarcity of forest products and services (possibly influenced by global environmental ideologies and by national political factors external to the forest sector); and smallholder-driven, labor intensive tree-based land use intensification.

These forest transitions frequently occur with a significant outsourcing of forest exploitation to neighboring countries via increased imports of wood and, sometimes, agricultural products—a process that is referred to as displacement of land use. Importing wood products is the economic equivalent of exporting ecological impacts. A general equilibrium model showed that forest conservation and environmental protection in countries with a significant forestry sector would lead to leakage—mainly to developing, tropical forest countries—for at least 65 percent of the timber stock being protected locally (Gan and McCarl 2007). The national-scale reforestation of Vietnam since 1992 was achieved by displacing forest extraction to other countries equivalent to about 40 percent of the regrowth of Vietnam's forests from 1987 to 2006 (Meyfroidt and Lambin 2009). About half of these wood imports were illegal. For most of the developing countries that recently experienced a forest transition, displacement of land use abroad accompanied the local reforestation. Additional global land use change embodied in their wood imports did offset 74 percent of their total reforested area, a figure that is reduced when taking into account their exports of agricultural goods (Meyfroidt et al. 2010). Economic globalization thus facilitates a forest transition in some countries through a displacement of demand overseas, but other countries absorb these demands and undergo large-scale agricultural expansion. For example, Brazil is currently facilitating forest regrowth elsewhere, by contributing massive quantities of agricultural products to national and global markets. This makes the 80 percent decrease in deforestation in the Brazilian Amazon in 2011, compared to 2004, even more remarkable.

The examples of forest transitions highlight the fact that global scale factors increasingly cause land changes, with a growing separation between the locations of production and consumption of land-based commodities. Land-use decisions related to these commodities are largely driven by factors in distant markets, mostly associated with wealthy urban consumers, and to a lesser extent by local-scale factors (DeFries et al. 2010). Consumers "outsource" their land use to other countries and a virtual land trade develops through the land use embodied in the international trade in agricultural and wood products (Anderson 2010). The distant factors affecting land use are not restricted to trade patterns. They also include remittances sent by migrants, the specific organization of global commodity value chains, channels of foreign investments in land, the transfer of market or technological information to producers via a diversity of networks (including producer associations, Internet resources, and delivery of information via cell phone), and the development and promotion of niche commodities that target narrow but wealthy market segments with high value commodities produced in limited quantities (Le Polain and Lambin 2012b).

Implications for Food Security

The current trend towards specialization and concentration of agricultural production offers a variety of challenges and opportunities for food security. Impacts on food security can be studied at three scales: (1) the microscale of households and firms; (2) the mesoscale of countries and regions; (3) and the macroscale of the globe.

Concerning the microscale, consumers of food around the world have benefitted from having more efficient and diverse supply chains that bring products to their tables. As the global population urbanizes, the urban poor increasingly have the potential to gain access to affordable food. Smallholders who remain in rural areas make up the majority of the food insecure population, and for them the outcomes of globalization are mixed. Their property rights are often loosely defined and insecure. Their land holdings are sometimes the target of large agribusiness enterprises which—with the complicity of weak states—acquire large tracts of productive lands for their farming operations.

In fact, the pressure on lands with good agro-ecological suitability is increasing. This is because the potential available cropland that is still accessible is limited globally, and existing smallholder farmland is more easily converted to grow non-staple crops than is land currently under natural ecosystems. These recent waves of land acquisitions or enclosure, whether for productive purposes (food or bioenergy crops, timber extraction, mining, hydraulic projects) or conservation and carbon sequestration projects (protected areas, carbon storage, afforestation/reforestation projects), create conflicts between smallholders and other customary resource managers and new capital interests (see Smith and Naylor, Chapter 8). They gravely affect the food security of smallholders by excluding them from their land (Cotula 2012; White et al. 2012).

Another trend may be observed, however. In some regions, global value chains have integrated smallholders through supply contracts, by turning farmers into laborers in larger farm operations or suppliers to agribusiness firms (Swinnen et al. 2010), or by opening access to higher-value, differentiated market segments. Recent evidence suggests that smallholders involved in contractual agriculture in high-value sectors, such as horticulture, and for differentiated products have access to improved technology, product quality, and sanitary standards, compared to smallholders not under contract (Maertens et al. 2012; Rueda and Lambin 2013b).

At the mesoscale, once-remote regions are more connected to one another, and thus local agricultural failures can be more easily compensated through imports. This reduces the vulnerability of local populations to food insecurity and famine and potentially opens new markets for higher return on investment. On the other hand, the global food system becomes more exposed to supply shocks with the concentration of food production. Socioeconomic or political events in one location can trigger price spikes with repercussions in distant places, which in turn can cause political and socioeconomic disruptions of their own. The cereal price increases experienced in 2008 were one of the factors spurring a series of protests and eventually contributing to regime and policy changes in a few countries, particularly in those highly dependent on imports to satisfy their domestic needs (Naylor and Falcon 2010). Furthermore, the dominance of single agro-ecological zones for the provision of specific commodities (such as the Latin American Southern Cone for soybean production) increases global vulnerability to climatic shocks, thus amplifying the effects of shortages. A looming scarcity of suitable land for agriculture that is not under forest or protection status, and increased competition between different land uses restrict future options to expand or shift food production zones to other world regions that are likely to have a more favorable future climate. Future cropland expansion will come at a high cost in terms of natural ecosystems and the ecosystem services they provide, and upon which societies depend.

At the global scale, the rapid conversion of natural ecosystems is fueled by agricultural growth and urban and suburban expansion. It severely threatens the provision of ecosystem services, particularly those that are critical for the success of agriculture and food production. Water supply, local climate regulation, and pollination, natural pest control, and other benefits of biodiversity, are services that forests and other natural habitats provide and that are key to agricultural productivity. It has been estimated, for example, that cross-pollination is key for at least 30 percent of the world's plants (Chavarria 1999). As agriculture and other land uses encroach on natural systems or abandon highly degraded lands, the capacity of local ecosystems to supply water, natural pest control, and other ecosystem services is highly compromised. The loss of these key ecosystem services, due to the conversion of natural ecosystems surrounding farmland, threatens the sustainability of food production. This is an additional stress on future food security.

Conclusion

The land-use constraints on food security are associated with increasing demand for land for non-staple crops, animal feed, biofuels, and carbon sequestration projects. This demand encroaches on the needs of the poor, and accounts for most of the negative environmental impacts of agricultural expansion. The geographical patterns of agricultural production over the 2000–2010 period show that, while production of staple crops has increased on existing agricultural land thanks to intensification gains, production of non-staple crops and livestock is expanding into natural ecosystems. The resulting ecological costs include the loss of significant ecosystem services. The products derived from non-staple crops, although mostly grown in the rural areas of developing countries, are for the most part consumed by urban populations living in rich countries and in emerging economies. Reshaping these trends to expand their benefits to the rural poor, while minimizing negative environmental externalities, is a challenge as corporations increase their reach and control over supply chains, diets become more homogeneous, and races to lock up remaining available agricultural land produce leakage of land conversion into countries with weaker states.

The same forces of globalization that threaten food security for the "bottom billion" also offer new opportunities. The final consumers of agricultural commodities, the corporations involved in their transformation and retailing, and civil society at large show a growing concern for sustainability, and are developing various private-public partnerships to promote sustainable land use (Dauvergne and Lister 2012). These actors are starting to express a preference for goods with supply chains that have been certified as meeting sustainability criteria, including environmental stewardship and fair trade benchmarks. Simultaneously, large agribusiness corporations are increasingly adopting sustainability standards, in response to consumer demand and societal pressure, and applying these to their suppliers. There is potential for the globalization of agriculture to lift small producers and the food insecure out of food insecurity and into more sustainable land-use practices. But doing so will require the engagement of all sectors, not just agriculture. The environmental and health costs of the dietary transition are too high and should not be borne by the weakest members of society, whether they live in developing countries or emerging and prosperous economies.

References

Anderson, K. 2010. Globalization's effect on world agricultural trade, 1960–2050. *Philosophical Transactions of the Royal Society B 365*(1554): 3007–3021.

Angel, S., S. C. Sheppard, D. L. Civco, R. Buckley, A. Chabaeva, L. Gitlin, A. Kraley, J. Parent, and M. Perlin. 2005. *The dynamics of global urban expansion*. Department of Transport and Urban Development. Washington, DC: The World Bank.

Angel, S., Parent, J., Civco, D.L., Blei, A., and Potere, D. 2011. The dimensions of global urban expansion: Estimates and projections for all countries, 2000–2050. *Progress in Planning 75*: 53–107.

Angelsen, A. 2010. Policies for reduced deforestation and their impact on agricultural production. *Proceedings of the National Academy of Sciences USA 107*:19639–19644.

Anseeuw, W., M. Boche, T. Breu, M. Giger, J. Lay, P. Messerli, and K. Nolte. 2012. Transnational Land Deals for Agriculture in the Global South. Analytical Report based on the Land Matrix Database. CDE/CIRAD/GIGA, Bern/Montpellier/Hamburg.

Arce, A., and T. K. Marsden. 1993. The social construction of international food—a new research agenda. *Economic Geography 69*:293–311.

Bleaney, M. 1993. Liberalisation and the terms of trade of developing countries: A cause for concern? *World Economy 16*:453–466.

Borras, S. M., Jr., J. C. Franco, S. Gomez, C. Kay, and M. Spoor. 2012. Land grabbing in Latin America and the Caribbean. *Journal of Peasant Studies 39*:845–872.

Brink, A. B., and H. D. Eva. 2009. Monitoring 25 years of land cover change dynamics in Africa: A sample-based remote sensing approach. *Applied Geography 29*: 501–512.

Carlson, K. M.. L.M. Curran, D. Ratnasari, A.M. Pittman, B.S. Soares-Filho, G.P. Asner, S.N. Trigg, D.A. Gaveau, D. Lawrence, H.O. Rodrigues. 2012. Committed carbon emissions, deforestation, and community land conversion from oil palm plantation expansion in West Kalimantan, Indonesia. *Proceedings of the National Academy of Sciences USA 109*: 7559–7564.

Chavarria, G. 1999. Pollinator conservation. *Renewable Resources Journal* (Winter 1999–2000).

Cordain, L. 1999. Cereal grains: Humanity's double-edged sword. World Review of Nutrition and Dietetics *84*: 19–73.

Cordain, L., S. B. Eaton, A. Sebastian, N. Mann, S. Lindeberg, B. A. Watkins, J. H. O'Keefe, and J. Brand-Miller. 2005. Origins and evolution of the Western diet: Health implications for the 21st century. *American Journal of Clinical Nutrition 81*: 341–354.

Cotula, L. 2012. The international political economy of the global land rush: A critical appraisal of trends, scale, geography and drivers. *Journal of Peasant Studies 39*:649–680.

Dauvergne, P., and J. Lister. 2012. Big brand sustainability: Governance prospects and environmental limits. *Global Environmental Change 22*: 36–45.

Daviron, B., and S. Ponte. 2005. The coffee paradox: Global markets, commodity trade and the elusive promise of development. Zed Books in association with the CTA, Ede, Netherlands.

DeFries, R. S., T. Rudel, M. Uriarte, and M. Hansen. 2010. Deforestation driven by urban population growth and agricultural trade in the twenty-first century. *Nature Geoscience 3*: 178–181.

Drewnowski, A., and B. M. Popkin. 1997. The nutrition transition: New trends in the global diet. *Nutrition Reviews 55*: 31–43.

FAO. 2006. Cattle ranching and deforestation. Rome: FAO.

———. 2010. FAO Statistical Yearbook. Rome: World Food and Agriculture Food and Agriculture Organization of the United Nations.

———. 2012. FAOSTAT. Food and Agriculture Organization of the United Nations.

Foley, J. A., et al. 2011. Solutions for a cultivated planet. *Nature 478*: 337–342.

Gan, J., and B. McCarl. 2007. Measuring transnational leakage of forest conservation. *Ecological Economics 64*: 423–432.

Gibbs, H. K., A. S. Ruesch, F. Achard, M. K. Clayton, P. Holmgren, N. Ramankutty, and J. A. Foley. 2010. Tropical forests were the primary sources of new agricultural land in the 1980s and 1990s. *Proceedings of the National Academy of Sciences USA 107*: 16732–16737.

Grau, H. R., N. I. Gasparri, and T. M. Aide. 2005. Agriculture expansion and deforestation in seasonally dry forests of North-West Argentina. *Environmental Conservation 32*: 140–148.

Greenaway, D., and O. Morrissey. 1993. Structural adjustment and liberalisation in developing countries: What lessons have we learned? *Kyklos 46*: 241–261.

Hallam, D. 2009. International investments in agricultural production. How to feed the World in 2050. Proceedings of a technical meeting of experts, Rome, Italy, 24–26 June 2009.

Hansen, M., et, al. 2008. Humid tropical forest clearing from 2000 to 2005 quantified by using multitemporal and multiresolution remotely sensed data. *Proceedings of the National Academy of Sciences USA 105*: 9439–9444.

Hawkes, C. 2005. The role of foreign direct investment in the nutrition transition. *Public Health Nutrition 8*, 4: 357–365.

Hawkes, C. 2006. Uneven dietary development: Linking the policies and processes of globalization with the nutrition transition, obesity and diet-related chronic diseases. *Globalization and Health 2*: 4. http://www.ncbi.nlm.nih.gov/pmc/articles/PMC1440852/.

IIASA/FAO. 2012. Global Agro-ecological Zones (GAEZ v3.0). Laxenburg, Austria: IIASA; Rome: FAO.

Jepson, W., C. Brannstrom, and A. Filippi. 2010. Access regimes and regional land change in the Brazilian Cerrado, 1972–2002. *Annals of the Association of American Geographers 100*: 87–111.

Koh, L. P., J. Miettinen, S. C. Liew, and J. Ghazoul. 2011. Remotely sensed evidence of tropical peatland conversion to oil palm. *Proceedings of the National Academy of Sciences USA 108*: 5127–5132.

Lambin, E. F., H. Gibbs, L. Ferraira, R. Grau, P. Mayaux, P. Meyfroidt, D. Morton, T. K. Rudel, I. Gasparri, and J. Munger. 2013. Estimating the world's potentially available cropland using a bottom-up approach. *Global Environmental Change 23*(5): 892–901.

Lambin, E. F., and P. Meyfroidt. 2011. Global land use change, economic globalization, and the looming land scarcity. *Proceedings of the National Academy of Sciences 108*: 3465–3472.

Le Polain, Y., and E. Lambin. 2012a. Niche commodities and rural poverty alleviation: Contextualizing the contribution of argan oil to rural livelihoods in Morocco. *Annals of Association of American Geographers.*

———. 2012b. Niche commodities and rural poverty alleviation: Contextualizing the contribution of argan oil to rural livelihoods in Morocco. *Annals of Association of American Geographers 103*(3): 589–607.

Macedo, M., R. DeFries, D. Morton, C. Stickler, G. Galford, and Y. E. Shimabukuro. 2012. Decoupling of deforestation and soy production in the southern Amazon during the late 2000s. *Proceedings of the National Academy of Sciences USA* 1341–1346.

Maertens, M., B. Minten, and J. Swinnen. 2012. Modern food supply chains and development: evidence from horticulture export sectors in Sub-Saharan Africa. *Development Policy Review 30*: 473–497.

Martinelli, L. A., R. Garrett, S. Ferraz, and R. Naylor. 2011. Sugar and ethanol production as a rural development strategy in Brazil: Evidence from the state of Sao Paulo. *Agricultural Systems 104*: 419–428.

Meyfroidt, P., and E. Lambin. 2009. Forest transition in Vietnam and displacement of deforestation abroad. *Proceedings of the National Academy of Sciences USA 106*: 16139–16144.

———. 2011. Global forest transition: Prospects for an end to deforestation. *Annual Review of Environment and Resources 36*: 343–371.

Meyfroidt, P., T. Rudel, and E. Lambin. 2010. Forest transitions, trade, and the global displacement of land use. *Proceedings of the National Academy of Sciences USA107*: 20917–20922.

Morton, D. C., R. S. DeFries, Y. E. Shimabukuro, L. O. Anderson, E. Arai, F. d. B. Espirito-Santo, R. Freitas, and J. Morisette. 2006. Cropland expansion changes deforestation dynamics in the southern Brazilian Amazon. *Proceedings of the National Academy of Sciences USA103*: 14637–14641.

Naylor, R. L., and W. P. Falcon. 2010. Food security in an era of economic volatility. *Population and Development Review 36*: 693–723–.

Nefedova, T. 2011. Agricultural land in Russia and its dynamics. *Regional Research of Russia 1*(3): 292–295.

Pingali, P. 2007. Westernization of Asian diets and the transformation of food systems: Implications for research and policy. *Food Policy 32*: 281–298.

Popkin, B. M. 2003. The nutrition transition in the developing world. *Development Policy Review 21*: 581–597.

Reardon, T., and C. B. Barrett. 2000. Agroindustrialization, globalization, and international development—An overview of issues, patterns, and determinants. *Agricultural Economics 23*: 195–205.

Rice, R. A. and R. Greenberg. 2000. Cacao cultivation and the conservation of biological diversity. *AMBIO: A Journal of the Human Environment 29*: 167–173.

Rudel, T. K. 2002. Paths of destruction and regeneration: Globalization and forests in the Tropics. *Rural Sociology 67*: 622–636.

Rudel, T. K., O. T. Coomes, E. Moran, F. Achard, A. Angelsen, J. C. Xu, and E. Lambin. 2005. Forest transitions: towards a global understanding of land use change. *Global Environmental Change—Human and Policy Dimensions 15*: 23–31.

Rueda, X. and E. Lambin. 2013a. Linking globalization to local land uses: how eco-consumers and gourmands are changing the Colombian coffee landscapes. *World Development 41*: 286–301.

Rueda, X. and E. F. Lambin. 2013b. Responding to Globalization: Impacts of Certification on Colombian Small-Scale Coffee Growers. *Ecology and Society 18* (3): 21. http://www.ecologyandsociety.org/vol18/iss3/art21/.

Saitone, T. L., and R. J. Sexton. 2010. Product differentiation and quality in food markets: Industrial organization implications, in *Annual Review of Resource Economics*, Vol. 2, ed. G. C. Rausser, V. K. Smith, and D. Zilberman, 341–368.. Palo Alto, CA: Annual Reviews.

Schnepf, R. D., E. Dohlman, and C. Bolling. 2001. Agriculture in Brazil and Argentina: Developments and Prospects for Major Field Crops. Washington, DC: USDA.

Seto, K. C., M. Fragkias, B. Güneralp, and M. K. Reilley. 2011. A meta-analysis of global urban land expansion. *PlosOne 6*(8): e23777. http://www.plosone.org/article/info%3Adoi%2F10.1371%2Fjournal.pone.0023777.

Seto, K. C., B. Güneralp, and L. R. Hutyra. 2012. Global forecasts of urban expansion to 2030 and direct impacts on biodiversity and carbon pools. *Proceedings of the National Academy of Sciences USA*. http://www.pnas.org/content/early/2012/09/11/1211658109.full.pdf+html.

Steinfeld, H., P. Gerber, T. Wassenaar, V. Castel, M. Rosales, and C. de Haan. 2006. Livestock's long shadow: Environmental issues and options. Rome: United Nations Food and Agriculture Organization.

Swinnen, J. F., M. A. Vandeplas, and M. Maertens. 2010. Liberalization, endogenous institutions, and growth: A comparative analysis of agricultural reforms in Africa, Asia, and Europe. *World Bank Economic Review 24*: 412–445.

Tilman, D. 1999. Global environmental impacts of agricultural expansion: The need for sustainable and efficient practices. *Proceedings of the National Academy of Sciences USA 96*: 5995–6000.

Tulet, J. C. 2010. Peru as a new major actor in Latin American coffee production. *Latin American Perspectives 37*: 133–141.

UN. 2012. Commodity Trade Statistics Database. http://comtrade.un.org.

UNCTAD. 2009. World Investment report: Transnational corporations, agricultural production and development. United Nations, Switzerland.

USDA. 2012a. Corn: Background. http://www.ers.usda.gov/topics/crops/corn/background.aspx#.U5tg8PldW_g.

USDA. 2012b. PSD Online. USDA. http://apps.fas.usda.gov/psdonline/psdhome.aspx.

White, B., S.M. Borras, R. Hall, I. Scoones, W. Wolford 2012. The new enclosure: critical perspectives on corporate land deals. *Journal of Peasant Studies 39*: 619–647.

World Bank. 2010. Rising global interest in farmland: Can it yield sustainable and equitable benefits? Washington, DC: The World Bank.

Young, A. 1999. Is there really spare land? A critique of estimates of available cultivable land in developing countries. *Environment, Development, and Sustainability 1*: 3–18.

PART FIVE

Food in a National and International Security Context

13

Food and Security

Stephen John Stedman

How does food security relate to security in a broader sense of the word? This is a much more vexing question than it first seems. The difficulty lies in lack of agreement on what constitutes security and what its referent should be: all of humanity; the state; the nation and its citizens; or the international system writ large. All of this book's authors agree that in a time of unprecedented global wealth it is an outrage that roughly a billion people lack adequate food and nutrition. They also agree that enough food is produced to feed everyone in the world, and therefore individual hunger and malnutrition are scourges that can and should be eliminated. The problem to date—or one of them, anyway—is the lack of incentives for governments to adopt agriculture, trade, and energy policies that could reduce global hunger. It is easy to believe that if food security is considered a broader security problem, it will better align incentives for governments to tackle global hunger, and thus make it a more tractable problem to solve. There is much at stake, therefore, in interrogating the relationship between food security and broader security, and understanding the strengths and weaknesses of connecting the two.

Food security—like development, like rule of law, like human rights, like basic order—depends on the actions and policies of states. In an ideal world, each state in the international system should be responsible for providing food security to its citizens. This seems straightforward but it poses at least four challenges for global food security. The first is that some states manifestly make their citizens food insecure: North Korea comes to mind, but it is only an extreme case of a category of states that place the interests and needs of rulers above the ruled. The second is that some states that do want to meet their citizens' food needs are unable to do so, because of lack of resources, inappropriate policy, flawed implementation, or political disincentives. The record of agricultural development and food security of the last 50 years has been hit or miss in large part because of this second challenge. As Walter Falcon describes in Chapter 2, Indonesia got it right, but there are scores of poor countries that continue to get it wrong. It is worth remembering that over three hundred million food insecure people reside in India, considered today to be a rising economic power.

The third challenge has to do with the troublesome definition of security's appropriate scope. Some states accept that food security is essential to broader security, but their referent for security is neither all of humanity nor the international system writ large. These states primarily treat food as a *national* security problem, and accordingly, the policies that they pursue address the needs of their own citizens exclusively. Problems arise when those policies have negative externalities that undermine the food security of others outside their borders. This latter outcome can also be a product of the fourth challenge: for some developed states, affluence and abundance have allowed them to transcend food as a national security problem, but now agriculture and food policies in these states are captured by powerful business interests. National policies that result from the third and fourth challenges—agricultural protection from earlier legislation, food export bans, national restrictions placed on foreign humanitarian aid, and more recent subsidies for crop-based biofuels—all contribute to the displacement of food insecurity outwards from developed and developing economies to the least-developed nations.

Any strategy for achieving global food security must address these four challenges posed by states and their policies in order to succeed. To put it more strongly, global food insecurity today is the direct result of inadequate answers to these challenges. One key question is, who should provide food security to those who are ignored by their state, and on what basis should they act? The answer of the last 50 years has been that in the worst cases of famine and near-famine, international relief organizations step in and feed the starving, acting on a humanitarian imperative. Such aid is crucial in addressing the visible suffering of acute food shortages, but it does not address the political causes of famine, create a basis for long-term recovery, or solve the more intractable problem of chronic hunger.

For those whose food security is threatened by bad or ineffectual state policy, the answer of the last 50 years has been that international financial institutions, governmental development agencies, and NGOs advise, cajole, and support states to do better—and step in as a safety net when those states do not. Traditional development assistance certainly has a humanitarian component. But a closer look at its history yields a murkier mix of motivations that includes the economic interest of corporations, the trade interest of states, and the security interests of powerful states. Moreover, development advice has not been consistent over time, and has been hostage to whatever ideas hold sway at the moment. For those whose food security has been impaired by the actions of others, whether because of national strategies for food security or through agribusiness capture of government subsidies, the remedies of the last 50 years have been manifestly insufficient. Ideally, those agricultural protections and subsidies that increase the food insecurity of others would be addressed through international trade negotiations. But as Johan Swinnen shows in Chapter 5, these have proved among the toughest issues to solve in recent world trade rounds.

These inadequate responses to such a widely recognized and lamented crisis strongly suggest that the motivations and incentives for action have been off target, and in part explains why some analysts believe that if states treat global food security as an issue for international security, better outcomes would be achieved.

Although food insecurity can impact international security by weakening states and stoking regional insecurity, and by potentially causing violent conflict between states in the future, this does not imply that governments will act differently to promote global food security. For that to happen, three things have to occur. First, powerful states would have to believe that international order is at such risk from global food insecurity that it changes their incentives away from narrow national policies that have damaging effects on food security outside their borders. Second, poorer, less-developed states would have to reciprocate and choose pro-agricultural policies. Third, the international security frame would have to successfully create a shared interest in providing global food security while not simultaneously distorting the very policies necessary to achieve such security.

A Guide to the Worlds of Security

For those whose background is in food, agricultural, and development issues, connecting food security to broader security may seem to be a straightforward proposition. But that ignores the significance of many decades of tradition, and evolution, within the security community. Understanding those traditions, in academia, the policy arena, and within the military and intelligence communities, is a necessary first step to making effective linkages between food security and other conceptions of security.

Understanding food security's relationship to other types of security is made particularly difficult by the fact that the meaning of the word "security" is opaque, imprecise, and ideologically loaded. Students of the environment, agriculture, and earth sciences should be aware of several fault lines in debates over security. These lines help explain why what from the outside seems like a clear path to better food security is in fact a complex terrain of intractable positions and vested interests. Scholars, practitioners, and activists disagree on the appropriate referent for security (human, national, or international), and the narrowness or breadth of its definition. Indeed, a literature has developed about the risks of "securitizing" social and economic issues: declaring "war" on poverty, drugs, and HIV/AIDS, for example, may help garner resources to advance solutions (Waever 1995). But those solutions—or the agencies implementing them—might not be the ones that experts in poverty, drug abuse, and deadly diseases would choose (Elbe 2006). Success in funding these so-called wars has rarely been associated with success in addressing the underlying issues.

As concepts, food security and broad security are remarkably different. Definitions of food security may vary, but they vary within a limited set of possibilities. However much they differ, each definition serves the purpose of enabling or clarifying measurement and data collection. Not so for security broadly conceived. Referents differ, scope differs, and indeed, in some cases the definitions become catch-alls of very different factors and processes. Definitions of national and international security reflect fad, fashion, and historical concern. More than 30 years ago Barry Buzan, a professor at the London School of Economics, wrote what is still one of the best books on security. He argued that the concept was undefined, underdeveloped, and simple-minded (Buzan 1983). He rather pointedly observed that in most usage the term constitutes a barrier to helping one think about how better to provide security. The same message still applies today.

Most think tanks with "security" in their title now address a broader range of threats and issues than they did 30 years ago, which is certainly an improvement. But there seems almost to be a tacit agreement that one should not specify what counts and what does not count under the security rubric. Anything goes, except when it doesn't. Many think tanks in the United States still conflate American national security with international security, for example, and assume implicitly that whatever is good for the United States is good for the rest of the world.

The position of U.S. think tanks reflects an ongoing confusion and blurring of levels of analysis between national and international systems. National security encompasses how best to protect a given nation's sovereignty, citizens, territory, and values from external threat. The goal of increased national security can be pursued through different strategies along a continuum, with "self-help" (unilateral pursuit of domestic interests) at one end and institutionalized cooperation at the other. Self-help strategies derive from the belief that the world is an inherently dangerous place, with death of the state or nation as a possibility. In a world of international anarchy, the reasoning goes, there can be no overarching authority capable of providing protection, so states must primarily rely on their own actions, capabilities, and devices to protect themselves. In the middle of the continuum, states can pursue national security through various degrees of cooperation. Such strategies derive from the belief that one's national security is interdependent with others'; that security is not zero-sum, and that cooperation can further one's own national security while also furthering other countries' national security. And at the other end of the continuum, furthest from self-help approaches, nations can commit to constrain their own freedom of action, with the goal of increasing national security for all, through international institutions such as regimes, conventions, and treaties.

Much early academic study in the field of national security in the 1940s and 1950s emphasized self-help and focused disproportionately on military power as a bulwark against threats (Buzan and Hansen 2009). International security as a field of study developed in reaction to this truncated conception of national security. As the name implies, "international security" takes the international system as referent. The international system consists of states as the basic unit of the system, the fundamental rule of sovereignty which shapes diplomacy and creates obligations and mutual expectations of regular behavior among states, and the norms and institutions that address difficult conflict-producing issues and provide order and some global public goods. Given that international disorder and upheaval threaten the ability of states and markets to function in ways that benefit their citizens, the international security framework promotes an interest in developing and protecting a robust international system. While acknowledging that states will compete with one another for power, status, and influence, the international security framework insists that such competition must be bounded so as not to undermine the international system as a whole. The framework insists that a robust international system is essential to national security; that is, the goal is not to protect the international system per se, but to do so for the synergistic benefit of all states.

Fundamental to the study of international security in the 1970s and 1980s was the quest to understand and overcome what is known as the security dilemma (Jervis 1978). Under this phenomenon, the actions taken by one state to make itself secure against the threats of others may actually trigger greater fear and insecurity among adversaries, leading them to take actions in turn that lead to still-greater fear and insecurity in a tragic cycle. The cycle can lead all states to become increasingly insecure and creates an international system more prone to risky confrontations, crises, hostilities, and war. Many scholars of international security still strive to make states more aware of the dangers of unbridled pursuit of national security and to inculcate in policymakers a greater understanding of strategic interaction and the inadvertent risks of unilateral actions.

While the move from a national security perspective to an international security outlook was for the good, the focus of international security studies in the 1970s and 1980s was mostly limited to the superpower competition between the United States and the Soviet Union. This focus was justified on the basis that, because of nuclear weapons, a war between these two powers had the potential to destroy the planet. A war between two countries in Africa may lead to hundreds of thousands of deaths, but even so it would not destroy the international system. Only if such a war somehow drew the superpowers into direct conflict might such violence, however devastating to its victims, become a concern of international security studies. Large parts of the world, therefore, were simply irrelevant 30 years ago as far as the field of international security was concerned.

What changed—and what provides the reason for this chapter—is that the Cold War ended and notions of international security became much more expansive. My own career is illustrative. My doctoral thesis, written at Stanford in the latter half of the 1980s, focused on international mediation in civil wars. In 1989 I submitted it to a prestigious university book series on international security and reviewers rejected it because they asserted that peacemaking in civil wars was irrelevant to international security. When the Cold War ended later that year and world attention became riveted on civil violence in the Balkans and the Horn of Africa in the early 1990s, my work on mediation and civil wars was suddenly deemed part of international security studies. In 1996 I was invited to spend a year at Stanford's Center for International Security and Arms Control (CISAC), and in 1997 was asked to build a program in peacekeeping, peace-building and prevention of violent conflicts. Founded in 1983, CISAC hosted its first visiting fellow from Africa in 1995, which is to say for more than a decade a premier institution on international security treated Africa as irrelevant to international security. Symbolizing the change in CISAC's new broader view of international security, it was renamed the Center for International Security and Cooperation in 1997.

The inclusion of civil violence in the international security agenda was long overdue. In the 1980 and 1990s, the most devastating critiques of traditional state-centered concepts of security stemmed from the observation that states, the fundamental unit of the international system and whose preservation therefore was central to concepts of both national and international security, were in many parts of the world the greatest threat to the lives and security of their citizens (Buzan 1983; Goldgeier and McFaul 1992). Cambodia may have been an extreme case, but it was one of tens of states during the Cold War era that practiced politicide—the mass murder of its own citizens for reasons of regime ideology (Harff and Gurr 1988). In many parts of the world well-regulated bureaucratic states with norms of national interest and capacity to regulate social order were absent (Jackson 1990). Many African leaders treated sovereignty, in the words of former Australian Foreign Minister Gareth Evans, as a "license to kill" their own citizens if they stood in the way of leader or authoritarian enrichment and control (Evans 2008). And where one could not speak properly of a state at all—for example, Lebanon throughout much of the 1970s and 1980s—security was also illusory.

Within several years, from 1989 to 1992, the field of international security changed dramatically to include civil wars and internal violence. By doing so, neglected regions of the world became newly relevant to security studies. But it was unclear what the joining thread was between civil war as security concern and superpower rivalry as security concern. Some scholars argued that large-scale violence was what unified the two concerns, implying that the proper study of international security was the prevention and management of violence (Brown 2003). Other scholars suggested that it was the spillover effects of civil wars to regional disorder and rivalry that connected civil war to the traditional concerns of international security (Stedman 1996).

In 1994 a new alternative to national and international security emerged: human security, or the idea that individuals, or humanity writ large, should be the referent for security. The United Nations Development Program (UNDP) devoted its Human Development Report that year to this new concept (UNDP 1994). It argued that the conception of security should be broadened far beyond questions of military force, diplomacy, and violence. The report included seven new dimensions: economic security, food security, health security, environmental security, personal security, community security, and political security. The approach combined the twin beliefs that security is both a natural entitlement of all people, and the key to reducing deadly threats of all sorts. It was partly an argument based on what constitutes human dignity, which in itself should be protected. And it was partly an argument that war and violence took many fewer lives annually compared to other threats, including disease and malnutrition. The human security concept found an enthusiastic audience among development scholars and practitioners.

By the time that UNDP put forward the concept of human security, the field of international security studies had broadened to include all parts of the world, and civil wars, ethnic violence, and genocide. Many of its scholars, however, opposed human security's all-encompassing purview (Paris 2001).

First, the range of threats was boundless and seemed categorically different from what international security specialists studied. Death alone, regardless of cause, seemed inadequate as a metric for insecurity. For example, many more people die of car accidents than ethnic violence every year. Is traffic safety therefore a security issue? In India each year tens of thousands of people die from snakebites (WHO 2007). Are snakes therefore a security issue?

Second, the human security concept seemed to many international security scholars to be development under another name. It is one thing to say that poverty is a security issue because it is a cause of civil war and violence, or to say that economic nationalism and beggar-your-neighbor trade policies help cause interstate war. It is quite another to say that all insecurity stems from deprivation and want. Economic fears are real, but so are the fears of government oppression, ethnic violence, international aggression, and absence of rule of law. Moreover, by treating security as a composite of seven different dimensions, proponents of the human security approach assumed that all dimensions were attainable together and that there are no trade-offs among them, and therefore no need for a strategy of prioritizing particular dimensions or sequencing their attainment. The importance of sequencing objectives and recognizing trade-offs is a theme that runs throughout this volume.

Third, the human security approach's greatest weakness was and is its normative and positive theory of action. In its weakest claims it asks simply that states everywhere act generously towards people beyond their borders. In this it is no different from humanitarian appeals, and it remains unclear why invoking human security should tug at the heartstrings of citizens and purse strings of governments where basic human empathy does not. In its strongest claims, developed in the late 1990s and early 2000s in the doctrine of "the responsibility to protect," there is a positive obligation of governments to risk blood and treasure to protect civilians in other countries whose security is in immediate, lethal, and large-scale danger (Evans 2009). Here, claims of human security may run counter to first-order moral obligations of governments to their own citizens.

Nonetheless, since UNDP's launch of the human security concept, nongovernmental organizations, international organizations, and even some governments have endorsed the approach. Japan sponsored a Commission on Human Security in 2001; more than 20 governments belong to a "friends of human security" group at the United Nations and supporters of the concept successfully urged the General Assembly in 2005 to endorse what is known as the Responsibility to Protect, a general obligation to protect civilians in cases of mass violence. Whatever theoretical shortcomings the concept may have, for many member states and bureaucrats within the United Nations, human security and its emphasis on development and human needs became a rallying cry against what they saw as a security agenda for only the rich and powerful.

Security has been and remains a deeply contested concept with different meanings and different constituencies. This is even more so in the world of policy, diplomacy, and budgets than in academia, where the field of security studies has remained strongly tethered to relationships leading to violence.

The High-Level Panel on Threats, Challenges, and Change

I experienced the cacophony surrounding security first-hand in 2003 when I was chosen by the then-Secretary General of the United Nations, Kofi Annan, to direct an "eminent persons group" to examine how the United Nations could be transformed into a collective security system relevant for the twenty-first century. "Collective security" is yet another variant meaning of security: it is what the United Nations is meant to provide and what its predecessor, the League of Nations, had failed at providing. In brief, collective security is an arrangement in which all members agree that a threat against one member is a threat against all and commit to come to the aid of any member that is threatened.

When the United Nations was formed in 1945 as a collective security organization, it was meant to counter the threat of international aggression by states. For collective security to work, all members must share a perception of threat: no agreement on threats, no collective security. This quandary has hobbled the United Nations throughout its history, when periodically members could not agree on whether a particular affront or behavior rose to the level of a threat against international peace and security. In 2003 emotions at the United Nations were extremely raw precisely because the Security Council rejected arguments by the United States that suspected weapons programs of Saddam Hussein's Iraqi regime and its ongoing lack of compliance with Security Council resolutions constituted immediate threats to international peace and security. As a result, the UN withheld its mandate for the United States invasion of Iraq.

Before I took the position, I heard vague rumblings of controversy and fractious disagreements within the Secretary General's office about what exactly constituted "security." In UN parlance of the time, traditional security—the prevention, management, or termination of wars within or between states, and the military and diplomatic means to achieve these goals—was referred to as "hard" security. Counterposed to this conception was "soft" security, where humans were the reference point of security and whose well-being and protection rested on freedom from want, illness, and deprivation. In essence, soft security was a different label for human security, and it was never clear why its proponents allowed it to be burdened with the moniker "soft." As was often said within the UN, the soft threats of poverty, deadly infectious disease, chronic malnutrition, and environmental degradation were as difficult to manage as the hard threats of international aggression.

Publicly, my group, named the High-Level Panel on Threats, Challenges and Change, had no instructions on which concept of security should guide us. *Privately*, the Secretary General instructed me that he wanted the Panel's advice on three issues, all in the purview of "hard" security:

1) Given new threats such as transnational terrorism, is preemptive or preventive force by states acceptable under the Charter of the United Nations, and should the Charter's language on use of force in self-defense be amended?
2) What should be done in cases of large-scale killing, when the Security Council refuses to act? How can we reconcile the use of military force that, while serving legitimate humanitarian purposes, might be illegal according to the UN Charter and international law?
3) How can the Security Council, the body within the United Nations with sole responsibility for responding to threats to peace and international security, be reformed to better reflect the realities of the twenty-first century?

These issues all made sense given the circumstances surrounding the creation of the Panel. The year 2003 was a potential turning point for the United Nations. The United States' decision to invade Iraq in March 2003 without Security Council authorization had created a terrible dilemma for the Secretary General and threatened to relegate the United Nations to complete irrelevance. In his speech to the opening of the General Assembly in September that year,

Kofi Annan clearly described the dilemma (UN 2003). On the one hand, he pointed out, the use of preventive force by the UN's most powerful member state had the potential to destroy one of the most basic tenets of international law—that force should only be used by individual states in self-defense. On the other hand, all member states of the UN needed to remind themselves that if the most powerful member state had no confidence in the United Nations and collective security, it would do whatever it believed it had to do to make its people secure. (In mid-2003 there was much speculation that the United States might actually withdraw from the United Nations.) The only way out of the dilemma was to make the United Nations effective against new transnational threats and to rebuild collective security to make it relevant to the challenges of this century.

From the outset then, I believed that the panel would view security through a narrow, traditional lens. My goal was to reach consensus on the first two tough questions, involving the use and authorization of force, and put forward a workable plan for reforming the Security Council. I knew that these were controversial issues and that consensus would be difficult to achieve, but at least our agenda was clearly defined.

Had we just answered those three questions, however, much of the world would have felt cheated. These questions were the concerns of the most powerful countries only, and bore little relation to the security concerns of poorer, less-developed countries.

The composition of the Panel itself guaranteed that we would take a broader look at security. Its 16 members included Sadaka Ogata, then High Commissioner for Refugees, who had just co-chaired a commission on human security; Robert Badinter, former French minister of justice, and a giant in the field of human rights, having almost single-handedly ended the death penalty in France; Enrique Iglesias, director of the Inter-American Development Bank; Nafis Sadik, former Executive Director of the United Nations Population Fund; and Gro Harlem Bruntland, former prime minister of Norway, former Director General of the World Health Organization, and the former chair of a major global commission on sustainability and the environment.

As feared, the very first meeting of the Panel threatened to fracture on the definition of security and how broad a view the Panel should take. I believed it was worthless to try to force the Panel into a premature agreement on the definition of security. It was clear to me though that the Panel could not simply endorse the perspective of human security, if only for the simple reason that the United Nations is an organization of states, and most states first and foremost think of security in national and occasionally international terms. And the fact was that several prominent members were very state-centric in security terms (these panelists included Yevgeny Primakov, the former Prime Minister of Russia; Qian Qichen, the former Foreign Minister of China; Amre Moussa, Secretary General of the Arab League; and Brent Scowcroft, the former National Security Adviser of the United States). Moreover, the Secretary General had wanted advice on some very traditional security matters. So we instead steered Panel members to put forward what they believed were the most salient threats to security, however defined. All in all, my staff counted 35 different threats put forward by our panelists.

One question kept running through my head: How could we promote collective security when there is no shared threat perception among members? We had a list of security threats that ran from severe malnutrition to HIV/AIDS, malaria, and tuberculosis, to youth unemployment, to organized crime, to migration, to genocide, to terrorism, to nuclear, biological, and chemical weapons (but at least no car accidents or snakes!).

Before our next Panel meeting, my staff and I began what would turn out to be almost two years of global consultations, traveling to every part of the world. We wanted to know from governments, nongovernmental organizations, leaders, politicians, and citizens what they perceived to be the biggest threats to their security. It was quickly apparent that, like our Panel, there was no agreement on what exactly the word security should mean, and therefore what the most pressing threats to security are. What also became clear was that both governments and individuals perceived significantly different threats depending on their power, privilege, and region. This confirmed what we feared: if our panel confined the definition of security narrowly to questions of military force or new threats from catastrophic terrorism, we would be guilty of conflating the threats to the security of the most powerful governments of the United Nations with threats to international security.

Based on our consultations, we defined international security as protection of the international system from "any event or process that leads to large scale death or diminution of life chances and undermines States as the basic unit of the international system" (UN 2004). At that time we identified six clusters of threats: (1) economic and social threats, including poverty, deadly infectious disease, and environmental degradation; (2) inter-state conflict; (3) internal conflict, including civil war, genocide, and other large-scale atrocities; (4) nuclear, radiological, biological, and chemical weapons; (5) terrorism; and (6) transnational organized crime.

Substantively this list included threats to human security, national security, and international security. We made clear that we included social and economic threats, not just because they endangered the lives of humanity, but because they also endangered states and therefore the international system.

In the process, we made several arguments that are relevant to the question of food and security. We insisted that in an age of transnational threats, states required the cooperation of other states to defend themselves against threats to their security. For collective security to work, states should cooperate with other states to help address those threats, and that cooperation should generate reciprocity in addressing threats to cooperating states. We also put forward an argument that these threats may be connected in ways not apparent at first glance. For example, if poverty, food insecurity, disease, and environmental degradation weaken states, they may become transit points for transnational organized crime and terrorism. Therefore it is in the direct interest of richer, more powerful states to assist poorer states in addressing social and economic threats—not just for humanitarian reasons, but for deeply pragmatic reasons, too. Finally, we argued that a collective security system for the twenty-first century had to get better at threat and conflict prevention. The best route for that to happen, we concluded, was through economic development of poor states.

Food Security and the Varieties of Security

FOOD AND HUMAN SECURITY

So then, how is food security related to security defined more broadly? Obviously access to adequate calories and nutrition is a component of human security, but it is not clear that anything is gained by saying so. Several of the chapters in this volume question the basic premise that all of the components of human security—for example, economic security, environmental security, political security, and food security—need go together. It may be desirable that food security be sustainable and environmentally sound, for example, but earlier chapters have shown that many countries have achieved food security at the expense of environmental security, not in conjunction with it. Economic security, achieved through rapid economic growth, may increase the challenges of maintaining food security; rising per capita caloric intake, changing diets with more meat consumption, and growing strain on land and water resources and the environment lead to greater difficulties in meeting food demands. Similarly, food security may be attained in circumstances where citizens do not enjoy all of their civil and political rights as in Indonesia in the 1970s and 1980s—and in some cases, such as India, citizens may be politically secure but food insecure.

FOOD AND NATIONAL SECURITY

Still, there are some clear linkages between food security and security as traditionally conceived. For poor, weak governments today, food insecurity can be a direct threat to national security. This was also the case for the countries of early modern Western Europe in the beginning stages of state-building. Dictators in 2008 who looked anxiously at rising prices of bread are just the historical descendants of Louis XVI, whose reign, like those of his predecessor French kings, was threatened by periodic food riots. Charles and Louise Tilly's research on food riots and the centrality of food supply in the rise of the nation-state make clear that food security can be elemental to statebuilding and national security (L. Tilly 1971; C. Tilly 1975).

Most civil wars today occur in weak states that are poor, rife with disease, and food insecure. Grievances of constituent groups and opportunities to rebel caused by weak states, high unemployment, and lootable resources abound. Political scientists and economists continue to wrestle with the relative contributions of geography, poverty, disease, hunger, youth unemployment, bad governance, and human rights abuses to civil wars and endemic state failure (Blattman and Miguel 2010). Food insecurity is clearly part of this complex of causes that hollow out states, however, and if states are to be set right then food insecurity must be overcome.

Massive food insecurity, especially when provoked by a crisis of short supply and high prices, may contribute to the mobilization of citizens against states that lack capacity and legitimacy. Depending on the response of authorities, grievances can be met with violence and beget more protest. This was certainly a part of the story in Tunisia and Egypt in 2010, as food insecurity was one of a multiple set of factors that facilitated the collective action to bring down authoritarian regimes and usher in transitional periods of uncertain outcomes.

But the story of food and national security also has a darker side. One of the earliest and most important textbooks in international relations, Hans J. Morganthau's *Politics Among Nations* (1954), actually identifies food production as an element of national power (which all states envy and pursue to assure survival under anarchy):

> Another relatively stable factor that exerts an important influence upon the power of a nation with respect to other nations is natural resources.
>
> To start with the most elemental of these resources, food, a country that is self-sufficient, or nearly self-sufficient, has a great advantage over a nation that is not and must be able to import the foodstuffs it does not grow, or else starve . . .
>
> A deficiency in home-grown food has thus been a permanent source of weakness for Great Britain and Germany which they must somehow overcome, or face the loss of their status as great powers. Countries enjoying self-sufficiency, such as the United States and Russia, need not divert their national energies and foreign policies from their primary objectives in order to make sure that their populations will not starve in war . . .
>
> Conversely, permanent scarcity of food is a source of permanent weakness in international politics. Of the truth of this observation, India is at present the prime example . . . Regardless of the other assets of national power which are at its disposal, the permanent deficiencies in food compel it to act in its foreign policy from weakness rather than from strength.

This is not distant history. Morgenthau wrote those words fewer than 70 years ago. One of the more unseemly episodes in American foreign policy came more recently still. It was during the Nixon administration when, under duress from OPEC's 1973 oil embargo, officials began to speak of America's "food weapon," which they rather uncritically asserted could break the oil cartel (Wallensteen 1977; Paarlberg 1978). Unseemly, not because the United States previously had not used food as a diplomatic instrument to earn compliance of poor, hungry nations—it had and did—but because an administration would talk so boldly about food in such blunt terms.

Humanitarianism has succeeded in changing contemporary moral sensitivities towards food, civilians in wartime, and the behavior of warring parties. But nearly 100 years ago, British strategy in World War I relied on naval blockade and the deprivation of food to Germany. This was simply an application to Europe of imperial military tactics used frequently to put down insurrections in Africa, the Middle East, and Asia (Porch 2000). Food becomes essential to national survival, let alone security, when its deprivation is a tool of war.

These are the legacies behind the search for food "self-sufficiency," whether in India, Israel, or other states. To be food insecure is to be weak in a way that others can exploit. To be food secure is perhaps an instrument to gain advantage over the insecure.

Thus, a nation's concept of its security requirements sometimes has implications for its food policies. Israel, for example, has a policy of food self-sufficiency, as befits a country born from genocide, inhabiting a small territory in an agriculturally challenging environment among hostile neighbors. All of these factors lead policymakers to emphasize self-suficiency and to be mistrustful of interdependence (Tal 2007). India similarly places enormous importance on self-determination and is fiercely protective of its freedom of action. Its food and agricultural policies have been motivated by a dream of self-sufficiency—that in the worst case of disorder, it should still be able to feed itself.

There is a deeply paradoxical aspect to the relationship of food and national security that emerges as states develop. During early stages of state-building, food security is a crucial element in regime stability and national security. But as states become stronger and richer, food transforms into a distributional issue in domestic politics. Affluence and abundance allow agribusiness to capture gains from protection and subsidies, whose security rationale has all but disappeared. In the United States, for example, agricultural protections and subsidies that were essential to farmer survival during the Great Depression accrue to large agribusiness today, and have little relationship to food security.

FOOD AND INTERNATIONAL SECURITY

There are several paths through which large-scale global food insecurity could trigger or magnify events and processes that could imperil international order and threaten states as the basic unit of the international system. Here I explore three of these paths.

The first and most direct path concerns food insecurity and the viability of many states. As previously mentioned, food insecurity is part of a complex, multifaceted set of causes of state weakness and civil violence. Food insecurity can be indicative of bad governance; it is certainly both a cause and an effect of failed economic development; and it can become a grievance that triggers collective action. As Scott Rozelle and colleagues argue in Chapter 3, food insecurity, in its nutritional aspects, can place limits on how much countries can grow economically. Several emerging economies, including China, potentially face a "middle income trap" that is created in no small part by chronic nutritional deficiencies. If international security is enhanced by functioning states and thriving economies, then food security matters.

The second path concerns food insecurity, regime collapse, and regional disorder. Returning to the recent cases of Egypt and Tunisia, where authoritarian regimes were brought down by civil violence triggered partially by food price spikes, it seems clear that these states per se were not in jeopardy of collapse. Nor was food insecurity in and of itself a fundamental cause of regime collapse, but it was a notable element in a complex causal chain set in motion by poor governance.

That these regimes were part of the Middle East—a regional insecurity complex with active fault lines—raised these countries, and indirectly their food security, to a problem of international security. These regimes had stable relations with the United States and Europe, and in the case of Egypt, it had partnered with the United States and Israel to create a framework of order for the region. While their conservatism worked to assuage fears of regional violence, their authoritarianism and deep unpopularity at home raised basic questions about their ability to sustain that regional cooperation should they fall. Regime collapse in these countries mattered for international security because of the possibility that replacement regimes might abandon or actively undermine the existing basis for regional stability.

The third path concerns food insecurity amidst future resource shortages, whether for land, water, energy, or minerals. Competition for resources prompts more powerful states to act aggressively outside their borders to secure access to and supply of goods needed to assure the food security of their own populations. Such competition, it is said, is fraught with the danger of potential violent escalation (Klare 2001). When climate-change induced stress is added to the picture, it is easy to conjure up scenarios of resource wars fought between states and a general spectre of unending scarcity and violent conflict.

It is not hard to find such prophecies by international security experts (Kaplan 1994). One small problem, however, is that while some of these prophecies are one or two decades old, they still have not been realized. Indeed, since 1994 when Robert Kaplan first wrote his influential article, "The Coming Anarchy," there has been a major decline in the number of civil wars around the world (Human Security Centre 2005). This is not to argue that one should be complacent, but to acknowledge that there may be powerful antidotes to any particular vision of an apocalyptic future. Indeed, it is worth noting that when the United States National Intelligence Council put forward its 2025 scenarios of future threats, resource-induced competition and associated conflict was but one of four scenarios (National Intelligence Council 2008). The purpose of such scenarios is to provoke policymakers to think creatively about how they could manage likely future challenges to avoid bad outcomes. Perhaps the safest approach to the question of food scarcity and major power conflict would be to acknowledge the potential for food insecurity to intersect and aggravate competition among powerful states in ways that could endanger international security, and to invest in rules, institutions, and policies to mitigate that potential.

Food Insecurity and Broader Security: How to Proceed?

Global food security will not be achieved by simply invoking its relationship to broader security. Calling something a security threat may bring greater governmental resources and attention to bear on a problem, but it also might bring traditional security bureaucracies, such as defense, the military, and intelligence communities to the table in ways that are unhelpful for solving the problems of hunger and malnutrition. Whatever else might be said about militaries, defense establishments, and intelligence communities, they are all deeply rooted in a national security mindset. If you tell them, for example, that food insecurity is a threat, their inclination will be to ensure that whatever happens in the world, their own nation's food supply will be protected. In doing so they will probably not spend a lot of time thinking about whether the measures to protect the nation's food supply are harmful or helpful to other nations, each trying to protect its own food supply.

Proponents of international security attempt to convince governments to limit their self-help solutions to their national security threats, so that they do not make others insecure in ways that may come back to eventually make everyone more insecure in the future. If food becomes scarce in the future, self-help solutions will aggravate conflicts, trigger violence, and jeopardize international order.

If the world is entering into a new era of food insecurity and resource shortages, strong states will have already considered the problem in terms of national security and thought about how to meet the needs of their citizens. To prevent food insecurity from endangering international security, rules, institutions, and policies must be developed to help soften the impacts of shortages, create equitable and predictable responses to scarcity, and ensure that resource competition does not endanger the overall international system. In short, what is needed is what is lacking: an international effort to build global food security. Such an effort will have to include basic issues concerning energy, climate, trade, and overall agricultural production.

Contemporary initiatives to increase global food security have yielded unimpressive results. The traditional powers of the United States, Europe and Japan, and the rising powers of China, India, Brazil, and Indonesia all acknowledge that global food insecurity is an affront to humanity. Since 2009, global food security has had a place on the agendas of the G8 and G20. In 2011, the French presidency of the G20 put together an impressive array of technical meetings, with the goal of mobilizing international organizations and G20 member states to strengthen global food security. France, however, was unable to push through the resulting set of integrated and far-reaching policy recommendations. The G20 instead settled for scatter-shot initiatives and half-measures.

This outcome would be predicted for an issue that some states treat as a humanitarian problem and others treat as a national security problem. At first glance, global food security should be an obvious candidate for G20 action. Roughly one billion people go hungry on a daily basis, sporadic outbreaks of famine kill tens of thousands, and food price volatility causes political unrest and threatens vulnerable civilians. Concerns about the future ramifications of population growth, resource shortages, and climate-induced stress raise basic doubts about the ability of future generations to feed themselves, and the larger security consequences that will follow from any shortfall. Despite its importance, the overall international governance of global food security is weak. In its absence, governments pursue food security on a national basis, and treat global food security largely as a humanitarian afterthought. Such a compartmentalization of the issue brings its own irony: in an interdependent world, the national pursuit of food security can have negative externalities that exacerbate food insecurity elsewhere, which is then treated as an issue of charity for those less fortunate.

Underneath the G20's shiny humanitarian veneer lie the food policies of its members—policies that have major negative repercussions on the food security of others. Agricultural subsidies, food export bans, and wrong-headed supports for biofuels are among the culprits. In a morally pure world, the individual G20 members would agree among themselves to each stop those policies that weaken the food security of others. But one should not hold one's breath waiting for the food and agricultural equivalent of unilateral disarmament.

As it stands, any solution to global food insecurity pursued by these 20 countries will be necessarily incomplete, and perceived to be largely self-serving. Most food-insecure countries are not at the G20 table, and will be skeptical of any steps taken by the G20 on their behalf. Whatever the G20 decides to do, the attention of other governments will be on what the individual member states of the G20 *refuse* to do: end subsidies, end food export bans, and end large overseas land grabs that include little or no protection for local populations.

The G20 is also an imperfect instrument for global food security for the simple reason that it will take more than G20 action to end global food insecurity. As mentioned at the beginning of this chapter, food security, like economic development, is a problem that requires national governments to implement good policies. International institutions can help create an enabling environment for good national policies to succeed, but the burden is on national governments to choose good policies and provide the right public goods for agricultural development and food security (Paarlberg 2002). Therefore, any solution to global food insecurity will depend on the actions of not just the G20 states, but also of governments of food-insecure countries around the world.

This conclusion suggests a different strategy for achieving global food security: a bargain between the G20 and poor, food-insecure countries. Such a bargain would first explicitly lay out what is expected of governments in terms of agricultural, food, and rural development policies. The commitment of governments of food-insecure countries to pursue such policies would trigger G20 commitments on trade, aid, and food governance. This would include eliminating internal policies that produce negative externalities for others, as well as increasing their assistance in creating effective, sustainable safety-net programs for the most food-insecure households in food-insecure states. Such an approach could ultimately provide more food security for the poor *and* for the G20 countries. A world where most nations actually pursue smart agricultural development policies is a world where big powerful countries are less likely to need to ban food exports or acquire vast arable land holdings on other continents, to name but two of the policies that individual G20 member states currently pursue to protect their own national food security.

The likelihood of such a bargain depends on rich, powerful countries accepting that they can be food secure in ways that do not diminish the food security of others. It also requires food-insecure countries to exercise their sovereignty in ways that provide the public goods and policies to enable food security for their own citizens. Arguments that tie food security to international security can help create a framework and rationale for such a bargain. But to truly work, those arguments must convince the powerful that national food security can be addressed through measures that do not harm the food security of others. And such arguments must convince weak and ineffective governments that their authority and strength—and yes, their national security—depend in part on the measures they take to make their own citizens food secure.

References

Blattman, C., and E. Miguel. 2010. Civil war. *Journal of Economic Literature*, *48*(1): 3–57.

Brown, M. E. 2003. Security problems and security policy in a grave new world. In *Grave New World: Security Challenges in the 21st Century*, ed. M. E. Brown. Washington, DC: Georgetown University Press.

Buzan, B. 1983. *People, states, and fear: The national security problem in international relations*. London: Harvester Wheatsheaf.

Buzan, B., and L. Hansen. 2009. *The evolution of international security studies*. Cambridge: Cambridge University Press.

Elbe, S. 2006. Should HIV/AIDS be securitized? The ethical dilemmas of linking HIV/AIDS and security. *International Studies Quarterly 50*(2): 119–144.

Evans, G. 2008. State sovereignty was a licence to kill. Interview with SEF News (Stiftun Entwicklung und Frieden) http://www.globalr2p.org/media/files/gareth-_state-sovereignty-was-a-licence-to-kill.pdf.

———. 2009. *The responsibility to protect: Ending mass atrocity crimes once and for all.* Washington, DC: Brookings Institution.

Goldgeier, J. M., and M. McFaul. 1992. A tale of two worlds: Core and periphery in the post-Cold War Era. *International Organization 46*(2): 467–491.

Harff, B., and T. Gurr. 1988. Toward empirical theory of genocides and politicides: Identification and measurement of cases since 1945. *International Studies Quarterly 32*(3): 359–371.

Human Security Centre. 2005. *Human Security Report 2005: War and Peace in the 21st Century*. New York: Oxford University Press.

Jackson, R. 1990. *Quasi-States: Sovereignty, international relations, and the Third World.* Cambridge: Cambridge University Press.

Jervis, R. 1978. Cooperation under the security dilemma. *World Politics 30*(2): 167–214.

Kaplan, R. 1994. The coming anarchy. *The Atlantic*, Feb. 1, 1994.

Klare, M. 2001. *Resource wars: The new landscape of global conflict*. New York: Holt.

Morgenthau, H. J. 1954. *Politics among nations: The struggle for power and peace*, 2nd ed. New York: Knopf.

National Intelligence Council. 2008. *Global Trends 2025: A transformed world*. Washington, DC: U.S. Government Printing Office.

Paarlberg, R. L. 1978. Food, oil and coercive power. *International Security 3*(2): 3–19.

———. 2002. Governance and Food Security in an Age of Globalization: Food, Agriculture and the Environment Discussion Paper 36. Washington DC: IFPRI.

Paris, R. 2001. Paradigm shift or hot air? *International Security 26*(2): 87–102.

Porch, D. 2000. Imperial wars. In *The Oxford history of modern war*, ed. C. Townshend. Oxford: Oxford University Press.

Stedman, S. J. 1996. Conflict and conciliation in sub-Saharan Africa. In *The international dimensions of internal conflict*, ed. M. E. Brown. Cambridge: MIT Press.

Tal, A. 2007. To make a desert \bBloom: The Israeli agricultural adventure and the quest for sustainability. *Agricultural History 81*(2): 228–258.

Tilly, C. 1975. Food supply and public order in modern Europe. In *The formation of national states in Western Europe*, ed. C. Tilly. Princeton, NJ: Princeton University Press.

Tilly, L. 1971. The food riot as a form of political conflict in France." *Journal of Interdisciplinary History 2*(1): 23–57.

UN. 2003. The Secretary-General Address to the General Assembly, New York, September 23, 2003. http://www.un.org/webcast/ga/58/statements/sg2eng030923.htm.

UN. 2004. A More Secure World: Our Shared Responsibility. New York: United Nations.

UNDP. 1994. *Human development report, 1994*. Oxford: Oxford University Press.

Waever, O. 1995. Securitization and descuritization. In *On Security*, ed. R. Lipschutz. New York: Columbia University Press.

Wallensteen, P. 1976. "Scarce goods as political weapons: The case of food. *Journal of Peace Research 13*(4): 277–298.

WHO. 2007. *Rabies and Envenomings: A Neglected Public Health Issue*. Report of a Consultative Meeting, World Health Organization, Geneva, January 10, 2007.

14

From Politics to Farm Plots

A FIELD PERSPECTIVE ON FOOD SECURITY

Rosamond L. Naylor

Many readers will have opened this book with the nagging question: Why, in a world of such impressive technological progress and economic advancement, are billions of people still food insecure? The answer revolves mainly around economic access to food—most people go hungry because they cannot afford a nutritious diet. Although this answer sounds straightforward, the dynamics of food insecurity are far more complicated.

Misguided policies and institutions in a number of developing countries that obstruct agricultural development and prevent sustained rural income growth are the main contributors to insufficient access. The spillovers from agricultural policies in rich countries to global food prices also reduce food security in many impoverished regions. The confounding roles of energy markets, infectious disease, water availability, land rights, climate and other biophysical conditions, and environmental management constitute yet another ring of explanations. And the historical focus of food security on obtaining calories, as opposed to full nutrition, is a problem that plagues countries throughout the world. Addressing the core need to produce and distribute sufficient food is necessary for solving hunger. But close examination of the roots of food insecurity in any given region reveals greater complexity, with several intersecting and overlapping causes. In short, food security is not a static concept—it is evolving over space and time.

As Stephen Stedman correctly points out in Chapter 13, ending hunger will require the collective participation of rich, middle, and poor countries alike. It will require the study, insight, and coordination of specialists from nearly every domain of academia, business and policy. And it will require sound governance on every level from local to global. Politics and policies are powerful in the world food economy, and it is often the political system, rather than a specific policy or program, that needs fixing in order to improve food security.

This volume provides several lenses through which to capture this expansive view of food security, from macro to micro, from past to present. What can one learn by shifting perspective from the rarefied world of high-level politics, to everyday life in the field, where individuals succeed or fail at meeting their food needs? At one of our research sites in Kenya in the summer of 2013, during the final stages of editing this volume, I was reminded of just how challenging it can be to improve food security for rural households at the bottom end of the income distribution.

A Kenyan graduate student and I were investigating a program launched by the Ministry of Fisheries in 2009 to create a viable pond-based aquaculture industry for tilapia and catfish. The initial intentions of the program were laudable: to spur rural income growth throughout the country; to enhance the production of farmed fish in order to offset the steady decline of Kenya's wild capture fisheries; and to provide greater supplies of animal protein for the country's rapidly growing population. Substantial government funds had been allocated to the program in 2009 as part of a larger economic stimulus program in response to the global financial crisis, and the money had to be spent during the first year in order for the program to survive politically. The funds covered the construction of ponds throughout the country, and the cost of supplying fingerlings and fish feeds to farmers. There were three main problems, however: very few rural Kenyans knew how to raise fish in ponds; a significant share of the ponds lacked reliable access to water and inputs; and extension services were inadequate to reach most farmers, especially in the first few years of the program. Many households had agreed to participate in the program because it was essentially free for them, and they had little to lose.

By 2013, many of the poorest farmers in the program had abandoned their ponds and were growing an assortment of crops in the sunken area that once held water. A large share of those farmers who continued to produce fish were dependent on government subsidies for break-even earnings. As a result, the government was increasingly turning its attention toward entrepreneurs who could succeed at raising fish, producing aquaculture feeds, and bringing these products to market at scale without extensive subsidies. After all, for a developing country with limited financial capital, a high return on public investment is critical. Although the program was focused on enhancing fish supplies (i.e., food availability) and protein intake (i.e., utilization), it was not specifically aimed at increasing incomes of poor households (i.e., economic access to food). Even if the program succeeds in launching an aquaculture industry, it remains unclear whether or not it will improve Kenya's food security over the long term. Nonetheless, the fact that the government chose to invest in aquaculture production as a key component of its national stimulus program, and that the Ministry of Agriculture was demonstrating adaptive management and policy design when farmers failed during the initial stages of the program are both positive signs for the country's rural economy.

Why is Kenya's aquaculture story so interesting in the context of this volume? First, it highlights some of the main challenges that governments face in reaching the bottom fifth of the income distribution. In many sub-Saharan African countries, private markets and information services are poorly developed. There is a need to provide extension services to farmers, and to connect farmers to markets, both for input purchases (including credit) and for sales. The importance of engaging the private sector early in the process of agricultural growth and supply chain development was emphasized in the cases of Indonesia (Chapter 2) and China (Chapter 3)—two of the world's most successful cases of poverty and hunger alleviation. For sub-Saharan Africa, investing in information technologies that can serve the poor is likely to be a valuable component of any country's development strategy. In addition, investing in energy technologies that can support irrigation, transportation, processing, and other critical components of the agricultural system is needed (Chapters 6, 9).

Understanding the risk perceptions of extremely poor households is also essential for promoting effective development strategies. Perceptions of risk are often influenced by a household's level of poverty, health, market access, education, and location (including water and land availability and climate conditions), as discussed in several chapters in this volume. In Kenya we found that some of the poorest farmers were willing to spend money on feeds to raise chickens in their yard but were not willing to pay for feed to raise tilapia in their pond, even though the latter could potentially yield much higher profits. From the farmers' point of view, they had raised chickens all their lives—they knew how to raise a healthy chicken and what it could earn in the market. Prior to the new aquaculture program, they had never raised a fish. They did not know if they could keep the fish alive, and if successful, whether they could sell the fish at a reasonable price. Very few of the poorest farmers had seen their neighbors raise a fish and successfully turn a profit. Emulating neighbors is a key variable in technology and system adoption in most rural economies.

For some experts within the Ministry of Fisheries, the frustration of not being able to translate the basic economic calculation of relative profits to the farmers often boiled down to a more rudimentary perception—that these farmers were poor because they had a mindset of being poor, and that little could be done to change this mindset. This perception, which is seen in many developing country contexts, is often a large part of the problem in reaching extremely poor households. It has some validity, not because of culture or basic intelligence, but because poor people often demonstrate a rational, and highly risk-averse, strategy toward functioning at the edge of life and death. For families who have suffered extreme and chronic malnutrition over long periods, it is possible that the mindset of being poor is also influenced partially by cognitive impairment from stunting (Chapters 3, 7). In either case, the solution is to address endemic conditions of malnutrition, not to walk away from the lowest income group.

If the poor are left behind in the development process, in Kenya or elsewhere, there is little hope of ending food insecurity. Each year the United Nations Food and Agricultural Organization (FAO) publishes a report on the *State of Food Insecurity in the World* (SOFI), and the 2012 report was particularly instructive.[1] A key lesson from this report is that when large inequities in income and asset distribution are present, reducing poverty and food insecurity through broad-based economic growth is extremely difficult. Several chapters in this volume discuss how inequities in income and resource distribution can perpetuate food insecurity. For example, property rights and land laws in sub-Saharan African countries often favor foreign investors or national elites over smallholder farmers, creating further income disparities (Chapter 8). The evolution of water institutions as countries move from less developed to more developed commonly favors large-scale, centralized water schemes over small-scale distributed delivery to poor households and farmers (Chapter 11). Income and asset disparities that result in food insecurity often have an insidious feedback to more permanent conditions of inequality. Despite rapid economic growth in several emerging economies (e.g., China, India, Brazil), hundreds of millions of people in these countries have remained malnourished, particularly with micronutrient deficiencies that perpetuate income disparities via low educational attainment and low labor productivity (Chapter 3). Preventing large income and asset inequities from occurring should be a mantra of the development community as it grapples with global food insecurity.

The problems of food insecurity are pervasive, and resolving them is not easy. Still, examples of solutions can be found almost everywhere one looks—if one looks closely, and with the right lenses. In 2012, I spent time with another graduate student at her field research site in Pará, one of Brazil's most heavily deforested states. On this particular trip, we drove for hours along dirt roads from Bélem to Paragominas, passing degraded pastures, burning forests, and massive oil palm plantations. We were traveling with an energetic scientist from The Nature Conservancy (TNC) who was also working with Cargill, a large multinational agribusiness company, to promote a program for sustainable sourcing of soybeans in the Amazon.[2] In Paragominas, we spent two days with a cattle producer, Mauro, and his family at their ranch in a forested area where they had lived since the mid-1980s. Mauro was intent on intensifying his cattle operation so he could remain in the cattle business despite the imposition of increasingly strict forest regulations. With ongoing revisions in Brazil's Forest Code, 80 percent of his land in the Amazon biome had to remain under forest cover as a "legal reserve," whereas earlier regulations had required that only

---

[1] See http://www.fao.org/docrep/016/i3027e/i3027e00.htm. Accessed September 25, 2013.

[2] For more information on the TNC-Cargill partnership, see http://www.cargill.com/news/viewpoints/what-can-cargill-and-tnc-learn-from-each-other/index.jsp. Accessed August 16, 2013.

50 percent of his land stay in legal reserve.[3] Mauro had worked diligently, often with the help of multinational companies, to improve his pasture productivity and to develop his secondary forests in order to harvest a wide range of indigenous food and pharmaceutical products. Through a combination of intensification and forest conservation, he found that his cattle were healthier and gaining weight more quickly, rainfall in the region had become more consistent, and the profits from his full operation were improving. He had essentially become a poster child for the TNC. He was a visionary rancher, yet his perspective remained clear and direct: "I am *not* an environmentalist," he told us. "I am a cattle rancher, trying to support my family, and trying to abide by the law."

Mauro's statement is testimony to the notion that it may be time to drop labels, such as "environmentalist" or "corporate agriculture," that emphasize differences and impede progress at the agriculture-environment interface. Environmental management is an essential part of Mauro's business plan, despite the fact that he is a large-scale rancher in the Amazon. If Mauro leaves his ranch, it is possible that the land would be purchased by a less responsible rancher or an oil palm producer who that would clear even more forestland to support Brazil's (bio)energy security. As discussed in Chapter 9, understanding the counterfactual to any agricultural development strategy is important for analyzing its potential consequences. As countries progress toward higher levels of economic development, resource conservation and environmental stewardship become more grounded in agricultural policymaking, as shown in the case of water institutions (Chapter 11) and policy objectives in the EU (Chapter 5). The U.S. Farm Bill also has sections on resource conservation and the environment, which date back to the Dust Bowl era of the 1930s (Chapter 4).

The future of the Kenyan aquaculture program is also intimately linked to issues of resources and the environment. Its success depends on adequate water to fill the ponds, and in turn could help alleviate stress on wild fisheries. A major theme of this volume is that food security, now and in the future, will depend on improved stewardship of resources and the environment. Given the complex interactions among agricultural systems and the environment described in various chapters, such improvement will require a suite of low-technology, high-technology, management, and institutional initiatives at local to regional scales. The twenty-first century is fundamentally different from the last century; there is virtually no disagreement among scientists that the world's climate is changing—as the fifth report of the Intergovernmental Panel on Climate Change makes clear[4]—and this change is likely to stress food production systems in all parts of the world in future decades (Chapter 9). In

[3] For more information on Brazil's Forest Code, see Garrett, R. D., E. F. Lambin, and R. L. Naylor, 2012. Land institutions and supply chain configurations as determinants of soybean planted area and yields in Brazil. *Land Use Policy.* http://dx.doi.org/10.1016/j.landusepol.2012.08.002.

[4] See http://www.ipcc.ch/index.htm. Accessed May 9, 2014.

addition, although population growth rates are tapering in most countries, global food systems will need to support up to three billion more people in this century. Can the world feed a growing population under conditions of rising resource scarcity and climate change? Yes, supply will effectively equal demand by definition. The more critical issues are: at what price, with what levels of nutrition, with what impact on ecosystems, and with how much inequality and chronic hunger?

In all important food-security cases that one can describe, solution strategies that are based on a single disciplinary perspective, or a silver bullet mentality, are bound to fail. For policymakers, entrepreneurs, and farmers, who must take responsibility for the decisions they make, the information they use does not come in tidy or narrow packages. That is why multiple perspectives, well grounded in strong analytic foundations, will be of increased importance to combat the serious food and agricultural challenges of the upcoming decades. Such an approach does not mean an integrated development strategy at all steps—history has shown a general failure of such strategies because agencies are typically siloed, and the cooperation needed to succeed with such a strategy is often lacking. Instead, our hope is that this volume will excite those interested in food security to delve deeper into disciplinary perspectives beyond their own and to remain rigorous, but also creative, in their search for solutions.

GLOSSARY OF SELECTED TERMS[1]

Anthropometry. Use of human body measurements to obtain information about nutritional status.

Dietary energy deficit. The difference between the average daily dietary energy (calorie) intake of an undernourished population and its average minimum energy requirement.

Dietary energy intake. The energy (calorie) content of food consumed.

Dietary energy requirement. The amount of dietary energy (calories) required by an individual to maintain body functions, health, and normal activity.

Dietary energy supply. Food available for human consumption, expressed in kilocalories per person per day (kcal/person/day). At the national level, it is calculated as the food remaining for human use after deduction of all non-food consumption (exports, animal feed, industrial use, seed, and wastage).

Food insecurity. A situation that exists when people lack secure access to sufficient amounts of safe and nutritious food for normal growth and development and an active and healthy life. It may be caused by the unavailability of food, insufficient purchasing power, inappropriate distribution, inability to absorb nutrition due to disease, or inadequate use of food at the household level. Poverty, poor conditions of health and sanitation, and inappropriate care and feeding practices are the major causes of poor nutritional status. Food insecurity may be chronic, seasonal, or transitory.

Food security. A situation that exists when all people, at all times, have physical, social, and economic access to sufficient, safe, and nutritious food that meets their dietary needs and food preferences for an active and healthy life.

Hidden hunger. Vitamin and mineral deficiencies, or micronutrient deficiencies, that compromise growth, immune function, cognitive development, and reproductive and work capacity. Those suffering from hidden hunger are malnourished, but may not sense hunger. Micronutrient deficiencies can also occur in people who are overweight or obese.

Kilocalorie (kcal). A unit of measurement of energy. One kilocalorie equals 1000 calories. In the International System of Units (ISU), the universal unit of energy is the joule (J). One kilocalorie = 4.184 kilojoules (kJ).

Macronutrients. The proteins, carbohydrates, and fats required by the body in large amounts and available to be used for energy (are measured in grams).

Malnutrition. An abnormal physiological condition caused by deficiencies, excesses, or imbalances in energy, protein, and/or other nutrients.

Micronutrients. Vitamins, minerals, and certain other substances required by the body in small amounts (measured in milligrams or micrograms).

---

[1] *Source*: FAO, WFP, and IFAD. 2012. The State of Food Insecurity in the World 2012. Rome: FAO.

Minimum dietary energy requirement. In a specified age/sex category, the minimum amount of dietary energy per person that is considered adequate to meet the energy (calorie) needs for light activity and good health. For an entire population, the weighted average of the minimum energy requirements of the different age/sex groups in the population (expressed as kilocalories per person per day).

Nutrition security. A situation that exists when secure access to an appropriately nutritious diet is coupled with a sanitary environment and adequate health services and care in order to ensure a healthy and active life for all household members. Nutrition security differs from food security in that it also considers the aspects of adequate caring practices, health, and hygiene in addition to dietary adequacy.

Nutritional status. The physiological state of an individual that results from the relationship between nutrient intake and requirements and from the body's ability to digest, absorb, and use these nutrients.

Overnourishment. Food intake that is continuously in excess of dietary energy requirements.

Overweight and obesity. Body weight that is above normal as a result of an excessive accumulation of fat, usually a manifestation of overnourishment measured by body mass index (BMI), a ratio of weight to mass and height. Overweight is defined as BMI ≥ 25–30 and obesity as BMI ≥ 30.

Stunting. Low height for age, reflecting a sustained past episode or episodes of undernutrition.

Undernourishment. Food intake that is insufficient to continuously meet dietary energy requirements. This term is used interchangeably with chronic hunger, or, in this case, hunger.

Undernutrition. The result of undernourishment, poor absorption, and/or poor biological use of nutrients consumed.

Underweight. Low weight for age in children, reflecting a current condition resulting from inadequate food intake, past episodes of undernutrition, or poor health conditions.

Wasting. Low weight for height, generally the result of weight loss associated with a recent period of starvation or disease.

INDEX

Note: Figures and tables are noted by page numbers in italics.